BIOLOGY OF DEPRESSIVE DISORDERS Part A

A Systems Perspective

THE DEPRESSIVE ILLNESS SERIES

Series Editor: J. John Mann, M.D.

Laboratories of Neuropharmacology
Department of Psychiatry
University of Pittsburgh School of Medicine
Pittsburgh, Pennsylvania

Volume 1 PHENOMENOLOGY OF DEPRESSIVE ILLNESS
Edited by J. John Mann, M. D.

Volume 2 MODELS OF DEPRESSIVE DISORDERS
Psychological, Biological, and Genetic Perspectives
Edited by J. John Mann, M. D.

Volume 3 BIOLOGY OF DEPRESSIVE DISORDERS
Part A: A Systems Perspective
Edited by J. John Mann, M. D.,
and David J. Kupfer, M. D.

Volume 4 BIOLOGY OF DEPRESSIVE DISORDERS
Part B: Subtypes of Depression and Comorbid Disorders
Edited by J. John Mann, M. D.,
and David J. Kupfer, M. D.

BIOLOGY OF DEPRESSIVE DISORDERS
Part A
A Systems Perspective

Edited by

J. John Mann, M.D.

and

David J. Kupfer, M.D.

University of Pittsburgh
Pittsburgh, Pennsylvania

SPRINGER SCIENCE+BUSINESS MEDIA, LLC

Library of Congress Cataloging in Publication Data

Biology of depressive disorders / edited by J. John Mann and David J. Kupfer.
p. cm.—(The Depressive illness series; v. 3-4)
Includes bibliographical references and index.
Contents: Pt. A. A systems perspective--pt. B. Subtypes of depression and comorbid disorders.
ISBN 978-1-4757-9500-4 ISBN 978-1-4757-9498-4 (eBook)
DOI 10.1007/978-1-4757-9498-4
1. Depression, Mental—Physiological aspects. I. Mann, J. John (Joseph John) II. Kupfer, David J. date. III. Series [DNLM: 1. Depressive Disorder—diagnosis. 2. Diagnosis, Differential. 3. Psychology, Clinical. W1 DE467p v. 3-4 / WM 171 B6153 1993]
RC537.B4995 1993
616.85'27—dc20
DNLM/DLC 92-48362
for Library of Congress CIP

ISBN 978-1-4757-9500-4

Originally published by Plenum Press, New York in 1993
Softcover reprint of the hardcover 1st edition 1993

CONTRIBUTORS

Wade H. Berrettini • Department of Psychiatry and Human Behavior, Jefferson Medical College, Philadelphia, Pennsylvania 19107

Daniel J. Buysse • Department of Psychiatry, University of Pittsburgh School of Medicine, Western Psychiatric Institute and Clinic, Pittsburgh, Pennsylvania 15213

Sanjay Dubé • Mood Disorders Module, Western Psychiatric Institute and Clinic, Pittsburgh, Pennsylvania 15213

Fred Grossman • Section on Clinical Pharmacology, Experimental Therapeutic Branch, National Institute of Mental Health, Bethesda, Maryland 20892

Roger F. Haskett • Department of Psychiatry, University of Pittsburgh School of Medicine, Western Psychiatric Institute and Clinic, Pittsburgh, Pennsylvania 15213-2593

William Hendrickse • Veterans Affairs Medical Center and Department of Psychiatry, University of Texas Southwestern Medical School, Dallas, Texas 75216

Gerald L. Kramer • Veterans Affairs Medical Center and Department of Psychiatry, University of Texas Southwestern Medical School, Dallas, Texas 75216

David J. Kupfer • Department of Psychiatry, University of Pittsburgh School of Medicine, Western Psychiatric Institute and Clinic, Pittsburgh, Pennsylvania 15213

Kevin Malone • Laboratories of Neuropharmacology, Western Psychiatric Institute and Clinic, Department of Psychiatry, University of Pittsburgh, Pittsburgh, Pennsylvania 15213

J. John Mann • Laboratories of Neuropharmacology, Western Psychiatric Institute and Clinic, Department of Psychiatry, University of Pittsburgh, Pittsburgh, Pennsylvania 15213

Timothy H. Monk • Human Chronobiology Program, University of Pittsburgh School of Medicine, Western Psychiatric Institute and Clinic, Pittsburgh, Pennsylvania 15213

Frederick Petty • Veterans Affairs Medical Center and Department of Psychiatry, University of Texas Southwestern Medical School, Dallas, Texas 75216

William Z. Potter • Section on Clinical Pharmacology, Experimental Therapeutic Branch, National Institute of Mental Health, Bethesda, Maryland 20892

Isak Prohovnik • Departments of Biological Psychiatry and Brain Imaging, New York State Psychiatric Institute and Department of Psychiatry, Columbia University, New York, New York 10032

Matthew V. Rudorfer • Section on Clinical Pharmacology, Experimental Therapeutic Branch, National Institute of Mental Health, Bethesda, Maryland 20892

Harold A. Sackeim • Departments of Biological Psychiatry and Brain Imaging, New York State Psychiatric Institute and Department of Psychiatry, Columbia University, New York, New York 10032

Andrew Winokur • Department of Psychiatry, University of Pennsylvania, Philadelphia, Pennsylvania 19104

PREFACE

This volume is the third in a series on depressive illness. The first volume, entitled *Phenomenology of Depressive Illness,* is devoted to a description of depressive illness from a variety of perspectives that include those of the patient, the clinician, and the psychiatric researcher. It describes the major subtypes of depressive illness and places them in the context of the life cycle.

The second volume in this series is entitled *Models of Depressive Disorders: Psychological, Biological, and Genetic Perspectives.* This volume describes several major models of depressive disorders, including genetic, cognitive, interpersonal, intrapsychic, and neurobiological models.

The third and fourth volumes deal with the biology of affective disorders in detail. These volumes are distinguished by a triaxial approach. In Volume III the biology of affective disorders is described from the perspective of individual transmitter systems and neurophysiologic and biologic processes. In Volume IV the biology of depression is addressed from the vantage point of symptom components of depression, and similarities and differences in the biology of depression are described compared to other psychiatric disorders with clinically overlapping features such as anxiety disorders or eating disorders. The effects on biology of comorbid conditions such as anxiety, personality disorders, alcoholism, and eating disorders are reviewed. These approaches incorporate issues of state versus trait. It is essential to distinguish features related to a constitutional predisposition to affective disorders from the pathophysiologic changes associated with affective disorders and from the secondary effects of treatment and illness. This ambitious approach has required the contributors to be forthcoming about areas where knowledge is lacking as well as introducing the reader systematically to currently available information. These volumes should therefore be of value to researchers, psychiatrists in training, and clinical practitioners who share a common interest in the scientific basis of psychiatry.

J. John Mann, M.D.
David J. Kupfer, M.D.

Pittsburgh, Pennsylvania

ACKNOWLEDGMENTS

This volume was made possible by the support of our wives and children, who permitted us the time to work on it. Ms. Jean Williams assisted in the preparation of the manuscript and the organization of the project. Our editor at Plenum Press, Norma Fox, showed great patience and was a valuable source of encouragement throughout the process of completing this volume. The contributing authors to this volume are really the reason for its value and quality, and to them go our thanks and acknowledgment.

CONTENTS

Chapter 1

NORADRENERGIC FUNCTION IN DEPRESSIVE DISORDERS

William Z. Potter, Fred Grossman, and Matthew V. Rudorfer

Introduction

Since the elaboration of the catecholamine hypothesis of affective disorders more than twenty years ago, researchers have attempted to test it mainly by measuring norepinephrine (NE) and 3-methoxy-4-hydroxyphenyl-glycol (MHPG), a major NE metabolite in urine, plasma, and cerebrospinal fluid (CSF). MHPG is the only quantitatively significant brain NE metabolite entering the circulation and therefore had originally been thought to reflect brain "activity" (Schildkraut, Orsulak, Schatzberg, Gudeman, Cole, Rohde, & LaBrie, 1978). It was subsequently demonstrated that the CNS contributes 20%–30% of NE metabolites to the circulation. Of this, only 50% of central MHPG appears as MHPG in the urine; the remainder is metabolized to other metabolites, of which 3-methoxy-4-hydroxyvanillyl-mandelic-acid (VMA) predominates. Peripherally released NE contributes the other 50% of circulating MHPG (Filser, Spira, Fischer, Gattaz, & Muller, 1988; Kopin, 1985; Kopin, 1984). Pathways of NE metabolism are shown in Figure 1. The extensive body of work leading to these conclusions and those on catecholamine

William Z. Potter, Fred Grossman, and Matthew V. Rudorfer • Section on Clinical Pharmacology, Experimental Therapeutic Branch, National Institute of Mental Health, Bethesda, Maryland 20892.

Biology of Depressive Disorders, Part A: A Systems Perspective, edited by J. John Mann and David J. Kupfer. Plenum Press, New York, 1993.

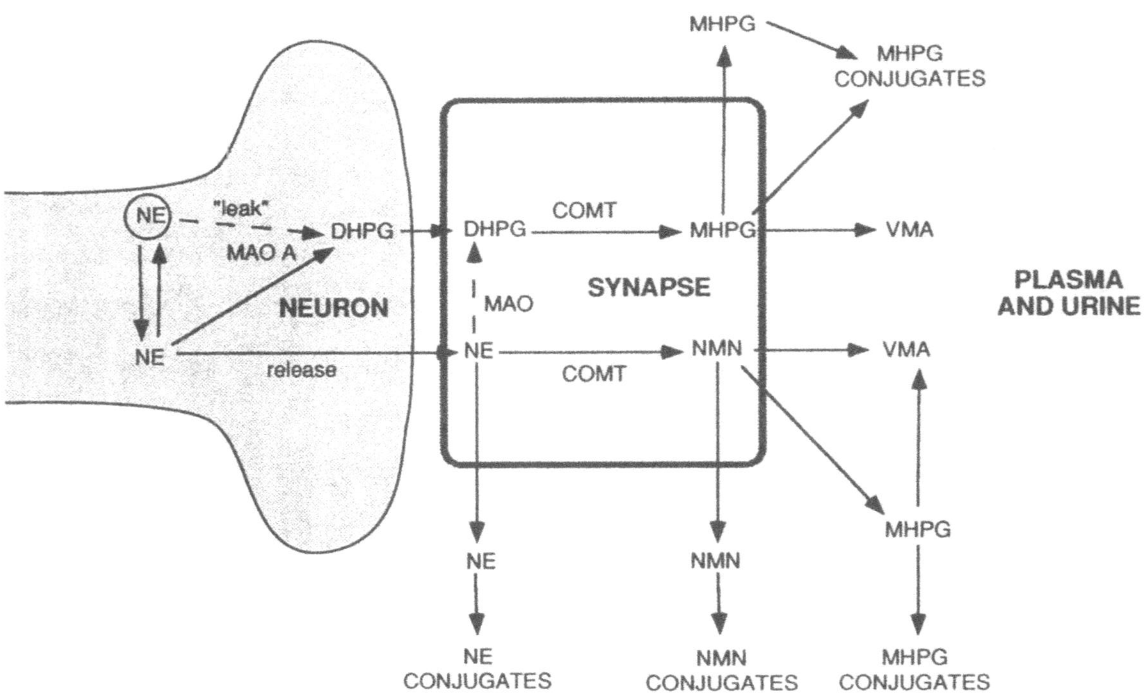

Figure 1. Biochemical pathways of intra- and extraneuronal release and metabolism of NE. MHPG-conjugates and VMA account for most of the major metabolites of NE found in plasma and urine. A small fraction of NMN and NE (which escapes reuptake) is released.

measures in affective illness have been reviewed in depth. We will only briefly summarize the traditional measures in urine, CSF, and plasma.

Current best evidence indicates that postganglionic sympathetic nerve endings are a major source of NE in CSF and that plasma is the major source of MHPG that readily crosses the blood brain–barrier (Kopin, 1984; Goldstein, Zimlichman, Kelly, Stull, Backer, & Keiser, 1987). Using the high correlation between plasma and CSF MHPG observed in comparisons of high and low catecholamine output states, an equation has been derived to "correct" for the contribution of plasma MHPG to that in CSF (Kopin, 1985; Goldstein et al., 1987). However, in the relatively narrow range of values observed in depressed patients and controls, the utility of such an equation to identify "brain MHPG" is questionable and, to date, has not revealed new insights in studies of depression (Linnoila, Guthrie, Lane, Karoum, Rudorfer, & Potter, 1986).

On the other hand, even if measures of MHPG in urine, plasma, and CSF do not directly reflect CNS NE activity or metabolism, these measures may still reflect an acute state or a stable heritable trait and therefore can be useful in studying depression (Roy, Jimerson, & Pickar, 1986). Moreover, as will be discussed below, novel measures and/or algorithms for combining more than one parameter may have relevant functional correlates.

Traditional Urinary, Plasma, and CSF Measures of Norepinephrine in Affective Disorders

Urinary Measures

Historically, measures of MHPG in urine of depressed patients versus healthy volunteers yielded differences that caught the attention of the field. At least in some groups of bipolar depressed patients, urinary output of MHPG has been found to be reduced (Davis, 1990). Studies in unipolar depressed patients yield completely inconsistent values for urinary MHPG.

Some investigators have argued that this variance can be used to discriminate patients within the unipolar group. For instance, one group described subgroups of low, normal, and high MHPG excretion in unipolar depressed patients (Schildkraut, Orsulak, Schatzberg, Cole, & Rosenbaum, 1978). Another group divided depressed patients into subsets having low versus normal or high MHPG relating this to treatment response (Goodwin, Cowdry, & Webster, 1978). Indeed, several studies supported the notion that in unipolar depression low urinary MHPG predicts favorable responses to antidepressants which primarily block NE reuptake (imipramine, desipramine, nortriptyline, maprotiline) (Mass, Fawcett, & Pekirmenjian, 1972; Maas, Kocsis, Bowden, Davis, Redmond, Hanen, & Robins, 1982). Others have failed to confirm this (Muscettola, Potter, Pickar, & Goodwin, 1984; Janicak, Davis, Chan, Altman, & Hedeker, 1986). Low levels of urinary MHPG have been reported to approach normal values with clinical improvement, suggesting that this measure may be state dependent in depressed patients (Pickar, Sweeney, Maas, & Heninger, 1978). In any event, urinary MHPG by itself has not proved a sufficiently robust and consistent measure to be generally accepted as a useful tool in diagnosis or prediction of treatment response.

Plasma Measures

Measuring NE and MHPG in plasma has the advantages of convenience, accessibility, and control. It allows frequent sampling under a variety of conditions such as controlled physical and mental stress. However, unlike 24-hour urine collection for 2–3 consecutive days, there is greater intraindividual variability over time, at least for plasma samples at single time points; studies have reported fluctuations of MHPG ranging from ±12% to ±43% over a 1-hour period (Leckman, Maas, & Heninger, 1981; Leckman, Maas, Redmond, & Heninger, 1980a). Therefore, usually encountered sampling intervals may not detect fluctuations of NE and MHPG. Some studies have measured "continuous" plasma levels, obtaining blood samples at multiple intervals (e.g., 3 hours); over a 24-hour period, however, given the very short half-life of NE (< 10 min) some fluctuations might still go undetected, and it is difficult to collect blood samples beyond 24 hours (DeMet, Halaris, Gwirtsman, Reno, & Becker, 1985; Halaris & DeMet, 1981).

Table 1 provides an updated summary of studies that measured plasma levels of NE while supine and/or following an orthostatic challenge in depressed patients. During an orthostatic challenge, blood samples are taken while lying and after standing for a period of time. The average NE increase during this procedure is approximately 100%, which is similar to that seen during a short sustained physical stress (Rudorfer, Ross, Linnoila, Sherer, & Potter, 1985). Many studies show elevated plasma NE in depressed patients versus healthy volunteers while in the supine position. A subgroup of bipolar depressed patients may actually show reduced supine NE (Rudorfer et al., 1985). In studies using an othostatic

Table 1. Plasma Norepinephrine Concentrations Across Studies

Study	Controls			Depressed patients		
	Norepinephrine (pmole/ml)			Norepinephrine (pmole/ml)		
	N	Supine	Upright	*N*	Supine	Upright
Wyatt et al. (1971)	22	1.21	—	1	2.18	—
Jones et al. (1979)	12	1.42	3.19	—	—	—
Esler et al. (1982)	17	1.39	—	11	2.56	—
Lake et al. (1982)	22	1.70	3.03	15.8	3.21	4.72
Veith et al. (1983)	8	2.31	3.45	14	1.8	3.4
Barnes et al. (1983)	—	—	—	9	2.24	4.24
Siever et al. (1984)	21	1.21	—	—	—	—
Cooper et al. (1985)	—	—	—	9	5.41	—
Rudorfer et al. (1985)	12	1.56	2.74	4	1.35	3.57
Roy et al. (1985)	10	0.96	1.89	17,16	2.05	4.06
de Villiers et al. (1986)	20	1.2	—	15	3.1	—
Rothschild et al. (1987)	6	1.87	—	22	2.5	—
Hauger et al. (1988)	—	—	—	6	1.14	—
Sevy et al. (1989)	14	0.51	—	14	0.61	—
Weighted means		1.34	2.88		2.34	3.91

challenge, however, the change in plasma NE in the upright position is consistently greater in depressed patients whether unipolar or bipolar. In our studies using an orthostatic challenge, despite exaggerated increases in plasma NE following assumption of the upright position in depressed patients, cardiovascular responses (i.e., heart rate and blood pressure) have not differed from those in healthy volunteers (Rudorfer et al, 1985). These observations suggest that the NE system is "inefficient" in some forms of depression since maintaining blood pressure and heart rate required a greater release of NE. Another approach has been the determination of NE "spillover" calculated by measuring the ratio of the infusion rate of tritiated NE to the specific activity of plasma NE with repeated blood sampling following an infusion of tritiated NE to a plateau concentration (Esler, Turbott, Schwartz, Leonard, Bobik, Skeurs, & Jackman, 1982). Using this more definitive technique which avoids potential artifacts from interindividual differences in NE clearance, both the rate of "spillover" and clearance of tritiated NE were elevated in depression (Esler et al., 1982). Another study found that endogenously, but not non-endogenously, depressed patients show a selective increase in NE "spillover" and no intergroup differences in clearance rate when compared to controls (Veith, Halter, Murburg, Barnes, Villacre, & Bachus, 1985).

Studies of plasma MHPG have yielded variable results and have not revealed significant overall differences between depressed patients and healthy volunteers. Plasma levels of MHPG tend to be lower in bipolar than unipolar depressed patients. This is comprehensively reviewed elsewhere (Goodwin, Jamison, 1990; Siever, 1987).

In summary, current studies of plasma NE and MHPG indicate that when compared to healthy volunteers, unipolar depressed patients have an average to elevated resting plasma level of both. Plasma NE shows significantly exaggerated responses in both bipolar and unipolar depression to the stimulus of going from a supine to standing position. In general, plasma MHPG levels have been less informative than those of plasma NE. We will turn later to the possible utility of measures such as the plasma DHPG/NE ratio which have not yet been applied in depression.

CSF Measures

For many years, measures of NE and its metabolites in cerebrospinal fluid (CSF) were thought to directly reflect brain NE "activity." As alluded to earlier, this assumption is problematic since correlations have been found between plasma and CSF NE and MHPG (Kopin, 1985; Goldstein et al., 1987). Pharmacologic studies in dogs reveal parallel changes between plasma and CSF which can be explained on the basis of sympathetic outflow determining concentrations in both compartments (Goldstein et al., 1987). Furthermore, plasma and CSF NE in humans are similarly affected by a variety of pathophysiological conditions (Schatzberg, Samson, Bloomingdale, Orsulak, Gerson, Kizuka, Cole, & Schildkraut, 1989). The extent to which an absolute blood–brain barrier exists for NE is unclear since patients with cerebral vasospasms have been shown to have elevated concentrations of CSF NE, suggesting arterial contributions (Ziegler, Milano, & Hull, 1985). As stated earlier, plasma MHPG readily crosses the blood–brain barrier. On the other hand, patients with pheochromocytoma having elevated concentrations of plasma NE do not show elevated concentrations in CSF (Ziegler et al., 1985). Thus, there are contributions of NE and MHPG to the CSF from other sources; however, their proportions under specific circumstances remain unclear.

Certain other methodological problems are unique to CSF measures. Standards for obtaining spinal fluid such as elapsed time between needle insertion and sample collection have not been established. Therefore, subjects may not have the same degree of accommodation to the stress of the needle stick. Moreover, sampling at a single point in time may not reflect the biochemical process of depression, but rather a state-dependent fluctuation from a recent external or internal stress.

Three of the six studies of CSF concentrations of NE which compare depressed patients and control subjects in the same study report elevations (Table 2), one a decrease, and the other two no significant change. One notes the lower reported values for both groups after investigators began to use HPLC with electrochemical detection in the 1980s. We have found lower concentrations of CSF NE in a small number of depressed patients, especially bipolars as compared to control subjects. CSF concentrations of NE have been reported to be elevated in depressed patients with an atypical presentation, higher scores for nurse-rated anxiety, and with a longer duration of hospitalization (Post, Jimerson, Ballenger, Lake, Uhde, & Goodwin, 1984). Perhaps the most novel finding has been that in dysphoric mania, defined by the coexistence of high depression ratings, NE in CSF correlates modestly but significantly ($r = \sim 0.5$) with ratings of dysphoria and anxiety but not with ratings of mania (Post, Rubinow, Uhde, Roy-Byrne, Linnoila, Rosoff, & Cowdry, 1989). Concentrations of CSF MHPG measured in depressed patients and control subjects yield variable results. Excluding studies where patients were free of medications for less than ten days does not alter the inconsistent results, although a longer period drug free is probably necessary to test these variables. Unipolar and bipolar differences were not found in most of these studies. A detailed tabulation of values has recently been provided by Goodwin and Jamison (Goodwin et al., 1990).

Interpretations of measurements of NE and MHPG in CSF remain uncertain. It seems likely that CSF measures reflect the activity of NE in the brain as well as the sympathetic nervous system, but any signifi-

Table 2. Cerebrospinal Fluid Norepinephrine Concentrations Across Studies[a]

Study	Controls		Depressed patients	
	N	pg/ml	*N*	pg/ml
			Pre-HPLC	
Dencker et al. (1966)	7	200	13 (depressed)	700
Post et al. (1978)	17	225	13 (depressed)	210
			Post-HPLC	
Christensen et al. (1980)	18	102	15 (unipolar and bipolar)	132
Linnoila et al. (1983)	—	—	16	95
Siever et al. (1984)	—	—	12	319
Berrettini et al. (1985)	23	67	12 10 (bipolar) 2 (unipolar)	76
Roy et al. (1988)	25	100	28	131
Post et al. (1989)	48	95	26	157
Weighted means		91		146

[a] modified from Davis, 1990

cance of these measures remains to be determined. Overall, one can view results from CSF studies of depression as qualitatively consistent with those in other compartments. To date, however, they have not added much to our knowledge on either a theoretical or practical level.

Alternative Approaches to the Study of Norepinephrine in Urine, Plasma, and CSF

Accounting for Diurnal Variation

The observation that NE and MHPG exhibit a diurnal variation in humans, as shown in postmortem brains as well as plasma and urine, suggested that there may be an underlying circadian rhythm, which can be altered as a function of depression (Linsell, Lightman, Mullen, Brown, & Causin, 1985; DeMet et al., 1982; Wehr, Muscettola, & Goodwin, 1980; Carlsson, Svennerholm, & Winblad, 1980). This would suggest a different approach from that taken in the studies summarized above. Thus, one could measure, for instance, urinary MHPG in 3-hour intervals around the clock and look for a shift in phase as characteristic of depression. Application of this design has led to the suggestion that depression is associated with a phase advance of MHPG. On the other hand, Sack, James, Doran, Sherer, Linnoila, and Wehr (1988) have demonstrated that diurnal variations in plasma MHPH are evoked by changes in physical activity, posture, or other external influences. They found that when patients and healthy controls were in a controlled environment with a constant routine, the fluctuations in plasma MHPG from patients participating in routine ward activities were absent. Thus, a true circadian rhythm has yet to be convincingly demonstrated.

Controlling for Variation in Clearance and "Leakage"

Other sources of variance in measuring NE and MHPG which might suggest alternate strategies are listed in Table 3. One of the less obvious but important sources stems from the dependence of the concentrations of NE and metabolites on volume of distribution, rate of entry (i.e., intraneuronal and extraneuronal "leakage" versus neuronal release), and rate of clearance (i.e., reuptake, metabolism, excretion) as opposed simply to production rate. As mentioned earlier, the question of variable clearance is partly controllable by infusing radiolabeled NE. Another approach is to leave interindividual variability unexplained

Table 3. Sources of Variance for Measuring NE and Its Metabolites in Various Compartments

1. Rate of entry and clearance
2. Volume of distribution
3. Possible circadian rhythm
4. Assay methods
5. Interindividual variability—age, race, gender
6. Diet
7. Level of stress
8. Collection technqiue[a]

[a] Linnoila and co-workers reported that at least two consecutive 24-hr samples are necessary to obtain reliable information about the output of NE and its metabolites (Linnoila et al., 1986).

and to minimize Table 3 variance using each subject as his or her own control. Obviously, this is only applicable to longitudinal and not to cross-sectional designs.

The question of variations in intraneuronal "leakage" of NE may partly be addressed by looking at the relative concentrations of NE and its deaminated metabolite(s) in plasma. Evidence to date supports the interpretation that under basal conditions plasma concentrations of the NE metabolite dihydroxyphenylglycol (DHPG) are largely determined by intraneuronal metabolism in sympathetic nerves of NE that "leaks" from storage vesicles (Goldstein, Eisenhofer, Stull, Folio, Keiser, & Kopin, 1988; Ludwig, Gerhardt, Halbrügge, Walter, & Graefe, 1988). It has been shown in humans that relative changes in plasma DHPG and NE (ΔDHPG/ΔNE) can be used to differentiate changes resulting from alterations of NE uptake, metabolism via MAO, and/or intraneuronal storage (Goldstein et al., 1988). Whether such a ratio has applications to comparing unmedicated psychiatric groups versus controls remains to be seen. It also needs to be established whether the use of plasma DHPG yields information different from that using MHPG, the only plasma metabolite measured so far in most psychiatric studies.

Urinary Constructs as an Integrated Measure of Metabolic and Physiologic Functions

Yet another strategy uses the relative amounts of urinary NE and metabolites within an individual as either a physiologic measure or a constant. One can calculate a measure that may better reflect average neuronal release of NE based on the model shown in Fig. 1. As can be seen, NE is deaminated intraneuronally by monoamine oxidase A (MAO A) to dihydroxyphenylglycol (DHPG) which is O-methylated (extraneuronally) by catechol-methyltransferase (COMT) to MHPG. MHPG can be either excreted unchanged, conjugated, or oxidized to vanillylmandelic acid (VMA), which in turn is excreted in urine. However, some of the MHPG is derived via DHPG from intraneuronal NE that is never released, but "leaks" from storage vesicles as described above and is oxidized by MAO A. On the other hand, normetanephrine (NMN) is formed exclusively from released or leaked NE that escapes reuptake. Much of the NMN is then converted to MHPG, leaving a moderate fraction that is excreted unchanged or as a conjugate. There is also a small fraction of NE that escapes reuptake and is released as free NE in urine. Therefore, NE and NMN should better reflect released (and leaked) NE, while MHPG and VMA more closely represent metabolism of NE. Changes in the ratio of NE and NMN to MHPG and VMA might also occur under circumstances of reduced firing rates of sympathetic neurons without inhibition of NE synthesis if there was a constant basal rate of extraneuronal leakage of NE from sympathetic nerves. This mechanism has been suggested as contributing to the maintenance of free plasma NE concentrations under circumstances in which spillover (see above) is acutely reduced following administration of the norepinephrine uptake inhibitor, desipramine (Kopin, 1985).

Concomitant measures of plasma NE and DHPG as described above can only address the question of whether intraneuronal leakage is altered. We know of no adequate control for the fraction of unchanged NE that leaks from the neuron in the absence of firing, although studies in volunteers with clonidine suggest that it is small and perhaps negligible in comparison to that accounted for by intraneuronal leakage from storage vehicles and metabolism via MAO to DHPG (Kopin, 1984). DHPG will, of course, readily diffuse across the membrane and be metabolized further to MHPG and then VMA.

Given the biochemical and physiological mechanisms discussed above, one can theoretically also determine daily whole body turnover of NE by adding the 24-hour excretion of urinary NE and its metabolites (NE + NMN + VMA + MHPG). Furthermore, since NE + NMN reflect released NE, the ratio of NE and NMN to whole body NE turnover (NE + NMN/NE + NMN + VMA + MHPG) provides an integrated measure of "released (or leaked) NE" which can potentially be used to differentiate properties of antidepressant treatments and characterize subtypes of affective disorders.

One application of the whole body NE turnover measure has been in a longitudinal study of patients under controlled conditions following chronic treatment with four biochemically distinct classes of antidepressants. All four antidepressant drugs reduced whole body NE turnover (Potter, Rudorfer, & Linnoila, 1988). Considered in light of the down-regulation of β_1-receptors in animal cortices following chronic antidepressants, this finding suggests that the noradrenergic system has become more efficient since physiological functions (i.e., cardiovascular parameters) and psychological functions (i.e., mood) are maintained or enhanced despite a reduction in total output of neurotransmitter and receptor sensitivity.

Some antidepressants have been shown to increase the proportion of NMN to MHPG and VMA. For example, chronic treatment with clorgyline in patients dramatically increases NMN while reducing total NE turnover, revealing effects on both metabolism and synthesis as might be expected following a monoamine oxidase inhibitor (Potter et al., 1988). We are currently exploring the properties of a putative new antidepressant, idazoxan, a potent and selective α_2-antagonist. Figure 2 illustrates an increase in the derived urinary measure for "released" NE following chronic oral treatment of depressed patients with idazoxan. Total turnover of NE and its metabolites, however, decreased. These data are compatible with a sustained increased release of NE by the sympathetic nervous system following α_2 (presynaptic) receptor antag-

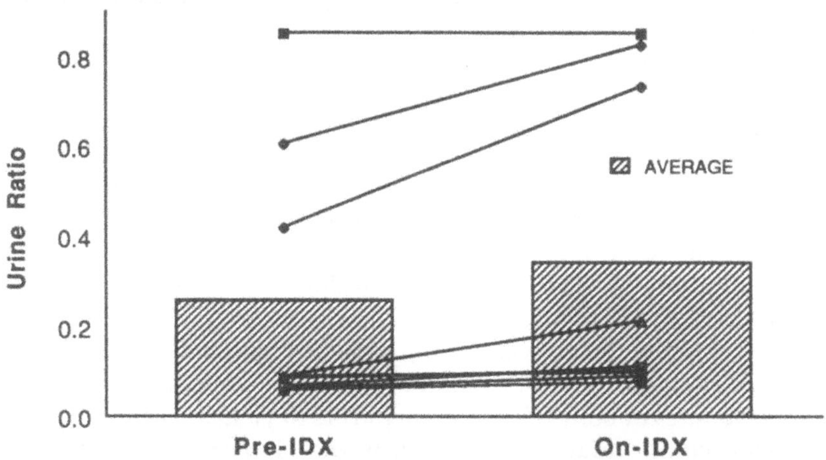

Figure 2. Mean urinary values for NE, NMN, MHPG, and VMA from depressed patients following oral treatment with idazoxan (120 mg/day) for 4–6 weeks were used to calculate the urine ratio. This ratio is an integrated measure of "NE release" determined from the ratio of NE and NMN to whole body NE turnover (NE + NMN/NE + NE + NMN + MHPG + VMA). In these pateints, whole body NE turnover decreased while the integrated measure of "NE release" increased following chronic treatment with idazoxan. Free and conjugated NE, NMN, MHPG, and VMA were averaged from three consecutive 24-hr urine samples using gas chromatography and mass spectometry.

onism and a presumably compensatory decrease in NE turnover. The finding of increased plasma NE after idazoxan suggests that decreased turnover does not fully reverse the effect of α_2 blockade.

Another line of evidence supports the possibility of functional significance of the derived measure of "NE release." Thus, as shown in Fig. 3, we observe that basal and stimulated adenylate cyclase activity in platelets and lymphocytes from humans is directly correlated with the average extraneuronal output of NE as well as "released NE" as measured in urine. The positive correlation with platelet adenylate cyclase activity is consistent with down-regulation (when blood NE is increased) or up-regulation (when blood NE is decreased) of the α_2 receptor complex. In lymphocytes the negative correlation between "released NE" and adenylate cyclase activity is consistent with down- or up-regulation of the β-receptor complex when circulating NE is, respectively, increased or decreased.

Current urinary measures of "NE release" and the whole body NE turnover in untreated depressed patients versus healthy volunteers (Fig. 4) show an increased "NE release" in depressed patients. This urinary finding is compatible with the plasma (and CSF) findings of a hyperresponsive and inefficient noradrenergic system discussed earlier in this chapter.

Discrimination Function Equations for Urinary Measures

Discriminant function analysis provides a very different type of approach to utilizing urinary measures of NE and its metabolites. Interestingly, in one center it has been possible to partially replicate the ability of a discriminant function equation to distinguish between bipolar and unipolar patients (Schildkraut, Orsulak, Schatzberg, Gudeman, Cole, Rohde, & LaBrie, 1978; Schatzberg, Samson, Bloomingdale, Orsulak, Gerson, Kizuka, Cole, & Schildkraut, 1989). The equation utilizes urinary metabolites of both NE and epinephrine to yield a "D-score" which provides a cutoff value below which is the bipolar range and above which is the unipolar range. Other groups have been unable to utilize this particular discriminant function equation since their absolute values of urinary parameters in patient populations differ from those used to calculate the original equation. These difficulties are discussed in detail elsewhere (Potter & Linnoila, 1989). Suffice it to note that without standardization of assays discriminant function equations cannot be generally applied. Furthermore, any equation derived simply on the basis of ability to discriminate between diagnostic groupings must be relatable

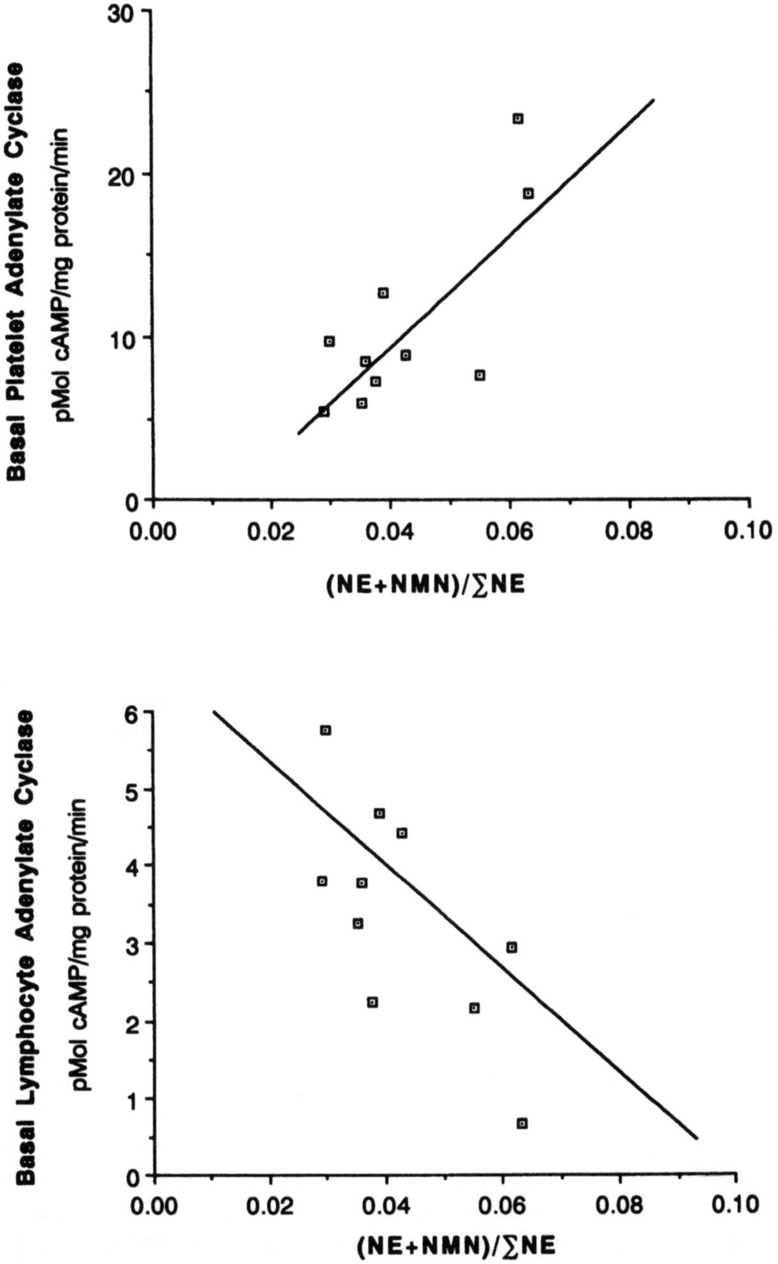

Figure 3. Adenylate cyclase concentrations were measured from platelets and lymphocytes of 10 healthy volunteers. Urinary NE, NMN, MHPG, and VMA were measured from two consecutive 24-hr urine samples to determine the integrated measure for "NE release" (NE + NMN/NE + NMN + MHPH + VMA). The integrated measure for "NE release" is significantly correlated positively with basal platelet adenylate cyclase activity ($R = 0.76$, $P = 0.01$) and negatively with basal lymphocyte adenylate cyclase activity ($R = -0.70$, $P = 0.02$).

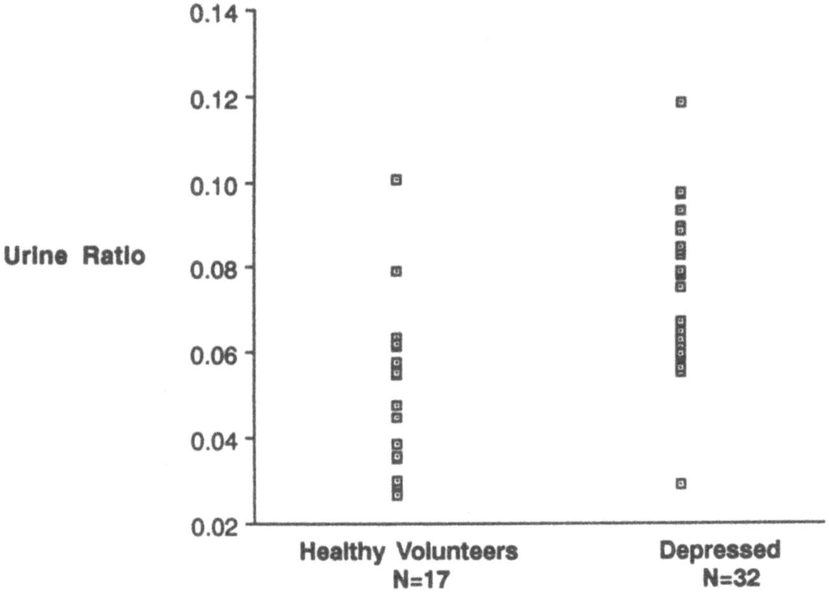

Figure 4. Mean urinary values for NE, NMN, MHPG, and VMA from three consecutive 24-hr urine samples from unmedicated depressed patients and healthy volunteers at rest. The integrated measure for "NE release" (urine ratio) is increased in depressed patients.

to some functional measures if it is to be utilized in asking questions about pathophysiology.

Adrenergic Receptors

Alpha Receptors

In recent years, biochemical and functional studies of depression have focused on receptor sensitivity and function. Investigators have utilized platelets to look for alterations in α_2 adrenoreceptors. While it is assumed that functions of brain α_2 adrenoreceptors are reflected by those of platelet receptors, this has never been proven. Indeed, it is highly unlikely that the standard approaches to study α_2-adrenoreceptors on platelets yield interpretable information on regulation in brain.

To begin with, α_2-adrenoreceptors are known to be tissue specific, both in density and response. For example, estrogen increases the number of α_2-adrenoreceptors in rabbit bladder and uterus, decreases those in platelets, and does not affect those in brain, spleen, and kidney

(Mishra, Hamilton, Jones, Leslie, & Reid, 1985; Mishra, Hamilton, Jones, & Reid, 1985; Roberts, Goldfien, Tsuchiya, Goldfien, & Insel, 1979). Other sources of variance include possible β_2-receptors on platelet membranes, which may interfere with radioligand binding studies (Wang & Brodde, 1985). Furthermore, there is now strong evidence of additional non-α_2-adrenoreceptor binding sites (imidazoline sites) to which drugs previously described as selective for α_2-adrenoreceptors bind with high affinity (Michel, Regan, Gerhardt, Neubig, Insel, & Motulsky, 1990). Therefore, considerable caution must be exercised in interpreting data from radioligand studies of human platelets as reflecting events in the brain. One can argue, however, that platelet receptors may be a stable state or trait parameter and therefore group differences would still be of interest. Whether such differences are established depends in large part on the methods used.

Radioligand studies across groups are confounded by methodological problems, such as varying patient populations, sex, age, clinical state, drug washout period, and assay technique. These methodological issues have been extensively reviewed by others (Piletz, Schubert, & Halaris, 1986). Figure 5 lists studies of radioligand binding in depressed patients and normal volunteers using the α_2 antagonist, yohimbine, and the α_2 agonist, clonidine. As can be seen, the range across all studies varies considerably, especially when yohimbine is used as a radioligand. Furthermore, studies using yohimbine yield no significant differences in the number of α-receptors (β_{max}) between depressed patients and healthy volunteers. Studies using clonidine have a relatively narrow range across most groups for α_2- receptor number (β_{max}), and show an increase in depressed patients.

There are, however, problems using clonidine as a radioligand that go beyond reproducibility of measures across groups. Recent studies suggest heterogeneity of α_2-adrenoreceptors and that clonidine binds with high affinity to nonadrenergic, catecholamine-insensitive sites (Meeley, Ernsberger, Granata, & Reis, 1986; Ernsberger, Meeley, Mann, & Reis, 1987; Bricca, Dontenwill, Molines, Feldman, Belcourt, & Bousquet, 1988; Bricca, Dontenwill, Molines, Feldman, Belcourt, & Bousquet, 1989; Boyajian & Leslie, 1987). These sites are referred to as imidazoline sites since ligands with an imidazoline structure (i.e., idazoxan, clonidine, *p*-amino-clonidine) bind to these sites. Therefore, rather than selectively labeling high-affinity α_2-adrenoreceptors on human platelets, clonidine also binds more extensively to nonadrenergic (imidazoline) sites. This makes it difficult to arrive at any meaningful inferences about α_2-adrenoreceptors on the basis of platelet clonidine binding studies.

Functional responses of α_2-receptors to the administration of clonid-

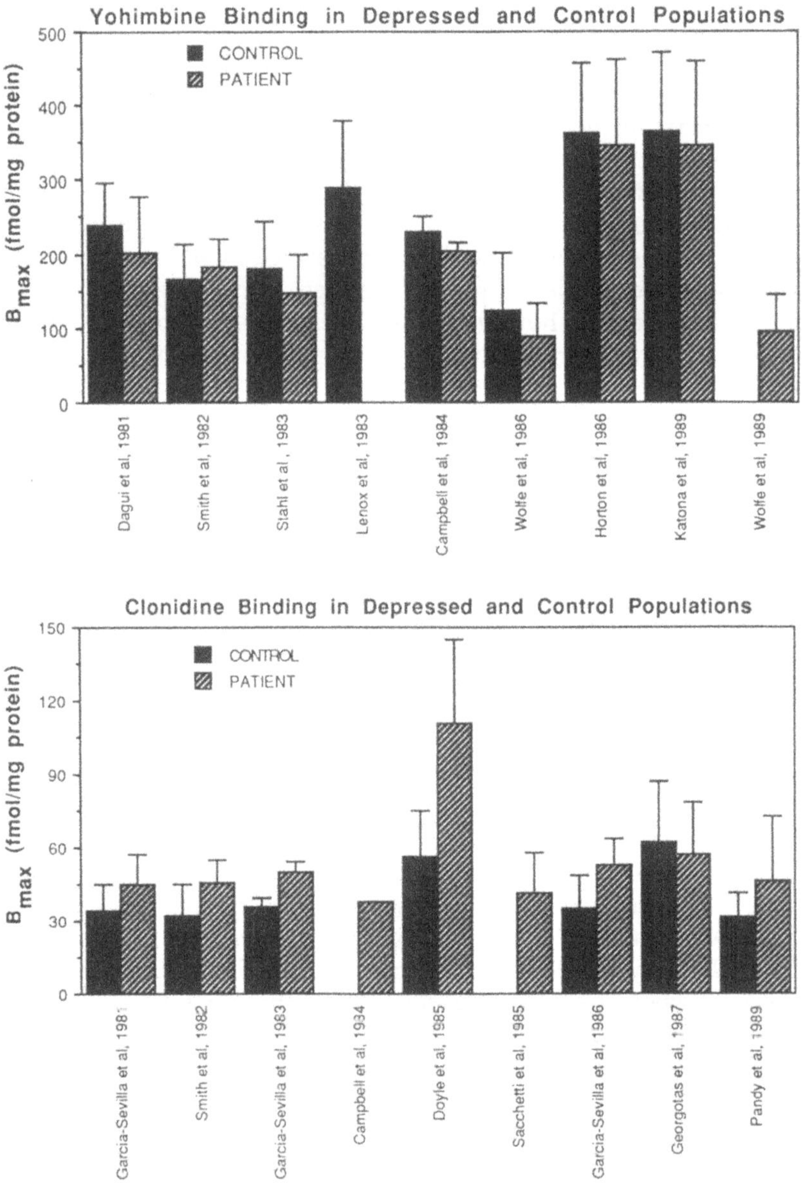

Figure 5. β_{max} for platelet α_2 receptors in depressed patients and healthy volunteers using ^{3}H-clonidine and ^{3}H-yohimbine as radioligands. Note the considerable range of values for β_{max} across studies, especially with ^{3}H-yohimbine as the radioligand.

ine, which might better reflect events in the brain, have also been measured in depressed patients. The most consistent and reproducible finding following acute intravenous clonidine is a blunted growth hormone response in depressed versus healthy subjects, suggesting α_2-adrenergic dysfunction (Charney, Heninger, Sternberg, Hafstad, Giddings, 1982; Siever, Uhde, Silverman, Jimerson, Aloi, Post, & Murphy, 1982; Amsterdam, Maislin, Skolnick, Berwish, & Winokur, 1989). However, blunted responses are not specific to depression and have been reported in patients with panic, general anxiety, and obsessive-compulsive disorder (Uhde, Vittone, Siever, Kaye, & Post, 1986; Charney & Heninger, 1986; Nutt, 1989; Siever et al., 1983). Additionally, it remains possible that alterations distal to the α_2-adrenoreceptor in the hypothalamus, such as in growth hormone releasing hormone, and/or in somatostatin, may be responsible for hormone response. The responses of adrenocorticotrophin hormone and cortisol to an acute challenge of intravenous clonidine in depressed patients vary with reports of increased, decreased, or unchanged levels of both (Siever, Kafka, Targum, & Lake, 1984). Taken together, patterns of functional responses to the administration of clonidine mediated through the central nervous system are inconsistent in depressed patients.

Measurements of cAMP and platelet aggregation in response to NE or epinephrine have been used to directly assess α_2-adrenergic function in platelets. Some, but not all, studies report a decreased NE induced inhibition of prostaglandin E_1- (PGE_1)-stimulated cAMP production in platelets of depressed patients (Garcia-Sevilla, Guimon, Garcia-Vallejo, & Fuster, 1986; Siever, Uhde, Jimerson, Post, Lake, & Murphy, 1984; Kafka, Nurnberger, Siever, Targum, Uhde, & Gehrson, 1986; Kafka & Paul, 1986). Some of these studies also report an increased number of platelet α_2-adrenoreceptors in depressed patients which might represent an attempt to compensate for a "subsensitive" α_2 response to NE. Others note that it is difficult to adequately interpret studies which use PGE_1 to stimulate adenylate cyclase as reflecting α_2-adrenoreceptor function since the PGE_1 response itself may be decreased (Garcia-Sevilla, Padro, Giralt, Guimon, & Areso, 1990). Furthermore, platelet aggregation to α_2-adrenoreceptor agonists may be increased in depressed patients, suggesting that increased platelet clonidine binding could reflect "supersensitive" rather than "subsensitive" α_2-adrenoreceptors.

There is no obvious way to reconcile the findings of increased platelet clonidine binding, decreased α_2 inhibition of platelet adenylate cyclase activity, increased platelet aggregation, and decreased plasma growth hormone response to clonidine. Further studies are necessary to clarify the significance of these findings. However, selective α_2-

adrenoreceptor agonists that do not bind to imidazoline sites as well as other methodological developments should precede further clinical studies since the available methods are not adequate to study such a complex system.

β-Receptors

One of the most consistent findings linking the noradrenergic system with depression is the decrease in β_1-receptor sensitivity in rodent forebrains following chronic, but not acute, administration of antidepressants and electroconvulsant shock (Sulser, Vetulani, & Mobley, 1978; Banergee, King, Riggi, & Chandra, 1977; Bergstrom & Kellar, 1979; Deakin, Owen, Cross, & Dashwood, 1981). The therapeutic significance of the finding, however, has not been clearly established. In suicide victims, higher densities of cortical β-receptors have been demonstrated when compared to controls (Mann, Stanley, McBride, & McEwen, 1986; Biegon & Israeli, 1988). In contrast, a recent study found no difference in cortical β-receptor densities between suicide victims, with a history of depression, and controls (DePaermentiez, Cheetham, Crompton, Katona, & Horton, 1989). Studies using humans are limited since peripheral cells, such as lymphocytes and leukocytes, which express β_2-receptors and exist in an environment very different from that of central neuronal cells, must be depended upon to measure the biochemical and functional properties of the β-receptor complex. Studies measuring β-receptor number in untransformed lymphocytes and leukocytes of untreated depressed patients have yielded conflicting results. Many report a decrease in β-receptor number (Extein, Tallman, Smith, & Goodwin, 1979; Wright, Crichton, Loudon, Morton, & Steel, 1984; Pandey & Davis, 1986; Carstens, Engelbrecht, Russell, Van Zyl, & Taljaard, 1988; Maglazzi, Gietzen, Maddock, Haack, Doran, Goodman, & Weiler, 1989), while others describe an increase or no change when compared to healthy volunteers (Pandey, Jamcak, & Davis, 1985; Sarai, Nakahara, Kanagana, & Yoshihara, 1982; Healy, Carney, & Leonard, 1983; Healy, Halloran, Carney, & Leonard, 1985; Mann, Brown, Halper, Sweeney, Kocsis, Stoker, & Bilezikina, 1985; Zohar, Bannet, Drummer, Fisch, Epstein, & Belmaker, 1983; Cooper, Kelly, & King, 1985). These studies and possible methodologic sources of their differences have been discussed in a very thorough critical appraisal (Werstiuk, Steiner, & Burns, 1990).

Most studies measuring β-receptor mediated cAMP increases following stimulation by an agonist report decreased responsiveness in depressed patients compared to healthy volunteers (Pandey, Dysken,

Garver, & Davis, 1979; Mann et al., 1985; Halper, Brown, Sweeney, Kocsis, Peters, & Mann, 1988; Ebstein, Lerer, Shapira, Shenesh, Moscourch, & Kindler, 1988). One study reported an increase in β-receptor mediated cAMP concentration in bipolar patients during an untreated depression when compared to euthymic patients treated with antidepressants (Klysner, Geisler, & Rosenberg, 1990). Another recent study using leukocytes to measure isoproterenol-stimulated cAMP from depressed patients and healthy volunteers found a resensitization of β-receptors following ECT similar to the levels found in the healthy volunteers (Mann, Mahler, Wilner, Halper, Brown, Johnson, Kocsis, & Chen, 1990). In any event, even using different methods, the common finding of decreased function during depression emerges (Werstiuk, Steiner, & Burns, 1990). The extent to which β-receptor sensitivity is dependent on circulatory catecholamines remains controversial. One study has revealed relatively elevated plasma NE concentrations and lymphocyte β-receptor coupling in certain subgroups of depressed patients (Buckholtz, Davies, Rudorfer, Golden, & Potter, 1988). However, plasma NE sampled at a single point during the day probably does not reflect the average steady state concentration to which blood elements are exposed. As described earlier, we find high inverse correlations between an integrated urinary measure of "NE release" and lymphocyte β-receptor sensitivity (Fig. 3), which are in the expected direction for an influence of catecholamines on function. As emphasized by Werstiuk et al. (1990), one should also consider the possible impact of elevated cortisol on β-receptor function in depressed patients. They provide a list of nine recommendations "for improved experimental design in future studies," to which the interested reader is referred. Most of their concerns also would apply to studies of α_2-adrenergic receptors on platelets once a more adequate ligand has been identified.

Conclusion

Efforts to study norepinephrine output and function in depression have evolved for almost two decades and yielded a cumulative data base pointing to some overall degree of abnormality. Studies to date utilize measures that may all reflect events in the sympathetic nervous system. The implicit assumption is that even if this is the case, the abnormal events in the sympathetic nervous system reflect altered central nervous system (CNS) regulation. Our models, however, are inadequate for specifying precise CNS sites or mechanisms that would explain all of the diverse peripheral findings.

How then can we move beyond our current strategies and still pursue the question of noradrenergic abnormalities in depression? Two extensions of measuring subtypes of noradrenergic receptors suggest themselves. The first would involve development of a PET or SPECT ligand that would at least test whether the density of a specific receptor was altered in some brain region. It is only a question of time until this obvious approach is implemented. Even in the brain, measures of receptor density, however, are unlikely to be that informative. A second application of receptor measures, which has already been tried in preliminary studies, is to use cell lines from groups of patients. There may be some genetically determined generalized abnormality of noradrenergic signal transduction which could be identified in a peripheral cell. If this were specific to, for instance, a single G protein, only a limited number of functions might be affected. Such an abnormality could be expressed both in the CNS and sympathetic nervous system. Although highly speculative, these possibilities remain of interest given the cumulative experiences from studies on norepinephrine in depression and the frequency with which the noradrenergic system is implicated in the action of antidepressant treatments.

References

Amsterdam JD, Maislin G, Skolnick GB, Berwish N, Winokur A: Multiple hormone responses to clonidine administration in depressed patients and healthy volunteers. *Biol Psychiat* 1989; 26: 265–278.

Banerjee SD, King LS, Riggi SS, Chandra SK: Development of beta-adrenergic receptor subsensitivity by antidepressants. *Nature* 1977; 268: 455–456.

Beckmann H, Goodwin FK: Urinary MHPG in subgroups of depressed patients and normal controls. *Neuropsychobiol* 1980; 6: 91–100.

Bergstrom DA, Kellar KJ: Effect of electroconvulsive shock on monoaminergic receptor binding sites in rat brain. *Nature* 1979; 278: 464–466.

Biegon A, Israeli M: Regionally selective increases in beta-adrenergic receptor density in the brains of suicide victims. *Brain Research* 1988; 442: 199–203.

Boyajian CL, Leslie FM: Pharmacological evidence for alpha$_2$ adrenoceptor heterogeneity: Differential binding properties of [^{3}H]rauwolscine and [^{3}H]idazoxan in rat brain. *J Pharmacol Exp Ther* 1987; 241: 1092–1098.

Bricca G, Dontenwill M, Molines A, Feldman J, Belcourt A, Bousquet P: Evidence for the existence of a homogeneous population of imidazoline receptors in the human brainstem. *Eur J Pharmacol* 1988; 150: 401–402.

Bricca G, Dontenwill M, Molines A, Feldman J, Belcourt A, Bousquet P: The imidazoline preferring receptor: Binding studies in bovine, rat, and human brain-stem. *Eur J Pharmacol* 1989; 162: 1–9.

Buckholtz NS, Davies, AO, Rudorfer MV, Golden RN, Potter WZ: Lymphocyte beta-adrenergic function versus catecholamines in depression. *Biol Psychiat* 1988; 24 451–457.

Campbell IC, McKernan RM, Checkley SA, Glass IB, Thompson C, Shur E: Characterization of platelet alpha$_2$ adrenoceptors and measurement in control and depressed subjects. *Psychiat Res* 1986; 14: 17–31.

Carlsson A, Svennerholm L, Winblad B: Seasonal and circadian monamine variations in human brains examined post mortem. *Picta Psychiat Scand* 1980; 61: 75–83.

Carstens ME, Engelbrecht AH, Russell VA, Van Zyl AM, Taljaard JJF: Biological markers in juvenile depression. *Psychiat Res* 1988; 23: 77–88.

Charney DS, Heninger GR: Abnormal regulation of noradrenergic function in panic disorders. *Arch Gen Psychiat* 1986; 43: 1041–1054.

Charney DS, Heninger GR, Sternberg DE, Hafstad KM, Giddings S, Landis H: Adrenergic receptor sensitivity in depression. *Arch Gen Psychiat* 1982; 39: 290–294.

Cooper SJ, Kelly JG, King DJ: Adrenergic receptors in depression effects of electroconvulsive therapy. *Br J Psychiat* 1985; 147: 23–29.

Daiguji M, Meltzer HY, Tong C, U'Prichard DC, Young M, Kravitz H: $\beta_2$adrenergic receptors in platelet membranes of depressed patients: No change in number or ^{3}H-yohimbine affinity. *Life Sci* 1981; 29: 2059–2064.

Davis BA: *Biogenic Monamines and their Metabolites in the Urine, Plasma, and Cerebrospinal Fluid of Normal, Psychiatric, and Neurological Subjects*. Boca Raton, FL: CRC Press, 1990, 179–312.

Deakin JFW, Owen F, Cross AJ, Dashwood MJ: Studies on possible mechanisms of action of electroconvulsive therapy: Effects of repeated electrically induced seizures on rat brain receptors for monoamines and other neurotransmitters. *Psychopharmacol* 1981; 73: 345–349.

Demet EM, Habarus AE, Gwirtsman HE, Reno RM, Becker PI: Effects of desipramine on diurnal rhythms of plasma 3-methoxy-4-hydroxyphenylglycol (MHPG) in depressed patients. *Psychopharmacol Bull* 1982; 18: 221–223.

Demet GM, Halaris AE, Gwirtsman HE, Reno RM, Becker PI: Diurnal rhythm of 3-methoxy-4-hydroxyphenylglycol (MHPG): Relationship between plasma and urinary levels. *Life Sci* 1985; 37: 1731–1741.

DePaermentiez F, Cheetham SC, Crompton MR, Katona CLE, Horton RW: Lower cortical β-adrenoceptor binding sites in post-mortem samples from depressed suicide victims. *Br J Pharmacol* 1989; 98: Supp., 818.

Doyle MC, George AJ, Ravindran AV, Philpott R: Platelet α_2-adrenoreceptor binding in elderly depressed patients. *Am J Psychiat* 1985; 142: 12.

Ebstein R, Lerer B, Shapira B, Shenesh Z, Moscouch DG, Kindler S: Cyclic AMP second messenger signal amplification in depression. *Br J Psychiat* 1988; 152: 665–669.

Ernsberger P, Meeley MP, Mann JJ, Reis DJ: Clonidine binds to imidazole binding sites as well as α_2-adrenoceptors in the ventrolateral medulla. *Eur Pharmacol* 1987; 134: 1–13.

Esler M, Turbott JT, Schwartz R, Leonard P, Bobik A, Skeurs H, Jackman G: The peripheral kinetics of norepinephrine in depressive illness. *Arch Gen Psychiat* 1982; 39: 295–300.

Extein I, Tallman J, Smith CC, Goodwin FK: Changes in lymphocyte beta-adrenergic receptors in depression and mania. *Psychiat Res* 1979; 1: 191–197.

Fawcett JA, Maas JW, Dekirmenjian H: Depression and MHPG excretion: Response to dextroamphetamines and tricycle antidepressants. *Arch Gen Psychiat* 1972; 26: 246–251.

Filser JG, Spira J, Fischer M, Gattaz WF, Müller WE: The evolution of 4-hydroxy-3-methoxy-phenylglycol sulfate as a possible marker of control norepinephrine turnover studies in healthy volunteers and depressed patients. *J Psychiat Res* 1988; 22: 171–181.

Garcia-Sevilla JA, Garcia-Vallejo P, Guimon J: Enhanced α_2-adrenoceptor-mediated platelet aggregation in patients with major depressive disorder. *Eur J Pharmacol* 1983; 94: 359–360.

Garcia-Sevilla JA, Guimon J, Garcia-Vallejo P, Fuster MJ: Biochemical and functional evidence of supersensitive platelet α_2-adrenoreceptors in major affective disorder: Effect of long-term carbonate treatment. *Arch Gen Psychiat* 1986; 43: 51–57.

Garcia-Sevilla JA, Hollingsworth PJ, Smith CB: Alpha$_2$-adrenoceptors on human platelets: Selective labelling by [^{3}H] clonidine and [^{3}H]yohimbine and competitive inhibition by antidepressant drugs. *Eur J Pharmacol* 1981; 74: 329–341.

Garcia-Sevilla JA, Padro D, Giralt MT, Guimon J, Areso P: α_2-adrenoceptor-mediated inhibition of platelet adenylate cyclase and induction of aggregation in major depression. Effect of long-term antidepressant drug treatment. *Arch Gen Psychiat* 1990; 47: 125–132.

Georgotas A, Schweitzer J, McCue RE, Armour M, Friedhoff AJ: Clinical and treatment effects on ^{3}H-clonidine and ^{3}H-imipramine binding in elderly depressed patients. *Life Sci* 1987; 40 2137–2143.

Goldstein DS, Eisenhofer G, Stull R, Folio CJ, Keiser HR, Kopin IJ: Plasma dihydroxyphenylglycol and the intraneuronal disposition of norepinephrine in humans. *J Clin Investigat* 1988; 81: 213–220.

Goldstein DS, Zimlichman R, Kelly GD, Stull R, Backer JD, Keiser HR: Effect of ganglion blockade on cerebrospinal fluid norepinephrine. *J Neurochem* 1987; 49: 1484–1490.

Goodwin FK, Cowdry RW, Webster MH: Predictors of drug response in the affective disorders: Toward an integrated approach. In M.A. Lipton, A.D. Maschio, & K.F. Killam (Eds.), *Psychopharmacology: A Generation of Progress.* New York: Raven Press, 1978, 1277–1288.

Goodwin FK, Jamison KR: Biochemical and pharmacologic studies. In: Goodwin FK, Jamison KR (Eds.), *Manic Depressive Illness,* New York: Oxford, 1990, 416–502. University Press.

Halaris AE, Demet EM: Parameters affecting norepinephrine metabolism in healthy and depressed populations (Abstract). *Proceedings of the 20th Annual Meeting of the American College of Neuropsychopharmacology,* 1981.

Halper JP, Brown RP, Sweeney JA, Kocsis JH, Peters A, Mann JJ: Blunted beta-adrenergic responsivity of peripheral blood mononuclear cells in endogenous depression. Isoproterenol dose–response studies. *Arch Gen Psychiat* 1988; 45: 241–244.

Healy D, Carney PA, Leonard BE: Monoamine-related markers of depression: Changes following treatment. *J Psychiat Res* 1983; 17: 251–260.

Healy D, Halloran AO, Carney PA, Leonard BE: Peripheral adrenoreceptors and serotonin receptors in depression: Changes associated with response to treatment with trazadone on amitriptyline. *J Affect Disord* 1985; 9: 285–296.

Hollister LE, Davis KL, Berger PA: Subtypes of depression based on excretion of MHPG and response to nortriptyline. *Arch Gen Psychiat* 1980; 37: 1107–1110.

Horton RW, Katona CLE, Theodorou AE, Hale AS, Davies SL, Tunnicliffe C, Yamaguchi Y, Paykel ES, Kelly JS: Platelet radioligand binding and neuroendocrine challenge tests in depression. Antidepressant and receptor function. *Ciba Found Sym* 1988; 123: 84–105.

Janicak PG, Davis JM, Chan C, Altman E, Hedeker D: Failure of urinary MHPG levels to predict treatment response in patients with unipolar depression. *Am J Psychiat* 1986; 143: 1398–1402.

Kafka MS, Nurnberger JI, Siever LJ, Targum S, Uhde TW, Gershon ES: Alpha$_2$-adrenergic receptor function in patients with unipolar and bipolar affective disorder. *J Affect Disord* 1986; 10: 163–169.

Kafka MS, Paul SM: Platelet α_2-adrenergic receptors in depression. *Arch Gen Psychiat* 1986; 43: 91–95.

Katona CLE, Theodorou AE, Davies SL, Hale AS, Kerry SM, Horton RW, Kelly JS, Paykel ES: [^{3}H]yohimbine binding to platelet α_2-adrenoreceptors in depression. *J Affect Disord* 1989; 17: 219–228.

Klysner R, Geisler A, Rosenberg R: Enhanced histamine and β-adrenoreceptor mediated cyclic AMP formation in leukocytes from patients with endogenous depression. *J Affect Disord* 1987; 13: 227–232.

Kopin IJ: Avenues of investigation for the role of catecholamines in anxiety. *Psychopathol* 1984; 17: Suppl. 1, 83–97.

Kopin IJ: Catecholamine metabolism: Basic aspects and clinical significance. *Pharmacol Rev.* 1985; 37: 333–364.

Lackmann JF, Maas JW, Heninger GR: Covariance of plasma free 3-methoxy-4-hydroxyphenethylene glycol and diastolic blood pressure. *Eur J Pharmacol* 70: 111–120.

Lackmann JF, Maas JW, Redmond Jr DE, Heninger GR: Effects of oral clonidine on plasma-3-methoxy-4-hydroxy phenethyleneglycol (MHPG) in man: Preliminary report. *Life Sci* 1980; 26: 2179–2185.

Lenox RH, Ellis JE, Van Riper DA, Ehrlich YH, Peyser JM, Shipley JE, Weaver LA: Platelet α_2-adrenergic receptor activity in clinical studies of depression. In Usdin E. (Ed.), *Frontiers in Neuropsychiatric Research*, London: Macmillan Press, 1983, 331–355.

Linnoila M, Guthrie S, Lane EA, Karoum F, Rudorfer M, Potter WZ: Clinical studies on norepinephrine metabolism: How to interpret the number. *Psychiat Res* 1986; 17: 229–239.

Linsell CR, Lightman SL, Mullen PE, Brown MS, Causin RC: Cicadian rhythms of epinephrine and norepinephrine in man. *J Clin Endocrinol Metab* 1985; 60: 1210–1215.

Ludwig J, Gerhardt T, Halbrügge T, Walter J, Graefe KH: Plasma concentrations of noradrenaline and 3, 4 dihydroxyphenylglycol under conditions of enhanced sympathetic activity. *Eur J Clin Pharmacol* 1988; 35: 261–267.

Maglazzi JR, Gietzen D, Maddock RJ, Haack D, Doran AR, Goodman T, Weiler PG: Lymphocyte beta-adrenoreceptor density in patients with unipolar depression and normal controls. *Biol Psychiat* 1989; 26: 15–25.

Mann JJ, Brown RP, Halper JP, Sweeney JA, Kocsis JH, Stocker PE, Bilezikian JP: Reduced sensitivity of lymphocyte beta-adrenergic receptors in patients with endogenous depression and psychomotor agitation. *New Eng J Med* 1985; 313: 715–720.

Mann JJ, Mahler JC, Wilner RJ, Halper JP, Brown RP, Johnson KS, Kocsis JH, Chen JS: Normalization of blunted lymphocyte N-adrenergic responsivity in melancholic inpatients by a course of electroconvulsive therapy. *Arch Gen Psychiat* 1990; 47: 461–464.

Mann JJ, Stanley M, McBride PA, McEwen BS: Increased serotonin2 and β-adrenergic receptor binding in the frontal cortices of suicide victims. *Arch Gen Psychiat* 1986; 43: 954–969.

Maas JW, Fawcett JA, Pekirmenjian H: Catecholamine metabolism, depressive illness, and drug response. *Arch Gen Psychiat* 1972; 26: 252–262.

Maas JW, Kocsis JH, Bowden CL, Davis JM, Redmond DE, Hanen I, Robins E: Pretreatment on neurotransmitter metabolites and response to imipramine or amitriptyline treatment. *Psychol Med* 1982; 12: 37–43.

Meeley MP, Ernsberger PR, Granata AR, Reis DJ: An endogenous clonidine-displacing substance from bovine brain: Receptor binding and hypotensive actions in the ventrolateral medulla. *Life Sci* 1986; 38: 1119–1126.

Michel MC, Regan JW, Gerhardt MA, Neubig RR, Insel PA, Motulsky HJ: Nonadrenergic [^{3}H]idazoxan binding sites are physically distinct from alpha$_2$-adrenergic receptors. *Mol Pharmacol* 1990; 37: 65–68.

Mishra N, Hamilton CA, Jones CR, Leslie C, Reid JL: Alpha adrenoceptor changes after estrogen treatment in platelets and other tissues in female rabbits. *Clin Sci* 1985; 69: 235–238.

Mishra N, Hamilton CA, Jones CR, Reid JL: The effects of sexual maturation on alpha adrenoceptors in female rabbits. *Eur J Pharmacol* 1985; 112: 243–247.

Muscettola G, Potter WZ, Pickar D, Goodwin F: Urinary 3-methoxy-4-hydroxyphenylglycol and major affective disorders. *Arch Gen Psychiat* 1984; 41: 337–342.

Nutt DJ: Altered central alpha$_2$-adrenoreceptor sensitivity in panic disorder. *Arch Gen Psychiat* 1989; 46: 165–169.

Pandey GN, Davis JM: Leucocyte beta-adrenergic receptors: A marker for central beta-adrenergic receptor function in depression. *Clin Neuropharmacol* 1986; 9: Suppl. 14: 353–355.

Pandey GN, Dysken MW, Garver DL, Davis JM: Beta-adrenergic receptor function in affective illness. *Am J Psychiat* 1979: 136: 675–678.

Pandey GN, Jamcak P, Davis JM: Studies of beta-adrenergic receptors in leukocytes of patients with affective illness and effects of antidepressant drugs. *Psychopharmacol Bull* 1985; 21: 603.

Pandey GN, Janicak PG, Javaid JI, Davis JM: Increased ^{3}H-clonidine binding in the platelets of patients with depressive and schizophrenic disorders. *Psychiat Res* 1989; 28: 73–88.

Pickar D, Sweeney DR, Maas JW, Heninger R: Primary affective disorder, clinical state change and MHPG excretion. A Longitudinal study. *Arch Gen Psychiat* 1978; 35: 1378–1383.

Piletz JE, Schubert DSP, Halaris A: Evaluation of studies on platelet $alpha_2$ adrenoreceptors in depressive illness. *Life Sci* 1986; 39: 1589–1616.

Post RM, Rubinow DR, Uhde TW, Roy-Byrne PP, Linnoila M, Rosoff A, Cowdry R: Dysphoric mania. *Arch Gen Psychiat* 1989; 46: 353–358.

Post RM, Jimerson DC, Ballenger JC, Lake CR, Uhde TW, Goodwin FK: In: Post RM, Ballenger JC (Eds.), *Neurology of Mood Disorders,* Baltimore: Williams and Wilkens, 1984, 539–553.

Potter WZ, Linnoila M: Biochemical classifications of diagnostic subgroups and D-type scores. *Arch Gen Psychiat* 1989; 46: 269–271.

Potter WZ, Rudorfer MV, Linnoila M: Effects of antidepressants in norepinephrine and its metabolites in cerebrospinal fluid, plasma, and urine. *Prog Catecholamine Res C: Clin Asp* 1988; 301–306.

Roberts JM, Goldfien RD, Tsuchiya AM, Goldfien A, Insel PA: Estrogen treatment decreases alpha adrenergic binding sites on rabbit platelets. *Endocrinol* 1979; 104: 722–728.

Rosenbaum AH, Schatzberg AF, Maruter T, Orsulak PJ, Cole JO, Grab EL, Schildkraut JJ: MHPG as a predictor of antidepressant responses to imipramine and maprotiline. *Am J Psychiat* 1980; 137: 1090–1092.

Roy A, Jimerson DC, Pickar D: Plasma MHPG in depressive disorders and relationship to the dexamethasone suppression test. *Am J Psychiat* 1986; 143(7): 846–851.

Rudorfer MV, Ross RS, Linnoila M, Sherer MA, Potter WZ: Exaggerated orthostatic responsivity of plasma norepinephrine in depression. *Arch Gen Psychiat* 42: 1186–1192.

Sacchetti E, Conte G, Pennati A, Vita A, Alciati A, Cazzullo CL: Platelet α_2-adrenoceptors in major depression: Relationship with urinary 4-hydroxy-3-methoxyphenylglycol and age at onset. *J Psychiat Res* 1985; 19: 579–586.

Sack DA, James SP, Doran AR, Sherer MA, Linnoila M, Wehr TA: The diurnal variations in plasma homovanillic acid level persists but the variation in 3-methoxy-4-hydroxyphenylglycol level is abolished under constant conditions. *Arch Gen Psychiat* 1988: 45: 162–166.

Sarai K, Nakahara T, Kanagana K, Yoshihara S: Lymphocyte beta-adrenergic receptor function in affective disorders. *Adv Biosci* 1982; 40: 161–165.

Schatzberg AF, Orsulak PJ, Rosenbaum AH, Kroger ER, Schildkraut JJ, Cole JO:

Catecholamine measures for diagnosis and treatment of patients with depressive disorders. *J Clin Psychiat* 1980; 41: 35–39.

Schatzberg AP, Samson JA, Bloomingdale KC, Orsulak PJ, Gerson B, Kizuka PP, Cole JO, Schildkraut JJ: Toward a biochemical classification of depressive disorders: Urinary catecholamines, their metabolites and D-type scores in subgroups of depressive disorders. *Arch Gen Psychiat* 1989; 46: 260–268.

Schildkraut JJ, Orsulak PJ, Schatzberg AF, Cole JO, Rosenbaum AH: Biochemical discrimination of subgroups of depression disorders based on differences in catecholamine metabolism. In: Usdin E, Hamin I (Eds.), *Biological Markers in Psychiatry and Neurology*, New York: Pergamon Press, 1978, 23–33.

Schildkraut JJ, Orsulak PJ, Schatzberg AF, Gudeman JE, Cole JO, Rohde WA, LaBrie RA: Toward a biochemical classification of depressive disorders: Differences in urinary excretion of MHPG and other catecholamine metabolites in clinically defined subtypes of depression. *Arch Gen Psychiat* 1978; 35: 1427–1433.

Siever LJ, Insel TR, Jimerson DC, Lake CR, Uhde TW, Aloi J, Murphy DL: Growth hormone response to clonidine in obsessive-compulsive patients. *Br J Psychiat* 1983; 142: 184–187.

Siever LJ, Kafka MS, Targum S, Lake CR: Platelet alpha-adrenergic binding and biochmeical responsiveness in depressed patients and controls. *Psychiat Res* 1984; 11: 287–302.

Siever LJ, Uhde TW, Jimerson DC, Lake CR, Silverman ER, Post ER, Murphy DL: Differential inhibitory noradrenergic responses to clonidine in 25 depressed patients and 25 normal control subjects. *Am J Psychiat* 1984; 141: 733.

Siever LJ, Uhde TW, Jimerson DC, Post RM, Lake CR, Murphy DL: Plasma cortisol responses to clonidine in depressed patients and controls. Evidence for possible alteration in noradrenergic-neuroendocrine relationships. *Arch Gen Psychiat* 1986; 41: 63–68.

Siever LJ: Role of noradrenergic mechanisms in the etiology of the affective disorders. In: Meltzer HY (Ed.), *Psychopharmacology: The Third Generation of Progress.* New York: Raven Press, 1987, 493–504.

Siever LJ, Uhde TW, Silberman EK, Jimerson DC, Aloi JA, Post RM, Murphy DL: Growth hormone response to clonidine as a probe of noradrenergic receptor responsiveness in affective disorder patients and controls. *Psychiat Res* 1982; 6: 171–183.

Smith CB, Hollingsworth PJ, Garcia-Sevilla JA, Zis AP: Platelet $alpha_2$ adrenoreceptors are decreased in number after antidepressant therapy. *Prog Neuropsychopharmacol Biol Psychiat* 1983; 7: 241–247.

Stahl SM, Lemoine PM, Ciaranello RD, Berger PA: Platelet $alpha_2$-adrenergic receptor sensitivity in major depressive disorder. *Psychiat Res* 1983; 10: 157–164.

Sulser F, Vetulani J, Mobley PL: Mode of action of antidepressant drugs. *Biochem Pharmacol* 1978; 27: 257–271.

Uhde TW, Vittone BJ, Siever LJ, Kaye WH, Post RM: Blunted growth hormone response to clonidine in panic disorder patients. *Biol Psychiat* 1986; 21: 1081–1085.

Veith RC, Halter JB, Murburg MM, Barnes RF, Villacre E, Bachus FI: Increased plasma NE appearance rate in dexamethasone resistant depression. *Fourth World Congress of Biological Psychiatry*, 1985, 186.

Wang, XL, Brodde O-E: Identification of a homogeneous class of β_2 adrenoceptors in human platelets by (^{125}I-iodopindolol binding. *J Cyclic Nucleotide Protein Phosphorous Res* 1985; 10: 439–450.

Wehr TA, Muscettola G, Goodwin FK: Urinary 3-methoxy-4-hydroxyphenylglycol circadian rhythms. *Arch Gen Psychiat* 1980; 37: 257–263.

Werstiuk ES, Steiner M, Burns T: Studies on leukocyte β-adrenergic receptors in depression: A critical appraisal. *Life Sci* 1990; 47: 85–105.

Wolfe N, Cohen BM, Gelenberg AJ: Alpha$_2$-adrenergic receptors in platelet membranes of depressed patients: Increased affinity for ^{3}H-yohimbine. *Psychiat Res* 1986; 20: 107–116.

Wolfe N, Gelenberg AJ, Lydiard RB: Alpha^2-adrenergic receptor sensitivity in depressed patients: Relation between ^{3}H-yohimbine binding to platelet membranes and clonidine-induced hypotension. *Biol Psychiat* 1989; 25: 382–392.

Wright AF, Crichton DN, Loudon JB, Morton JEN, Steel CM: β-adrenoreceptor binding defects in cell lines from families with manic-depressive disorder. *Ann Human Genet* 1984; 48: 210–214.

Ziegler MG, Milano AJ, Hull E: In Lake R, Ziegler MG (Eds.) *The Catecholamines in Psychiatric and Neurological Disorders.* Boston: Butterworth Publishers, 1985, 37–53.

Zohar J, Bannet J, Drummer D, Fisch R, Epstein RP, Belmaker RH: The responses of lymphocyte beta-adrenergic receptors to chronic propranolol treatment in depressed patients, schizophrenic patients, and normal controls. *Biol Psychiat* 1983; 18: 553–560.

Chapter 2

SEROTONIN AND MAJOR DEPRESSION

Kevin Malone and J. John Mann

Introduction

In 1948, a serum tonic factor released from platelets during blood clotting was first isolated, identified as a monoamine, 5-hydroxytryptamine, synthesized by Rappaport, Green, and Page (1968), and named serotonin (Page, 1968). By 1953, serotonin had been isolated from gut smooth muscle and from extracts of brain. In the 1960s a role for serotonin in the etiology of major depression was proposed. Reserpine, a drug that triggered depression in some patients, was found to deplete serotonin in brain (Pletscher, Shore & Brodie, 1955). Monoamine oxidase inhibitors (MAOI) increased serotonin content and were found to be antidepressants (Pletscher, 1991). The serotonin hypothesis of depression was developed, (Coppen, 1969; Lapin & Oxenkrug, 1969; Coppen, Prange, & Whybrow, 1972), and proposed that an increased vulnerability to either depression or mania was related to decreased serotonergic activity. Over the last 25 years, studies on blood platelets, plasma, and cerebrospinal fluid (CSF) of depressed patients reported serotonergic abnormalities. The last decade of clinical research on serotonin dysfunction in mental illness has utilized advances in neuroimaging, neuroendocrinology, neuropsychopharmacology, and molecular biology, as well as refinements in the classification of mental

Kevin Malone and J. John Mann • Laboratories of Neuropharmacology, Western Psychiatric Institute and Clinic, Department of Psychiatry, University of Pittsburgh, Pittsburgh, Pennsylvania 15213.

Biology of Depressive Disorders, Part A: A Systems Perspective, edited by J. John Mann and David J. Kupfer. Plenum Press, New York, 1993.

disorders. There has been a rapid increase in knowledge about the serotonergic system, which requires a reconsideration of the serotonin hypothesis of depression.

Serotonin has a significant but by no means exclusive central nervous system (CNS) role in the regulation of a variety of core biological functions such as sleep, appetite, circadian rhythm, and the regulation of psychological and behavioral dimensions such as affect, aggression, and anxiety (Meltzer & Lowy, 1987). A role for serotonin has been proposed in affective disorders, schizophrenia, anxiety disorders, eating disorders, and obsessive-compulsive disorder. State versus trait dependent neurobiological correlates, methods for studying serotonin in major depression, and the clinical implications of serotonergic system research receive special attention in this chapter.

Serotonergic system

Overview

Serotonin is one of several neurotransmitters within the CNS whose molecular structure, mechanism of action, and psychobiological effects have been intensely studied over many decades. Techniques in molecular biology, neuropharmacology, and physiology have been applied to the research of serotonin. Multiple serotonin receptors have already been recognized and cloned, and the likelihood is that there are other serotonin receptors that have yet to be discovered. Primary dysfunction of the serotonergic system, or changes secondary to dysfunction of other neurotransmitter systems, can be involved with multiple psychiatric disorders. How one transmitter system can be involved with multiple psychiatric disorders, and multiple transmitter systems appear involved with one psychiatric disorder such as depression, is a question that requires the effort of basic scientific and clinical researchers in tandem, and not separately. The multiplicity of serotonin receptors and the innervation by serotonin nerve terminals of numerous brain regions provides the biochemical specificity and the neuroanatomical basis for such a wide range of functions for serotonin as a neurotransmitter. At the same time there is also the possibility for a variety of dysfunctional or pathological states involving some aspect of the serotonin system.

Neuroanatomy of the Serotonergic System

The serotonin system consists of a morphologically diverse group of neurons (for review see Törk, 1990). These cell bodies are located in the brainstem raphe nuclei and regions of the reticular formation, and have complex axonal systems that innervate virtually all regions of the cortical nervous system. These axonal systems are particularly dense in the cerebral cortex, limbic structures, basal ganglia, many regions of the brain stem, and the gray matter of the spinal cord.

The serotonergic system consists of two distinct subdivisions: a rostral division with cell bodies in the midbrain and rostral pons, providing projections to the forebrain; and a caudal division located primarily in the medulla oblongata with major descending projections directed principally to the spinal cord. Advances in immunohistochemical analysis have yielded a greater understanding of the neuroanatomy of the serotonergic system. Nonetheless few data are still available on some of the areas of serotonergic cell bodies and their projections in the brain. The forebrain appears to be innervated by three types of serotonergic axons. The most common is thin, with small (less than 1-mm diameter) varicosities that are fusiform in shape (D-fibers). These axons are by far the most numerous and extend profusely over large areas of the forebrain. A second type of fiber has large round or oval varicosities which are on very thin axons. A third fiber, which is relatively thick (about 1 mm) and nonvaricose (M-fiber), is also found in smaller numbers in or close to the major tract systems from which serotonergic fibers emerge. The fine varicose axons systems (D-system) arises from the dorsal raphe nucleus with fibers that branch profusely in their target areas. The second system (M-fibers) along with the D-fiber system provides the dual serotonergic system innervating the forebrain. This second system—also called basket axons (M-fibers)—arises from the median raphe nucleus with thick nonvaricose axons, giving rise to branches with the characteristic beaded varicose axons.

The spinal cord receives a strong serotonergic innervation. The strongest serotonergic innervation is in the region of the dorsal horn especially lamina I, the ventral horn motor nuclei (laminae VIII and IX) and in the thoracic cord, the intermediolateral column. A second major target of the serotonergic projections in the spinal cord are the motor neurons, and a high density of serotonergic innervation can be observed in regions of lamina IX. The serotonergic axons innervating the ventral horn are the termination of a second major descending system, which arises primarily from the raphe obscuras and raphe pallidus nuclei. Many neurons of these raphe nuclei have been shown to colocalize serotonin with substance P or thyrotropin-releasing hormone. Coexistence of serotonin with peptides in the ventral horn can also be observed.

Notwithstanding that the most significant cells of origin of serotonergic effects to the spinal cord are in the medulla raphe nuclei, additional minor projections arise from scattered cells of the B9 group in the pontomesencephalic reticular formation and from cells located within the medial longitudinal fasciculus.

The Serotonin Receptor Family

Multiple receptor subtypes were described by Peroutka and Snyder (1979) and defined by radioligand binding studies in brain membranes. As a result of considerable subsequent research, the serotonin family of receptors can thus far be divided into three main classes: 5-HT_1, 5-HT_2, and 5-HT_3. 5-HT_1 and 5-HT_2 classes include receptor subtypes: 5-HT_{1A}, 5-HT_{1B}, 5-HT_{1D}, and 5-HT_{1C}, 5-HT_{2A}, 5-HT_{2B}, respectively. The precise role of each of these receptor subtypes in the etiology of psychiatric disorders remains unclear.

5-HT_{1A} Receptors. The development of agents active at the 5-HT_{1A} site, ranging from selective agonists and partial agonists to selective antagonists, has allowed study of 5-HT_{1A} receptors in the pathophysiology, and potential treatment of several psychiatric disorders. Evidence supporting the role for 5-HT_{1A} receptors in the pathophysiology of anxiety comes from the observation that benzodiazepines decrease serotonergic neurotransmission (Wise, Berger & Stein, 1972; Practt, Jenner, & Marsden, 1985; Trulson, Preussler, and Howell, 1982; Laurent, Margold, & Hunkel, 1983), and that selective 5-HT_{1A} agonists such as buspirone and other partial agonists have anxiolytic effects. In addition, gepirone, ipsapirone SR-3997 (Feighner & Boyer, 1989) and (^{3}H-8-hydroxy-2-[*n*-dipropylamino] tetralin, ^{3}H-8-OH-DPAT) 8-OH-DPAT have also been evaluated as anxiolytics in clinical trials. Autoradiographic studies reveal a heterogeneous distribution of binding in the brain. The highest densities of 5-HT_{1A} sites are found in the CA1 region and dentate gyrus of the hippocampus and in the raphe nuclei (Hoyer, Pazos, & Palacios, 1986). Multiple correlations have been established between the pharmacological characteristics of this binding site and specific effects of serotonin (Peroutka, 1988). Electrophysiologic studies have suggested that the serotonin autoreceptor located on serotonergic cell bodies (somatodendritic) in the dorsal raphe is a 5-HT_{1A} site, and that this receptor may partly mediate the inhibitory effect of serotonin on its own release.

A role for 5-HT_{1A} receptors in depression is suggested by the following observations. Chronic antidepressant treatment in rats results in a reduction in the number of 5-HT_{1A} receptors as well as 5-HT_2 receptors. Placebo-controlled treatment trials in patients with depression have found the 5-HT_{1A} agonists buspirone and gepirone more effective than placebo in reducing scores on the Hamilton Rating Scales for depression. In addition to the treatment of depression and anxiety disorders, 5-HT_{1A} receptor agonists may also be beneficial in the management of alcoholism (Tollefson, 1989; Bruno, 1989) and the regulation of appetite and food intake, (Doods, Boddeke, & Kalkman, 1988; Di Francesco, Petty, & Fozard, 1988; Marsden, Sleight, & Fone, 1989; Dourish, Hutson, & Curson, 1985; Gilbert & Dourish, 1987).

5-HT_{1B} Receptors. The 5-HT_{1B} receptor is species specific and is not found in human brain. In rat and mouse brain this receptor site has high affinity for serotonin, and has relatively low affinity for *d*-lysergic acid diethylamide and 8-OH-DPAT. 5-HT_{1B} activation may increase or decrease anxiety in rat models of anxiety (Broekkamp, Berendsen, & Jenk, 1989).

5-HT_{1D} Receptors. The 5-HT_{1D} site was described in 1987, and is found in human brain as well as bovine brain, but not in rat brain. It displays nanomolar affinity for 5-HT and 5 carboxyamidotryptamine (5CT). It has been shown (Peroutka, Schmidt, Sleight, & Harrington, 1990; Waeber, Dietl, & Hoyer, 1989) to be located throughout the human brain. Its presence on or near cerebral blood vessels has implicated the 5-HT_{1D} receptor in the etiology of migraine. Sumatriptan, a potent and selective 5-HT_{1D} receptor agonist, has been shown to cause vasoconstriction of cerebral arteries. Sumatriptan has also been shown to be effective in the acute treatment of migraine and to cause minimal side effects possibly owing to its selectivity for 5-HT_{1D} receptors (Ferrari, Wilkinson, Hirt, Lataste, Notter, & ICS 205-930 Migraine Study Group, 1991; Peroutka & McCarthy, 1989).

5-HT_{1C} Receptors. Because of the 5-HT_{1C} receptor's similarities to the 5-HT_2 receptor, the two receptor types are generally grouped together. It is related to the 5-HT_2 receptor, particularly in terms of a common second messenger system involving phophoinositide hydrolysis as well as molecular similarities. In contrast, other 5-HT_1 receptor subtypes are linked to the adenylate cyclase second messenger system. No 5-HT_{1C} selective agents are available at present, nor is any role currently identified for 5-HT_{1C} in psychiatric disorders. 5-HT_{1C} receptors are concentrated on choroid plexus tissue, and probably play a role in CSF formation. Lower density 5-HT_{1C} binding has also been found in other brain regions.

5-HT_2 Receptors. Radioligands including buspirone, LSD, mianserin, ketanserin, mesulergine and *N*-methyl-LSD have been used to label 5-

HT_2 sites (Pritchett, Bach, & Wozny, 1988). 5-HT_2 subtypes have been suggested on the basis of the binding of the radiolabeled derivative of the serotonin agonist DOB (4-bromo -2,5, dimethoxy amphetamine). The 5-HT_{2A} site exists in human and rat brain and has high affinity for ^3H ketanserin. The lower affinity component of ^3H ketanserin binding in the rat brain has been designated the 5-HT_{2B} site, and can also be detected in human and bovine brains (Pierce & Peroutka, 1989).

A role has been suggested for the 5-HT_2 receptor in the pathogenesis of major depression. Tricyclic antidepressant medication decreases the number of 5-HT_2 binding sites in the rat cerebral cortex (Peroutka & Snyder, 1980), and yet repeated electroconvulsive shock (ECS) as a model for ECT, increases the number of 5-HT_2 receptors. Nevertheless, the net effect of both antidepressant drugs and ECT may be the same, namely, an increase in overall serotonergic transmission, but the mechanisms involved may differ (Charney, Menkes, & Heninger, 1981; Brown & Mann, 1985). Both acute and chronic treatment with the 5-HT_2 antagonist mianserin paradoxically decrease 5-HT_2 receptor density in rat brain, especially in the frontal cortex (Blackshear & Sanders-Bush, 1982). Antidepressants may work presynaptically by increasing serotonin availability to postsynaptic receptors to such a degree that they produce a countertherapeutic compensatory downregulation of 5-HT_2 receptors. In contrast, ECS, and perhaps ECT, increase 5-HT_2 receptors possibly because of a direct postsynaptic effect.

5-HT_3 Receptors. The 5-HT_3 family of receptors is largely located in peripheral (non-CNS) tissue (Peroutka & Hamik, 1988; Watling, Aspley, & Swain). Within the CNS, these receptors have been shown to cause release of dopamine from rat striated slices (Barnes, Barnes, & Costall, 1989) and an inhibition of acetylcholine release from rat cortical slices (Bladina, Goldarb, & Green, 1988). Ondansetron, a 5-HT_3 receptor antagonist, inhibits dopamine-mediated hyperactivity in rats (Costall, Domeney, & Naylor, 1987). Thus, among other effects, 5-HT_3 receptors are involved in modulating other neurotransmitter systems.

Serotonergic Abnormalities in Major Depression

Serotonergic dysfunction has been linked to depression at a syndromal level as well as to several key components of depression such as appetite (Blundell, 1984), mood, anxiety, suicide risk, sexual function, sleep, and circadian rhythm changes (Meltzer et al., 1987). Serotonergic abnormalities in blood platelets, plasma, and CSF of depressed patients, as well as postmortem brain specimens from suicides and nonsuicides who had been depressed, have been reviewed in detail in the past decade (Mann, McBride, & Stanley, 1986; Mann, Arango, Marzuk, Theccanat, & Reis, 1989; Mann, Arango, & Underwood, 1990). It appears likely that diminished 5-HT function may be characteristic of at least a subgroup of patients with Major Depressive Disorders, and that a tonic serotonergic effect in harmony with other neurotransmitter systems may be necessary to maintain normal mood. It is worth focusing here on the results of studies utilizing techniques currently available for investigating central nervous system serotonergic activity in depressed patients.

Studies of 5-Hydroxyindoleacetic Acid in Lumbar Cerebrospinal Fluid

The metabolite of serotonin, 5-hydroxyindoleacetic acid (5-HIAA), has been studied in lumbar cerebrospinal fluid (CSF) as an index of central serotonin turnover in patients with major depression (Åsberg, Thorén, and Träskman 1976; Stanley, Träskman-Bendz, & Dorovnitis, 1985). Variables influencing lumbar CSF 5-HIAA concentration such as age, sex, body height, circadian and seasonal rhythms, drug intake, medical conditions, diet, physical activity, lumbar puncture (LP) intervertebral space location, subject position, amount of CSF drawn, handling and storage of CSF, analytic methods, and within-subject variables such as mental state and characterological variables have confounded interpretation of data from CSF studies (Åsberg, 1989). The hypothesis that presynaptic central 5-HT activity is diminished in a subgroup of patients with major depression because of reduced CSF 5-HIAA concentration is supported by several but not all studies. Ågren (1980) and Åsberg (1976, 1984) have reported that subgroups of drug-free depressed patients have reduced CSF 5-HIAA concentrations. Gibbons and Davis (1986) analyzed data on depressed patients from five research groups and found that CSF 5-HIAA levels corresponded to a bimodal distribution. In two large independent samples of depressed patients, approximately 35% of depressed patients in either sample fell into a "low" CSF 5-HIAA mode and had comparable levels of CSF 5-HIAA. Significantly more violent suicides or suicide attempts were found in the low CSF 5-HIAA group; a finding that has not been reported for bipolar disorder (Åsberg, Schalling, & Träskman-Bendz, 1987). Diminished CSF 5-HIAA values are not specific for depressive disorders or suicidal behavior. Studies have also found a correlation between externally directed aggressive/impulsive behavior and CSF 5-HIAA (Linnoila, Virkunen, & Scheinin, 1983; Lidberg, Åsberg, & Sunquist-Stensman, 1984; Lidberg, Tuck, & Åsberg, 1985).

Serotonin-Specific Neuroendocrine Tests

Manipulating levels of monoamines, such as serotonin, which are involved in the regulation of the secretion of hypothalamic factors such as prolactin-releasing factor (PRF) that in turn control the release of the hormone prolactin, offers the potential advantage of examining the physiologic responsivity of the brain, or at least the hypothalamic serotonin system (McBride, Tierney, DeMeo, Chen, & Mann, 1990).

Both intravenously administered tryptophan (a 5-HT precursor), and orally administered fenfluramine (a 5-HT releaser/uptake inhibitor), stimulate prolactin release, and have been used in the study of patients with major depression. These agents both assess the "net" (pre- and postsynaptic) aspects of central 5-HT activity in the brain but differ because fenfluramine releases endogenous stores of serotonin, in contrast to tryptophan, which stimulates 5-HT synthesis.

Blunted prolactin responses to these central serotonergic probes have been reported in patients with major depression (Heninger, Charney, & Sternberg, 1984; Meltzer, Perline, Tricou, Lowy, & Robertson, 1984; Siever, Uhde, Jimerson, Lake, Siberman, Post, & Murphy, 1984). There is evidence of an increased cortisol response to 5-hydroxyhyptophan (5-HTP) (Meltzer, Nash, & Ohmori, 1987) that appears to correlate with a history of suicide attempts, and has been suggested as being due to an increase in receptor number, based on findings in the brain of suicide victims. Reduced prolactin responses do not appear to correlate with severity of the depressive state, raising the possibility that these measures may be trait dependent, or correlate with other aspects of depression such as associated psychopathology, for example, suicide risk (Coccaro, Siever, Owen, & Davis, 1990; Mann, 1991). Recently, attenuated growth hormone (GH) and prolactin responses to intravenous L-tryptophan (LTP) in depression, which normalized with clinical recovery, have been reported (Upadhyaya, Pennell, Cowen, & Deakin, 1991). However, results were different for males and females and the number of patients was small.

Cortisol hypersecretion (Sachar, Hellman, Fukushima, & Gallagher, 1970) and resistance to dexamethasone suppression (Carroll, Feinberg, Greden, Tarika, Albala, Haskett, James, Kronfol, Lohr, Steiner, de Vigne, & Young, 1981) are the best established state-dependent biological changes in depression. Unfortunately, these and other neuroendocrine abnormalities in depression, such as blunted response of thyroid-stimulating hormone (TSH) to thyrotropic-releasing hormone (TRH), cannot be readily linked to a specific neurotransmitter system, since several neurotransmitters are involved in their regulation. Current research is directed toward elucidating specific monoamine abnormalities of depression. Thus far, the conclusion from neuroendocrine and CSF studies in depressed patients (Stanley et al., 1985) supports the hypothesis that diminished central serotonin function may be characteristic of only a subgroup of patients with major depression.

Postmortem Studies

Postmortem studies have measured concentrations of serotonin, 5-HIAA (Mann, Arango, Marzuk, Theccanat, & Reis) and serotonin receptors (Stanley & Mann, 1983; Mann, Stanley, & McBride, 1986; Mann, Arango, Underwood, Baird, & McBride, 1990) in suicide victims, and regardless of psychiatric diagnosis, have found reduced serotonin and/or 5-HIAA in the brain stem but not cerebral cortex. Ferrier, McKeith, Cross, Perry, Candy, and Perry (1986) reported a trend toward significant increase in 5-HT_2 receptors and a decrease in 5-HIAA particularly in those patients with evidence of a major depressive episode in the month before death compared with controls. Therefore, suicidality, as a symptom of either depression, a characterological disorder, or schizophrenia, appears to be associated with dysfunction of serotonergic activity.

Measurement of Peripheral Serotonergic Indices in Major Depression: Plasma Tryptophan

The recent report by Delgado et al. (Delgado, Charney, Price, Aghajanian, Andes, & Heninger, 1990) and earlier reports by Young, Smith, Pihl, and Ervin (1985) that some patients may suffer depressive relapse following a tryptophan (TRP) deficient diet, has lent further support to the hypothesis that mood states can be significantly altered by experimentally induced decreases in 5-HT function. This possibility had originally been suggested by Shopsin (Shopsin, Friedman, & Gershon, 1976), who reported that *p*-chlorophenylalanine (PCPA), which inhibits 5-HT synthesis, rapidly induced a recurrence of depressive symptoms in antidepressant-treated, recovered depressed patients. There is conflicting evidence concerning the ratio of plasma TRP to large neutral amino acids (LNAA), which compete with TRP for entry into the brain. Some studies have reported a decrease in the plasma TRP/LNAA ratio in depressed patients (Cowen, Charig, Fraser, & Elliott, 1987; Moller, Honore, & Lavsen, 1983; DeMyer, Shea, Hendrie, & Yoshimara, 1981; Joseph, Breweton, Reus, & Stebbus, 1984; Maes, DeRuyer, Claes, Bosma, & Surg, 1987), while others have not found this (Moller, Odiem, Kirk,

Bjerre, Fog-Moller, & Knudsen, 1985). Other investigators reported that a low TRP/LNAA ratio predicts clinical antidepressant response to treatment with L-TRP (Moller, Kirk, & Honore 1980) and amitriptyline (Moller, Honore, & Larsen, 1983) but not nortriptyline (Moller et al., 1985).

Plasma TRP studies following ingestion of large oral doses of TRP have also yielded conflicting results in depressed patients. Some have reported significantly lower TRP concentrations in major depression than in controls (Smith & Stromgren, 1981) which persisted after treatment, while other studies have failed to replicate this (Hoes, Loeffen, & Vree, 1981). A recent positron emission tomography study reported a significantly lower uptake of [^{11}C] 5-HTP across the blood–brain barrier in depressed patients compared with controls, irrespective of phase of illness (Ågren, Reiboring, Hartuig, Tedroff, Bjurling, Hornfeldt, Andresson, Lundquist, & Langstrom, 1991). These findings support earlier suggestions that the transport of 5-HTP across the blood–brain barrier is compromised in major depression.

Administration of TRP alone to depressed patients occasionally has an antidepressant effect. More striking is the facilitatory antidepressant effect of TRP in association with MAO inhibitors, which again would lend support to the serotonin hypothesis (Glassman & Platman, 1969). These findings need to be replicated. Lithium augmentation of an antidepressant response to antidepressant drugs such as imipramine has been suggested by DeMontigny (1984) to be the result of enhanced serotonin turnover. Such studies indicate that normal serotonergic function is a necessary although not sufficient condition for antidepressant efficacy for some antidepressant drugs.

Platelet Serotonin Uptake and Tritiated Imipramine Binding in Major Depression. Paasonen (1973) proposed the platelet as a model for the serotonin neuron because it appeared to share many, but not all of the characteristics of central serotonergic neurons. There have been many studies in the past two decades that have examined indices of platelet serotonin function in depression. Meltzer et al., (1987) reviewed reduced platelet serotonin uptake reports from studies of depressed medication-free patients. Several studies reported reduced platelet serotonin uptake with the uptake being most reduced in bipolar depressives (Tuomisto 1979; Kaplan & Mann, 1982; Meltzer, Arora, Baber, & Tricou, 1981). These studies also reported that reduced serotonin uptake into platelets seemed to be related to reduced numbers of 5-HT uptake sites (V_{max}), rather than reduced affinity of the recognition site (K_m). Furthermore, V_{max} does not change upon recovery in unipolar depressed patients (Coppen, Swade, & Wood, 1978), (Scott, Reading, & Loudon, 1979), or manic patients (Meltzer et al., 1981). The V_{max} does not appear to correlate with severity of depression. Normalization of V_{max} reported in manic patients (Goodnick & Meltzer, 1984) may be due to a persistent effect of chronic lithium treatment which can be associated with an increase in V_{max} values (Coppen et al., 1980).

Several research groups have reported a significant increase in the numbers of platelet 5-HT_2 binding sites in depressed patients who had made serious suicidal attempts (McBride, Brown, DeMeo, Keilp, Stanley, & Mann, 1987; Pandey, Pandey Janicak, Marks, & Davis, 1990; Biegon, Srinspoon, Blumenfeld, Bleich, Apter, & Mester, 1990) although this finding has not been replicated in another study (Arora & Meltzer, 1989).

Studies of platelet tritiated-imipramine binding have also yielded varying results (review in Briley, 1985) with some studies reporting reduced platelet tritiated-imipramine binding in depressed patients and others reporting negative findings. It is likely that the discrepancies between platelet study results are related to many factors: differences in subtypes and severity of depression, varying washout periods of previous drug treatments, duration of illness, circadian, and seasonal variations, technical variations in methods of platelet isolation, and determination of uptake and binding parameters.

Reduced imipramine binding may be characteristic of bipolar depressive patients, or of unipolar depressives with a first-degree family history of depression (Lewis & McChesney, 1985). Unlike the reduced V_{max} for platelet serotonin uptake, the number of imipramine binding sites (B_{max}) appears to increase with recovery from depression, as well as to increase after chronic treatment with some antidepressant drugs (Arora & Meltzer, 1988).

State- Versus Trait-Dependent Biological Correlates of Depression

With the large body of biological research that has explored the serotonergic system, it is disappointing how few data are available regarding the distinction between state- and trait-dependent serotonergic abnormalities in major depression. These differences have been alluded to above and are summarized in Table 1. Identification of neurochemical traits may facilitate genetic and family studies as well as detection of at risk individuals who are euthymic. Overall, most serotonin correlates of depression are trait dependent. Because they are generally altered by antidepressant treatment, data from untreated euthymic patients are essential in order to validate this conclusion.

Table 1. State- and Trait-Dependent Correlates of Serotonergic Dysfunction in Major Depression

	State	Trait
CSF		
5-HIAA[a]	+	+
Neuroendocrine challenge		
dl fenfluramine[b]		+
IV TRP[c,d]	?+	+
Peripheral parameters		
Platelet serotonin uptake[e]		+
^{3}H imipramine binding[f]	+	+
Plasma tryptophan[g]		?+

[a] Träskman-Bendz et al. (1984).
[b] Siever et al. (1984).
[c] Heninger et al. (1984).
[d] Upadhyaya et al. (1991).
[e] McBride et al. (1987).
[f] Arora et al. (1988).
[g] Smith et al. (1981).

In conceptualizing a "serotonin trait" Depue and Spoont (1986) drew on both animal and human data in serotonin studies to offer an elaborate framework to examine behavioral facilitation and inhibition systems, and the behavioral response to serotonin manipulation. They concluded that the serotonin influence is not specific to any particular response system.

It is not known how psychosocial factors, such as loss through bereavement, or other significant life events or stress may adversely affect human brain neurochemistry. Subtle and transient alterations in neuropeptide neurotransmitter activity with the CNS in response to significant life events may have a cascade affect in several neurotransmitter systems, and in vulnerable/neurochemical trait-deficient individuals, may precipitate significant alteration in behavior and mood, which may develop into a depressive disorder. Serotonin may have the role of a vulnerability factor or trait, where deficiency or dysfunction would not manifest psychological symptomatology, except as part of a neurochemical cascade in response to environmental or other stressors that may crystallize into overt symptomatology such as aggressive or a depressive episode, or suicidal behavior complicating a psychiatric illness. Ehlers, Frank, and Kupfer (1988) have proposed that social zeitgebers may alter biological rhythms and this may prove a way forward in understanding the interdependence of brain neurochemistry, biological rhythms, and the environment, where serotonin may have a significant role.

Given that serotonergic activity may represent a biological trait, research continues into the role of serotonin in behavior traits such as aggressivity and impulsivity, to identify correlations between biological measurements and personality traits. Such behavioral traits may be the phenotypic consequence of constitutional differences in levels of serotonin activity.

ANTIDEPRESSANT TREATMENT AND THE SEROTONERGIC SYSTEM IN MAJOR DEPRESSION

Whether the serotonergic dysfunction hypothesis is modified or indeed rejected by future research, the framework has been valuable for the development of antidepressant medications. Various classes of antidepressant treatments have been shown to enhance 5-HT transmission. (Blier, deMontigny, & Chaput, 1990; outlined in Table 2). The effect of these treatments is not confined to serotonin neurotransmission, nor are their mechanisms of enhancement of the serotonin system similar. However, the time scale of serotonin enhancement is often similar and in some cases correlates with the time course of therapeutic effect. However, depressed patients do not always respond to antidepressant treatment despite increased central serotonin neurotransmission (Charney, Heninger, & Sternberg, 1984), suggesting that this effect may be a necessary, but not sufficient, effect for an antidepressant response. The fact that a subgroup of clinically effective antidepressant drugs selectively influence only the serotonin system is further evidence for a role for the

Table 2. Effects of Antidepressant Treatments on Serotonin Neuronal Responsiveness in Human Subjects[a]

Treatment	Acute treatment (1–2 days)	Short-term treatment (10–14 days)	Long-term treatment (over 14 days)
Serotonin reuptake blockers (clomipramine, fluvoxamine, fluoxetine)	↓	Normalize	↑
MAOIs (tranylcypromine, phenelzine)	↓	Normalize	ì
Tricyclic antidepressants (desipramine, amitriptyline)	No change	No change	↑
$5\text{-}HT_{1A}$ Receptors agonists (buspirone gepirone)	↓	Normalize	ì
Electroconvulsive shock	Not determined	Normalize	↑

[a] Adapted from Blier et al. (1990).

serotonin system in the pathogenesis of depression. There is no conclusive evidence that any of the selective serotonergic agents are more effective than the less selective agents in treating depression, other than in those patients who cannot tolerate some of the side effects of the less specific antidepressants. Therefore, while there is a clinical role for pharmacologically manipulating central serotonergic activity in the treatment of depression, precisely which patients will benefit from this treatment, and why, remains unanswered.

Conclusion

Studies of the serotonergic system in depressive disorders are identifying a new series of research goals. Incorporating new data on serotonin receptors may partly explain how one transmitter system can have a multitude of effects. Determining the relative contribution of the serotonin system to depressive disorders, as opposed to behavioral components such as suicide risk, anxiety, appetite disorders, and sleep, may clarify a subgroup of serotonin depression. The relationship of serotonin function to behavioral traits or personality disorders may prove to be as important as its relationship to a depressive episode. Finally, understanding how antidepressant drugs and ECT affect the serotonin system may permit the development of more specific and effective antidepressant medications.

Acknowledgments

This work was supported by a Fogarty Fellowship Award No. 1FO5TW04538-01 to Dr. Malone and MH46745 to Dr. Mann. Ms. Jean Williams and Ms. Sally Evans provided secretarial assistance in the preparation of the manuscript.

References

Arora RC, Meltzer HY: Effect of desipramine treatment on ^{3}H-imipramine binding in the blood platelets of depressed patients. *Biol Psychiat* 1988; 23: 397–404.

Arora RC, Meltzer HY: Serotonergic measures in the brains of suicide victims: 5-HT2 binding sites in the frontal cortex of suicide victims and control subjects. *Am J Psychiat* 1989; 146: 730–736.

Ågren H: Symptom patterns in unipolar and bipolar depression correlating with monoamine metabolites in the cerebrospinal fluid. II. Suicide. *Psychiat Res* 1980; 3: 225–236.

Ågren H, Reibring L, Hartvig P, Tedroff J, Bjurling P, Hörnfeldt K, Andersson Y, Lundqvist H, Långström B: Low brain uptake of L-[^{11}C]5-hydroxytryptophan in major depression: A positron emission tomography study on patients and health volunteers. *Acta Psychiat Scand* 1991; 83: 449–455.

Åsberg M: Neurotransmitter monoamine metabolites in the cerebrospinal fluid as risk factors for suicidal behavior. In: Davidson L, Linnoila M (Eds.), *Report of the Secretary's Task Force On Youth Suicide,* Washington: U.S. Health & Human Services, 1989, 2-193–2-212.

Åsberg M, Bertilsson L, Martensson B, Scalia Tomba GP, Thorén P, Träskman-Bendz L: CSF monoamine metabolites in melancholia. *Acta Psychiat Scand* 1984; 69: 201–219.

Åsberg M. Schalling D, Träskman-Bendz L, Wägner A: Psychobiology of suicide, impulsivity, and related phenomena. In: Meltzer HY (Ed.), *Psychopharmacology: The Third Generation of Progress,* New York: Raven Press, 1987, 655–668.

Åsberg M, Thorén P, Träskman-Bendz L, Bertilsson L, Ringberger V: "Serotonin depression"—A biochemical subgroup within the affective disorders? *Science,* 1976; 191: 478–480.

Barnes JM, Barnes NM, Costall B, Naylor RJ, Tyers MB: 5-HT_3 receptors mediate inhibition of acetylcholine release in cortical tissue. *Nature* 1989; 338: 762–763.

Biegon A, Grinspoon A, Blumenfeld B, Bleich A, Apter A, Mester R: Increased serotonin 5-HT2 receptor binding on blood platelets of suicidal men. *Psychopharmacol (Berlin)* 1990; 100: 165–167.

Blackshear MA, Sanders-Bush E: Serotonin receptor sensitivity after acute and chronic treatment with mianserin. *J Pharmacol Exp Therapeut* 1982; 221: 303–308.

Bladina P, Goldfarb J, Green JP: Activation of a 5-HT_3 receptor releases dopamine from rat striatal slice. *Eur J Pharmacol* 1988; 155: 349–350.

Blier P, De Montigny C, Chaput Y: A role for the serotonin system in the mechanism of action of antidepressant treatments: Preclinical evidence. *J Clin Psychiat* 1990; 51: 4, 14–21.

Blundell JE: Serotonin and appetite. *Neuropharmacology* 1984; 23: 12B, 1537–1551.

Briley M: Imipramine binding: Its relationship with serotonin uptake and depression. In: Green AR (Ed.), *Neuropharmacology of Serotonin,* Oxford, UK: Oxford University Press, 1985, 50–78.

Broekkamp CLE, Berendsen HHG, Jenck F, Van Delft AML: Animal models for anxiety and response to serotonergic drugs. *Psychopathology* 1989; 22: 1, 2–12.

Brown RP, Mann JJ: A clinical perspective on the role of neurotransmitters in mental disorders. *Hosp Commun Psychiat* 1985; 36: 141–150.

Bruno F: Buspirone in the treatment of alcoholic patients. *Psychopathology* 1989; 22: 1, 49–59.

Carroll BJ, Feinberg M, Greden JF, Tarika J, Albala AA, Haskett RF, James N McI, Fronfol Z, Lohr N, Steiner M, de Vigne JP, Young E: A specific laboratory test for the diagnosis of melancholia. Standardization, validation, and clinical utility. *Arch Gen Psychiat* 1981; 38: 15–22.

Charney DS, Heninger GR, Sternberg DE: Serotonin function and mechanism of action of antidepressant treatment. *Arch Gen Psychiat* 1984; 41: 359–365.

Charney DS, Menkes DB, Heninger GR: Receptor sensitivity and the mechanism of action of antidepressant treatment. Implications for the etiology and therapy of depression. *Arch Gen Psychiat* 1981; 38: 1160–1180.

Coccaro EF, Siever LJ, Owen KR, Davis KL: Serotonin in mood and personality disorders. In: Coccaro EF, Murphy DL (Eds.), *Serotonin in Major Psychiatric Disorders.* Washington: American Psychiatric Press, 1990, 71–97.

Coppen A, Prange AJ Jr, Hill C, Whybrow PC, Noguera R: Abnormalities of indoleamines in affective disorders. *Arch Gen Psychiat* 1972; 26: 474–478.

Coppen A, Swade C, Wood K: Lithium restores abnormal platelet 5-HT transport in patients with affective disorders. *Br J Psychiat* 1980; 136: 235–238.

Coppen AJ: Biochemical aspects of depression. *Int. Psychiat Clin* 1969; 6: 53–81.

Costall B, Domeney AM, Naylor RJ, Tyers MB: Effects of the 5-HT_3 receptor antagonist, GR38032F, on raised dopaminergic activity in the mesolimbic system of the rat and marmoset brain. *Br J Pharmacol* 1987; 92: 881–894.

Cowen PJ, Charig EM, Fraser S, Elliott JM: Platelet 5-HT receptor binding during depressive illness and tricyclic antidepressant treatment. *J Affect Disord* 1987; 13: 45–50.

De Montigny C, Blier P, Chaput Y: Electrophysiologically identified serotonin receptors in the rat CNS. Effect of antidepressant treatment. *Neuropharmacology* 1984; 23: 12B, 1511–1520.

Delgado PL, Charney DS, Price LH, Aghajanian GK, Landis H, Heniger GR: Serotonin function and the mechanism of antidepressant action. Reversal of antidepressant-induced remission by rapid depletion of plasma tryptophan. *Arch Gen Psychiat* 1990; 47: 411–418.

DeMyer MK, Shea PA, Hendrie HC, Yoshimura NN: Plasma tryptophan and five other amino acids in depressed and normal subjects. *Arch Gen Psychiat* 1981; 38: 642–646.

Depue RA, Spoont MR: Conceptualizing a serotonin trait. A behavioral dimension of constraint. In: Mann JJ, Stanley M (Eds.), *Psychobiology of Suicidal Behavior*, New York: Annals of New York Academy of Science, 1986, 47–62.

Di Francesco GF, Petty MA, Fozard JR: Antihypertensive effects of 8-hydroxy-2-(di-*n*-propylamino)tetralin (8-PH-DPAT) in conscious dogs. *Eur J Pharmacol* 1988; 147: 287–290.

Doods HN, Boddeke HWGM, Kalkman HO, Hoyer D, Mathy M-J, Van Zwieten PA: Central 5-HT_{1A} receptors and the mechanism of the central hypotensive effect of (+)8-OH-DPAT, DP-5-CT. R28935, and urapidil. *J Cardiovascular Pharmacol* 1988; 11: 432–437.

Dourish CT, Hutson PH, Curzon G: Low doses of the putative serotonin agonist 8-hydroxy-20(di-*n*-propylamino)tetralin(8-OH-DPAT) elicit feeding in the rat. *Psychopharmacology* 1985; 86: 197–204.

Ehlers CL, Frank E, Kupfer DJ: Social zeitgebers and biological rhythms. A unified approach to understanding the etiology of depression. *Arch Gen Psychiat* 1988; 45: 948–952.

Feigner JP, Boyer WF: Serotonin-1A anxiolytics: An overview. *Psychopathology* 1989; 22: 1, 21–26.

Ferrari MD, Wilkinson M, Hirt D, Lataste X, Notter M, ICS 205-930 Migraine Study Group: Efficacy of ICS 205-930, a novel 5-hydroxytryptamine$_3$ (5-HT$_3$) receptor antagonist, in the prevention of migraine attacks. A complex answer to a simple question. *Pain* 1991; 45: 283–291.

Ferrier IN, McKeith IG, Cross AJ, Perry EK, Candy JM, Perry RH: Postmortem neurochemical studies in depression. *Ann NY Acad Sci* 1986; 487: 128–142,

Gibbons RD, Davis JM: Consistent evidence for a biological subtype of depression characterized by low CSF monoamine levels. *Acta Psychiat Scand* 1986; 74: 8–12.

Gilbert F, Dourish CT: Effects of the novel anxiolytics gepirone, buspirone and ipsapirone on free feeding and on feeding induced by 8-OH-DPAT. *Psychopharmacology* 1987; 93: 349–352.

Gillepsie, DD, Manier DH, Sanders-Bush E, Sulser F: The serotonin/norepinephrine-link in brain. II. Role of serotonin in the regulation of *beta* adrenoceptors in the low agonist affinity conformation. *J Pharmacol Exp Therapeut* 1988; 244: 154–159.

Glassman AH, Platman SR: Potentiation of a monoamine oxidase inhibitor by tryptophan. *J Psychiat Res* 1969; 7: 83–88.

Goodnick PJ, Arora RC, Jackman H, Meltzer HY: Neurochemical changes during discontinuation of lithium prophylaxis. II. Alterations in platelet serotonin function. *Biol Psychiat* 1984; 19: 6, 891–898.

Heninger GR, Charney DS, Sternberg DE: Serotonergic function in depression. *Arch Gen Psychiat* 1984; 41: 398–402.

Hoes MJAJM, Loeffen T, Vree TB: Kinetics of L-tryptophan in depressive patients: A possible correlation between the plasma concentrations of L-tryptophan and some psychiatric rating scales. *Psychopharmacology* 1981; 75: 350–353.

Hoyer D, Srivatsa S, Pazos A, Engel G, Palacios JM: [^{125}I]LSD labels 5-HT$_{1C}$ recognition sites in pig choroid plexus membranes. Comparison with [^{3}H] mesulergine and [^{3}]5-HT binding. *Neurosci Lett* 1986; 69: 269–274.

Joseph MS, Brewerton TD, Reus VI, Stebbins GT: Plasma L-tryptophan/neutral amino acid ratio and dexamethasone suppression in depression. *Psychiat Res* 1984; 11: 185–192.

Kaplan RD, Mann JJ: Altered platelet serotonin uptake kinetics in schizophrenia and depression. *Life Sci* 1982; 3: 583–588.

Lapin IP, Oxenkrug GF: Intensification of the central serotoninergic processes as a possible determinant of the thymoleptic effect. *Lancet* 1969; i: 132–136.

Laurent JP, Mangold M, Humbel U, Haefely W: Reduction by two benzodiazepines and pentobarbitone of the multiunit activity in substantia nigra, hippocampus, nucleus locus coeruleus and nucleus raphé dorsalis of *encéphale isolé* rats. *Neuropharmacology* 1983; 22: 4, 501–511.

Lewis DA, McChesney C: Tritiated imipramine binding distinguishes among subtypes of depression. *Arch Gen Psychiat* 1985; 42: 485–488.

Lidberg L, Åsberg M, Sundqvist-Stensman UB: 5-hydroxyindoleacetic acid levels in attempted suicides who have killed their children. *Lancet* 1984; i: 928.

Lidberg L, Tuck JR, Åsberg M, Scalia Tomba GP, Bertilsson L: Homicide, suicide and CSF 5-HIAA. *Acta Psychiat Scand* 1985; 71: 230–236.

Linnoila M, Virkkunen M, Scheinin M, Nuutila A, Rimond R, Goodwin FK: Low cerebrospinal fluid 5-hydroxyindoleacetic acid concentration differentiates impulsive from non-impulsive violent behavior. *Life Sci* 1983; 33: 2609–2614.

Maes M, De Ruter M, Claes R, Bosma G, Suy E: The cortisol responses to 5-hydroxytryptophan, orally, in depressive inpatients. *J Affect Disord* 1987; 13: 23–30.

Mann JJ: Integration of neurobiology and psychopathology in a unified model of suicidal behavior. *Excerpta Med Int Cong Ser* 968, 1(1991), 114–117.

Mann JJ, Arango V, Marzuk P, Tierney H, Ernsberger P, Reis DJ: Postmortem studies of receptors in human brain of suicides. *Neurochemical Correlates of Affective Disorders and their Treatment,* Symposium, Copenhagen (Abstract), 1988.

Mann JJ, Arango V, Marzuk PM, Theccanat S, Reis DJ: Evidence for the 5-HT hypothesis of suicide: A review of post-mortem studies. *Br J Psychiat* 1989; 155 (Suppl.): 7–14.

Mann JJ, Arango V, Underwood MD: Serotonin and suicidal behavior. *Ann NY Acad Sci* 1990; 600: 476–485.

Mann JJ, Arango V, Underwood MD, Baird F, McBride PA: Neurochemical correlates of suicidal behavior: Involvement of serotonergic and non-serotonergic systems. *Pharmacol Toxicol* 1990; 66: 37–60.

Mann JJ, McBride PA, Stanley M: Aminergic receptor binding correlates of suicide: Methodological issues. *Psychopharmacol Bull* 1986; 22: 741–743.

Mann JJ, Stanley M, McBride PA, McEwen BS: Increased serotonin$_2$ and β-adrenergic receptor binding in the frontal cortices of suicide victims. *Arch Gen Psychiat* 1986; 43: 954–959.

Marsden CA, Sleight AJ, Fone KCF, Jonson JV, Crespi F, Martin KF, Garrett JC, Bennett GW: Functional identification of 5HT receptor subtypes. *Comp Biochem Physiol[A]: Comp Physiol* 1989; 93A: 1, 107–114.

McAlpine DE, Panser LA, Swanson DW, O'Fallon WM, Melton LJ, III: Suicide in Olmsted County, Minnesota, 1951 through 1985. *Mayo Clin Proc* 1990; 65: 13–22.

McBride PA, Brown RP, DeMeo M, Keilp J, Stanley M, Mann JJ, Polly M: Platelet 5-HT$_2$ receptors: Depression and suicide. *American Psychiatric Association, New Research,* Chicago, ILL, NR 157 (Abstract), 1987.

McBride PA, Tierney H, DeMeo M, Chen J-S, Mann JJ: Effects of age and gender on CNS serotonergic responsivity in normal adults. *Biol Psychiat* 1990; 27: 1143–1155.

Meltzer HY, Arora RC, Baber R, Tricou B-J: Serotonin uptake in blood platelets of psychiatric patients. *Arch Gen Psychiat* 1981; 38: 1322–1326.

Meltzer HY, Lowy MT: The serotonin hypothesis of depression. In: Meltzer HY (Ed.), *Psychopharmacology: The Third Generation of Progress*, New York: Raven Press, 1987, 513–526.

Meltzer HY, Nash JF, Ohmori T, Kregel L, Arora R: Neuroendocrine and biochemical studies of serotonin and dopamine in depression and suicide. *International Conference on New Directions in Affective Disorders*, S60 (Abstract), 1987.

Meltzer HY, Perline R, Tricou BJ, Lowy M, Robertson A: Effect of 5-hydroxytryptophan on serum cortisol levels in major affective disorders. II. Relation to suicide, psychosis and depressive symptoms. *Arch Gen Psychiat* 1984; 41: 379–387.

Moller SE, Honoré P, Larsen OB: Tryptophan and tyrosine ratios to neutral amino acids in endogenous depression. Relation to antidepressant response to amitriptyline and lithium + L-tryptophan. *J Affect Disord* 1983; 5: 67–79.

Moller SE, Odum K, Kirk L, Bjerre M, Fog-Moller F, Knudsen A: Plasma tyrosine/neutral amino acid ratio correlated with clinical response to nortriptyline in endogenously depressed patients. *J Affect Disord* 1985; 9: 223–229.

Moller SE, Kirk L, Honoré P: Relationship between plasma ratio of tryptophan to competing amino acids and the response to $_{L}$tryptophan treatment in endogenously depressed patients. *J Affect Disord* 1980; 2: 47–59.

Paasonen MK: Blood Platelets as a model for aminergic neurons. *Proc 5th Int Cong Pharmacol* 1973; 4: 328–342.

Page IH: *Serotonin*, Chicago: Chicago Year Book Medical Publishers, 1968.

Pandey GN, Pandey SC, Janicak PG, Marks RC, Davis JM: Platelet serotonin-2 receptor binding sites in depression and suicide. *Biol Psychiat* 1990; 28: 215–222.

Pazos A, Cortes R, Palacios JM: Quantitative autoradiographic mapping of serotonin receptors in the rat brain. II. Serotonin-2 receptors. *Brain Res* 1985; 346: 231–249.

Peroutka SJ: 5-Hydroxytryptamine receptor subtypes. *Ann Rev Neurosci* 1988; 11: 45–60.

Peroutka SJ, Hamik A: [^{3}H]Quipazine labels 5-HT_3 recognition sites in rat cortical membranes. *Eur J Pharmacol* 1988; 148: 297–299.

Peroutka SJ, McCarthy BG: Sumatriptan (GR 43175) interacts selectively with 5-HT_{1B} and 5-HT_{1D} binding sites. *Eur J Pharmacol* 1989; 163: 133–136.

Peroutka SJ, Schmidt AW, Sleight AJ, Harrington MA: Serotonin receptor "families" in the central nervous system: An overview. *Ann NY Acad Sci* 1990; 600: 104–113.

Peroutka SJ, Snyder SH: Multiple serotonin receptors: differential binding to ^{3}H-5-hydroxytryptamine, ^{3}H-lysergic acid diethylamide and ^{3}H-spiroperidol. *Mol Pharmacol* 1979; 16: 687–699.

Peroutka SJ, Snyder SH: Long-term antidepressant treatment decreases spiroperidol-labeled serotonin receptor binding. *Science* 1980; 210: 88–90.

Pierce PA, Peroutka SJ: Evidence for distinct 5-hydroxytryptamine$_2$ binding site subtypes in cortical membrane preparations. *J Neurochem* 1989; 52: 2, 656–658.

Pletscher A: The discovery of antidepressants: A winding path. *Experientia* 1991; 47: 4–8.

Pletscher A, Shore PA, Brodie BB: Serotonin release as a possible mechanism of reserpine action. *Science* 1955; 122: 374–375.

Pratt JA, Jenner P, Marsden CD: Comparison of the effects of benzodiazepines and other anticonvulsant drugs on synthesis and utilization of 5-HT in mouse brain. *Neuropharmacology* 1985; 24: 1, 59–68.

Pritchett DB, Bach AWJ, Wozny M, Taleb O, Toso RD, Shih JC, Seeburg PH: Structure and functional expression of cloned rat serotonin 5HT-2 receptor. *EMBO J* 1988; 7: 13, 4135–4140.

Sachar EJ, Hellman L, Fukushima DK, Gallagher TF: Cortisol production in depressive illness. *Arch Gen Psychiat* 1970; 23: 289–298.

Sadzot B, Frost JJ, Wagner HN,Jr: In vivo labeling of central benzodiazepine receptors with the partial inverse agonist [^{3}H]Ro 15-4513. *Brain Res* 1989; 491: 128–135.

Scott M, Reading HW, Loudon JB: Studies on human blood platelets in affective disorders. *Psychopharmacol (Berlin)* 1979; 60: 131–135.

Shopsin B, Friedman E, Gershon S: Parachlorophenylalanine reversal of tranylcypromine effects in depressed patients. *Arch Gen Psychiat* 1976; 33: 811–819.

Siever LJ, Murphy DL, Slater S, de la Vega E, Lipper S: Plasma prolactin changes following fenfluramine in depressed patients compared to controls: An evaluation of central serotonergic responsivity in depression. *Life Sci* 1984; 34: 1029–1039.

Siever LJ, Uhde TW, Jimerson DC, Lake CR, Silberman ER, Poast RM, Murphy DL: Differential inhibitory noradrenergic responses to clonidine in 25 depressed patients and 25 normal control subjects. *Am J Psychiat* 1984; 141: 733–741.

Smith DF, Strömgren LS: Influence of unilateral ECT on tryptophan metabolism in endogenous depression. *Pharmacopsychiatry* 1981; 14: 135–138.

Stanley M, Mann JJ: Increased serotonin-2 binding sites in frontal cortex of suicide victims. *Lancet* 1983; i: 214–216.

Stanley M, Träskman-Bendz L, Dorovini-Zis K: Correlations between aminergic metabolites simultaneously obtained from human CSF and brain. *Life Sci* 1985; 37: 1279–1286.

Stockmeier CA, Kellar KJ: Serotonin depletion unmasks serotonergic component of [^{3}H]dihydroalprenolol binding in rat brain. *Mol Pharmacol* 1989; 36: 903–911.

Tollefson GD: Serotonin and alcohol: Interrelationships. *Psychopathology* 1989; 221, 37–48.

Törk I: Anatomy of the serotonergic system. *Ann NY Acad Sci* 1990; 600: 9–35.

Träskman-Bendz L, Åsberg M, Bertilsson L, Thorén P: CSF monoamine metabolites of depressed patients during illness and after recovery. *Acta Psychiat Scand* 1984; 69: 333–342.

Tuomisto J, Tukiainen E, Ahlfors UG: Decreased uptake of 5-hydroxytryptamine in blood platelets from patients with endogenous depression. *Psychopharmacology* 1979; 65: 141–147.

Upadhyaya AK, Pennell I, Cowen PJ, Deakin JFW: Blunted growth hormone and prolactin responses to L-tryptophan in depression; A state-dependent abnormality. *J Affect Disord* 1991; 21: 213–218.

Waeber C, Dietl MM, Hoyer D, Palacios JM: 5-HT_1 receptors in the vertebrate brain. Regional distribution examined by autoradiography. *Naunyn Schmiedebergs Arch Pharmacol* 1989; 340: 486–494.

Watling KJ, Aspley S, Swain CJ, Saunders J: [^{3}H]Quaternised ICS 205-930 labels 5-HT_3 receptor binding sites in rat brain. *Eur J Pharmacol* 1988; 149: 397–398.

Wise CD, Berger BD, Stein L: Benzodiazepines: Anxiety-reducing activity by reduction of serotonin turnover in the brain. *Science* 1972; 177: 180–183.

Young SN, Smith SE, Pihl RO, Ervin FR: Tryptophan depletion causes a rapid lowering of mood normal males. *Psychopharmacology* 1985; 87: 173–177.

Chapter 3

CHOLINERGIC SUPERSENSITIVITY IN AFFECTIVE DISORDERS

Sanjay Dubé

Introduction

Disturbance of cholinergic regulation is implicated in the aetiology of affective disorders, with some indication that the defect may lie at the level of the central muscarinic cholinergic receptors. Although direct experimental evidence in support of this hypothesis is lacking because of the absence of an "ideal" model to study *m*-cholinoceptor sensitivity in psychiatric patients, there is a large body of indirect data that points toward a cholinergic supersensitivity in mood disorders. Some of the controversies in support of this theory may be better understood with the identification and cloning of the muscarinic cholinergic receptor subtypes (M_1-M_5), each with a unique pharmacological profile and distribution in different regions of the brain.

In 1972, Janowsky, Davis, and colleagues proposed the adrenergic-cholinergic hypothesis of affective disorders. According to this hypothesis, mania is a disorder of adrenergic predominance, relative to acetylcholine activity, and depression is considered to be the converse phenomenon. The above approach, however, is reductionistic, and an "interaction" between several neurotransmitter systems may be paramount to the understanding of basic mechanisms involved in the patho-

Sanjay Dubé • Mood Disorders Module, Western Psychiatric Institute and Clinic, Pittsburgh, Pennsylvania 15213.

Biology of Depressive Disorders, Part A: A Systems Perspective, edited by J. John Mann and David J. Kupfer. Plenum Press, New York, 1993.

physiology affective disorders, the basic premise being that neurotransmitter systems do not, and indeed cannot, exist in isolation.

This chapter presents data supporting the cholinergic hypothesis of affective disorders and its interaction with other neurochemical systems. Pharmacological perturbation strategies for modeling affective disorders through muscarinic cholinergic networks on through manipulations of aminergic systems are presented. The role of genetic determinants of these behaviors is discussed in the context of different models, and applications to newer technology such as brain imaging are highlighted.

Effects of Cholinergic Agonists in Man

Some of the most convincing evidence that central cholinergic mechanisms affect affective states comes from descriptions of the psychological and behavioral effects of central cholinesterase inhibitors in man. Rowntree et al. (1950), studying normals, depressives, and manics; Gershon and Shaw (1961), observing insecticide (organophosphates)-poisoned normals; and Bowers et al. (1964), observing normals, reported that cholinesterase inhibitors exert anergic-inhibitory effects, intensify and induce depressive symptoms, and antagonize manic symptoms. Janowsky et al. (1973a,b) studied the pharmacological and behavioral effects of the centrally active cholinesterase inhibitor physostigmine and of neostigmine (a similar cholinesterase inhibitor which does not cross the blood–brain barrier), given intravenously in doses up to 4 mg. In methscopolamine (a non-centrally acting anticholinergic agent) pretreated subjects, physostigmine in contrast to neostigmine or placebo rapidly exerted inhibitory effects in virtually all patients who received it. These effects are outlined in Table 1. Behavioral changes occurred within 5–15 minutes of physostigmine administration, lasted between 20 and 90 minutes (corresponding to physostigmine's pharmacological profile), and

Table 1. Effects of Centrally Active Cholinesterase Inhibitors; Anergic-Inhibitory Syndrome

Lethargic	Drained
Slow thoughts	Hypoactive
Wants to say nothing	Lacks thoughts
Withdrawn	Motor retarded
Apathetic	Emotionally withdrawn
Lacks energy	

were rapidly antagonized by small doses of centrally active antimuscarinic drug atropine. The reversal of physostigmine's obvious behavioral effects by atropine suggests a mediation via a central cholinergic mechanism.

While some studies report mood depressant and antimanic effects of physostigmine (Janowsky et al., 1972; 1973b; and, Modestin et al., 1973 a,b), others (Carroll et al., 1973; Shopsin et al., 1975), demonstrated that physostigmine reduced only the mood and psychomotor symptoms of mania, but not necessarily grandiosity and manic thought content. Reversal of psychostimulant (methylphenidate) induced behavioral activation (a potential model for mania) by physostigmine, i.e., reversal of increased talkativeness, cheerfulness, increased interactions, friendliness, and euphoria was reported by Janowsky et al. (1973c). Furthermore, physostigmine has been reported to cause depression in the majority of a group of euthymic bipolars maintained on lithium (Oppenheimer et al. 1979). El Yousef et al. (1973) studied the effects of physostigmine on two chronic cannabis abusers. The anticholinesterase counteracted the "high state" and induced profound depression in both subjects. The depressive syndrome included feelings of uselessness, hopelessness, apathy, sadness, decreased thoughts and verbalizations, dysphoria, suicidal ideation, and extreme psychomotor retardation, which were relieved by intravenous atropine sulfate in one subject. This phenomenon is attributed to "cholinergic upregulation" or supersensitivity induced by Δ^9-THC (tetrahydrocannabinol), which causes a decrease in the quantal release of acetylcholine at neuromuscular junctions. This is consistent with increased physostigmine toxicity in marijuana-treated rats due to an increase in striatal and amygdaloid acetylcholine content following Δ^9-THC treatment (Rosenblatt et al., 1972; Kumbarachi & Nastok, 1980; Yashimura et al., 1974). Hence, centrally acting anticholinesterases significantly affect positive mood symptoms and motoric components of manic illness.

Specific Versus Nonspecific Effects of Cholinergic Agents

The anergic, dysphoric, and neuroendocrine effects of physostigmine (discussed below) could be attributed to the nausea syndrome. Evidence to the contrary was provided by studies which negative affect, neuroendocrine changes (increased prolactin and cortisol), and increased heart rate and blood pressure occurred without any nausea or vomiting whatsoever. Furthermore, arecoline (muscarinic cholinergic agonist) given to normals and affective disorder patients induced affective changes, as well as increased ACTH, cortisol, and β-endorphin without development of nausea or subjects' ability to distinguish between the arecoline and placebo infusions (Risch, 1982; Risch et al., 1984). Nevertheless, it is possible that physostigmine intoxication may cause a subjective experience of stress that is more pronounced in depressed patients and/or set in motion the neurochemistry and endocrinology of stress, with stress being regulated by cholinergic mechanisms.

Cholinergic Mechanisms in Stress and Affective Disorders

The traditional dichotomy of depression being either a biological or a psychological entity is simplistic and inadequate. The classical "endogenous" episode may occur as a "bolt from the blue" without a trigger or precipitant. On the other hand, an episode of melancholia may be strongly associated with a significant life event at a different point during the course of the same illness. Hence, both the biological and psychological determinants of depression may coexist. Weiss et al (1981) reported that an uncontrollable stressor (reductionistically conceived as being "psychological in nature"), produced a fall in brain biogenic amines, a feature of "biological" depression. A psychologically appearing illness may have a biological substrate. Several studies of life events and onset of depression (Lloyd, 1980), manic illness (Kennedy et al., 1983), and separation and loss induced depression (Rasmussen & Reite, 1982) suggest that affective disorder subjects are at increased risk of an episode in face of stress due to their neurobiological abnormalities. Janowsky et al. (1983a, 1984, 1985) proposed that acute and chronic stress is partially mediated by central muscarinic mechanisms with capacity to activate the adrenergic networks. Acetylcholine uniquely causes simultaneous effects on behavioral, cardiovascular, neuroendocrine, and noradrenergic systems typifying stress. This model proposes that an underlying cholinergic supersensitivity renders susceptible individuals vulnerable to manic and depressive episodes, depression being characterized by a "cholinergic overdrive" and mania by a "hypocholinergic-hypermonoaminergic" state. However, the cholinergic systems are unusually perturbable in every phase in bipolar patients, with depression being characterized by a drive toward down-regulation and subsensitivity of cholinergic and upregulation and supersensitivity of monoaminergic systems, and the reverse characterizing mania. This model highlights the dysfunction of homeostatic mechanisms, with an imbalance of the cholinergic and monoaminergic systems serving as pathogenic forces. Stressors cause the overly driven cholinergic systems to result in mania. That is, the more effective the cholinergic systems are in provoking compensatory or restorative responses in cholinergic systems (down-regulation and subsensitivity), the greater the probability of a "paradoxical" mania occurring. Furthermore, a hypercholinergic state may terminate in a hyperadrenergic state. The following observations support this position: the fact that muscarinic agonists release catecholamines from the adrenal medulla and cardiac tissue (Taylor, 1980); the induction of panic attacks by norepinephrine as well as methacholine (Lindemann & Finesinger, 1940); and a common diathesis between major depression and panic disorder as suggested by family studies (Leckman et al., 1983).

Evidence to Support the Cholinergic Hypothesis

Since the original observations of Rowntree et al. (1950) and Janowsky et al. (1983b) that cholinergic overdrive states produce a phenomenological depressive syndrome, several reports indicate that drugs activating the central muscarinic networks produce neuroendocrine (Dilsaver, 1986a,b; Risch et al., 1981a,b; Janowsky et al., 1980), polysomnographic (Sitaram, 1980), and psychological (Dilsaver, 1986b,d; Dilsaver & Greden, 1984; Janowsky et al., 1972, 1973a,b,c; Risch, 1982) abnormalities characteristic of endogenous depression in normal subjects. Affective disorder probands show a supersensitive response to these agents. Following a physostigmine challenge, affective disorder patients show a greater increase in dysphoria, behavioral inhibition, tension-anxiety, hostility, and fatigue as compared to controls and "miscellaneous" psychiatric diagnoses (Janowsky et al., 1973a,b,c), as well as greater increases in concentration of plasma β-endorphin compared to normals (Risch et al., 1981a,b). Physostigmine also induces nonsuppression of cortisol following dexamethasone (Carroll et al., 1980; Doerr & Berger, 1983) which is consistent with the possibility that dysfunction of a muscarinic mechanism may contribute to the dysregulation of the LHP Axis. Polysomnographic studies have shown a disturbance in sleep continuity and efficiency, reduction in delta wave sleep, decreased REM latency, and increased REM density in major depression (Kupfer & Foster, 1972). These effects are induced by arecoline and physostigmine infusions (McCarley, 1982; Sitaram & Gillin, 1980) and reversed by antimuscarinic agents (McCarley, 1982; Sagelas, 1969). Sitaram and Gillin (1980) found an accelerated induction of REM sleep following arecoline challenge in euthymic primary affective disorder patients, suggesting that muscarinic cholinergic supersensitivity is a trait marker of depressive illness.

Pharmacological Perturbation Strategies

These strategies involve measurement of change in a targeted neuronal system from a pretreatment baseline, for the purpose of modeling a pathophysiological state in the following areas: (1) behavioral and psychological, (2) physiological, (3) biochemical, and (4) receptor binding variables. Dependent variables that may be driven by a cholinergically driven up-regulation in depression include clinical (mood, affect, and psychomotor function), physiological (measures of LHPA function, polysomnographic parameters such as REM latency, arecoline-induced REM induction, REM density, and pupillometric data), biochemical (phospholipid or cyclic GMP accumulation), and in vivo *m*-Ach binding. Several classes of drugs up-regulate the muscarinic cholinergic systems, viz., opiates, cannabinoids, ethanol, and barbiturates (Dilsaver, 1987). Tricyclic antidepressants resemble scopolamine by producing postsynaptic mAchR blockade, thereby up-regulating/supersensitizing muscarinic mechanisms.

Sitaram et al. (1979) and Gillin et al. (1979) gave scopolamine at 9:00 a.m. to normal subjects for three successive mornings and were able to induce polysomnographic characteristics resembling major depression, viz, reduced REM latency and increased REM density. Withdrawal of TCAs has similarly been used to study possible mechanisms underlying LHPA dysregulation. Some of these studies are summarized in Table 2.

Desipramine (DMI) induced muscarinic receptor up-regulation may be related to (1) cholinoceptive denervation by postsynaptic mAchR blockade or (2) adrenergic inhibition of Ach release (Nomura et al., 1982, 1983). Mianserin's inability to increase the miotic response to pilocarpine is related to its weak affinity to mAchR (18.6% that of DMI in the caudate nuclei) as well as weak inhibition of norepinephrine uptake (Shur et al., 1983). Several other investigators have reported a hypothermic response to cholinomimetics and anticholinesterases (Freidman & Jaffe, 1969; Jaffe & Sharpless, 1968; Lomax & Jendeu, 1966). Conversely *chronic* anticholinesterase treatment blunts the hypothermic response to cholinergic challenge (Overstreet et al., 1973). Avery et al. (1982) studied 9 affective disorder patients and concluded that reduced REM latency, duration of first REM period, and increased REM density and body temperature retain their usual relationship during a depressive episode.

Table 2. Pertubation Strategies: Treatment Effects

	Sample	Strategy	Marker	Results
Dilsaver and Greden (1985)	14 Endog. Dep (E) 9 Drug-free Endog. Dep (C)	2-wk TCA withdrawal	Post DST Cortisol	12/14 patients had ↑ cortisol post-withdrawal. 71.5% (E) and 22.2% (C) had positive DST during withdrawal
Shur and Checkley (1982)	Depressed	Desipramine	Pupillary miosis	↑ miosis after 21 days
Shur et al. (1983)		Mianserin	Pupillary miosis	NSC
Dilsaver et al. (1987)	15 ♂ Sprague Dawley rats	Amitryptyline (AMI) 10 mg i.p. BID × 16d	Hypothermic response to oxotremorin (OXO)	a. OXO ↓ temperature prior to AMI b. AMI ↑ hypothermic response to OXO. Scopolamine but not methscopolamine blocks a)
Dilsaver and Hariharan (1989)	11 ♂ Sprague Dawley rats	Licl 160 mg/kg i.p. BID × 14d	Hypothermic Licl: response to nicotine	↑↑ hypothermic response to nicotine; saline: hyperthermic response
Dilsaver et al. (1988)	♂ Sprague Dawley rats	AMI 15 mg i.p. BID × 7 days	Hypothermic response to nicotine	↑ hypothermic response post-Tt compared to pre-Tt

A simultaneous phase advance of REM sleep and temperature cycles in man suggests a strong implication for a cholinergic mechanism in affective disorder.

Physiologic Hypersensitivity to Cholinomimetics

Polysomnographic abnormalities associated with depressive disorder include shortened REM latency, increased REM density and activity, reduction in slow wave sleep, and diminished sleep efficiency. These abnormalities of primary or endogenous depression can be produced by cholinergic up-regulation (Hobson et al., 1974, 1975; Hobson & McCarley, 1972; Hobson, 1981; Jouvet, 1975; Karczmar et al., 1976; McCarley, 1982; Sitaram et al., 1977) and suppressed by cholinoceptor blockade (Tollefson et al., 1982). Sitaram et al. (1980, 1981) reported a physiologic hyperresponsiveness to cholinomimetics in affective disorder patients. REM latency, an acetylcholine sensitive sleep parameter, is shortened by acetylcholine and prolonged by adrenergic agents and is frequently found significantly shortened in depression. Furthermore, infusion of 0.5 mg of arecoline 25 minutes after the end of the first REM period (second NREM period) produced a significant shortening of the time from the end of the infusion to the onset of the second REM period in patients with primary affective disorder. The cholinergic REM induction test (CRIT) has since then been replicated by several other groups (Gillin et al., 1988a; Kreig & Berger, 1987; Wager et al., 1988; Jones et al., 1985). Kreig and Berger (1987) studied the effect of an oral muscarinic agonist RS86 on sleep and HP Axis of 8 depressives and 8 controls. RS86 produced a significant shortening of REM latency and increase in REM sleep % in the depressed group as compared to controls. These findings are consistent with the role of muscarinic supersensitivity in regulation of REM sleep and in depression. Conversely, scopolamine is a muscarinic antagonist and was found to differentially affect REM sleep measures in depressives compared with normals, i.e., it prolonged REM latency significantly less in patients than in controls on the first night. Following three nights of scopolamine, on the first recovery night REM density increased significantly more in patients than in controls (Gillin et al., 1988b). This is the first demonstration that depressed patients differ more than controls in this respect and supports the earlier findings of Gillin et al. (1979), namely, that REM density increases after three nights of scopolamine administration in normal subjects. Whether there is a greater propensity to "up-regulate" muscarinic receptors in patients than controls is an interesting area of research. Similar to the CRIT, Berger and Lund (1982) found increased dreaming in depressives compared to controls following infusion of physostigmine 0.5 mg, 5 minutes after sleep onset. An interesting model to understand the pathophysiology of depression was the development of the Flinders Sensitive Line (FSL) of rats which were selectively bred for supersensitive responsivity to an acetylcholinesterase inhibitor di-isopropylfluorophosphate (DFP) (Overstreet, 1986a). Compared to randomly bred controls, FSL rats show greater physiologic sensitivity to cholinergic agonists (Overstreet et al., 1979, 1986b, 1986c; Overstreet & Russell, 1982, 1984). With regard to sleep findings Shiromani et al. (1988) found an increase REM%, reduced REM latency and shorter REM–REM cycle in the FSL than in the Flinders Resistant Line of rats. The changes in REM sleep parameters in the FSL strain are consistent with the amassed evidence that central cholinergic activity promotes REM sleep, a feature that is one of the most well established biologic abnormalities in depression (Gillin, 1982; Coble et al., 1981; Gillin et al., 1979b, 1984; Kupfer, 1976; Kupfer et al.,

1976; Sitaram et al., 1982; Spiker et al., 1978). Polysomnographic changes induced by cholinomimetics produce a phase advance resembling the derangement of the same ultradian rhythm that is observed in depression.

Euthymic patients (in remission) exhibit a more rapid arecoline- or physostigmine-induced onset of REM sleep than normal controls (Sitaram et al., 1977), and this finding suggests that cholinergic supersensitivity may be a "trait-marker" of affective illness. Kelwala (1984) provided further evidence by reporting an increased pupillary miotic response to 0.125% pilocarpine instillilation in depressed patients as compared to normals and nonaffective controls. Whether these subjects would have exhibited supersensitivity prior to their first affective episode, i.e., trait marker versus effect of illness, however, is yet unknown. Nonetheless, there is some evidence to suggest that some polysomnographic variables have a genetic (trait) component.

Genetic Basis of Cholinergic Supersensitivity in Depression

Webb and Campbell (1983) obtained one-night polysomnographs on 14 identical and fraternal adult twin pairs. Sleep onset latency, number of awakenings during sleep, number of shifts in sleep stages, and amount of REM sleep was significantly correlated in identical but not fraternal twins. Nurnberger et al. (1983) found concordance for sensitivity to REM sleep induction by arecoline ($i = 0.69$, $p < 0.02$) in seven monozygotic twin pairs. The REM_1–REM_2 interval was concordant for zygosity following infusion of arecoline but not placebo. These data suggest that supersensitive polysomnographic responses to arecoline may be heritable, i.e., a trait. Chouvet et al. (1983) showed similarity in structure of occurrence of rapid eye movement in the paradoxical sleep of monozygotic twins. Sitaram et al. (1987) compared the putative vulnerability marker/risk factor (rapid cholinergic REM sleep induction by arecoline) in 35 affectively "ill" and 31 "well" first-degree relatives of 34 probands chosen on the basis of presence of primary affective illness and positive vulnerability marker status. 66% of "ill" relatives had a supersensitive REM-induction response as compared to 22% of the "well" relatives. Effects of prior depressive episodes and treatment on cholinergic supersensitivity, however, could not be ruled out. Dubé and Sitaram (1988) found that affectively "ill" probands and "ill" relatives have a rapid (supersensitive) cholinergic REM-induction response, higher density of eye movements in the first REM period, lesser stage 2 and delta sleep, greater REM% sleep,and a shorter latency of the first REM period. Furthermore, ANOVA revealed significant main effects for family membership and diagnostic status interaction with respect to two sleep parameters, viz., total REM% and arecoline-induced REM2 time. In other words, the REM% and arecoline-induced REM2 induction time of individuals within families were significantly heritable and were influenced by diagnostic (ill/well) status as well. The above findings are supported by animal data that have shown that muscarinic receptor densities (Baumgold, 1987; Overstreet, 1984, 1989) and paradoxical sleep characteristics (Valatx et al., 1980; Chouvet, 1981; Aalto and Hilakivi, 1986; Daszuta and Gambarelli, 1985; Freidman, 1974; and Rosenberg et al, 1987) may have significant genetic component. The promising early report of increased mAchR binding on fibroblasts from human depressives and their ill relatives has not been confirmed (Kelsoe et al., 1985; Lenox et al., 1985).

Neuroendocrine Hypersensitivity to Cholinomimetics

Animal evidence indicates that cholinergic mechanisms may regulate prolactin (Gibbs et al., 1979; Grandison & Meites, 1976a; Subramanium & Gala, 1976), growth hormone (Mendelson et al., 1978; Bruni & Meites, 1978), and ACTH secretion (Bradbury et al., 1974; Edwardson & Benett, 1974; Hillhouse et al., 1975). Central cholinergic mechanisms are also implicated in the circadian variation of cortisol secretion (Hill & Wynber, 1974; Cozanitis, 1974; Ferrari et al., 1977).

The finding that depressed patients display an increased responsiveness of the hypothalamic pituitary-adrenocortical (HPA) axis to cholinergic stimulus provides evidence for the cholinergic supersensitivity hypothesis (Carroll et al., 1980).

In vivo studies indicate a muscarinic mechanism in the release of corticotropin releasing factor (CRF) and a nonadrenergic mechanism in the inhibition of CRF and ACTH (Ganong, 1980; Hedge et al., Kneko, 1977; Makara & Stark, 1976; Naumenko, 1967; Palkovitz, 1977). Risch and associates (Risch et al., 1983) have demonstrated that centrally active cholinergic agonists increase plasma concentrations of ACTH and β-endorphin (Risch et al., 1980, 1981b, 1982). Physostigmine, relative to placebo, causes a significant increase in plasma concentrations of ACTH and β-endorphin immunoreactivity in subjects with major affective disorder and schizoaffective disorder compared to normal or control subjects without affective disorders (Risch et al., 1983). A differential responsivity of ACTH and cortisol release secondary to cholinergic stimulation has been reported by the same group. In another study, physostigmine infusion caused a significantly greater elevation of plasma prolactin in 14 affective disorder patients than in 15 nonaffective psychiatric patients (Janowsky et al., 1982b). The increases correlated with a cholinergic overdrive and with increases in negative affect when the entire patient population was pooled. Negative affect was correlated with plasma β-endorphin concentration in control subjects. The findings suggest that correlation of increases in negative affect with prolactin and plasma β-endorphin elevations following physostigmine infusion point to simultaneous involvement of central neurochemical mechanisms regulating affect, β-endorphin, and prolactin secretion. These phenomena may reflect a single system or two cholinergic systems running in parallel. Cholinergic stimulation has been reported to produce cortisol nonsuppression (Carroll et al., 1980; Doerr & Berger, 1983) following dexamethasone, a state marker for depression (Carroll et al., 1981). Davis and Davis (1981) concluded that nausea (stress response) and not direct cholinergic activity nonspecifically elevated the peripheral cortisol, prolactin, and growth hormone concentrations. Contrary evidence was provided by Carroll et al. (1980), who showed that an emetic (ipecacuanha) failed to cause nonsuppression of cortisol following the dexamethasone challenge. Furthermore, physostigmine produces a major rise in plasma cortisol before producing nausea and argues against the "nonspecific stress" hypothesis. In a related study, they demonstrated that administration of methscopolamine (to block the peripheral side effects of physostigmine), did not block the rise in plasma cortisol. On the other hand, atropine pretreatment prevented the physostigmine-induced nonsuppression of cortisol following dexamethasone. In a recent study, however (Krieg & Berger, 1989), an oral muscarinic agonist RS86 was unable to produce an activating effect on the HPA axis (cortisol system). Thus, a role for mediation of the HPA axis through the nicotinic cholinergic neurons (Wilkins et al., 1982) or a nonspecific stress response has been suggested.

Hence, the neuroendocrine data from cortisol nonsuppression, physostigmine-induced β-endorphin, ACTH, and prolactin rise suggest a strong but not definitive role of central muscarinic cholinergic overdrive in affective states. A dose–response study of cortisol, ACTH, and β-endorphin immunoreactivity to physostigmine challenge in affectively ill patients and controls is indicated. If the dose–response curves in affectively ill patients fell to the left of those in normal controls, the cholinergic supersensitivity hypothesis would be confirmed.

An anomalous growth hormone (GH) response to thyrotropin-releasing hormone (TRH) in depression has been attributed to a muscarinic cholinergic mechanism (Coiro et al., 1987). Twelve male patients with unipolar major depression were given TRH (200 mg IV) with and without pretreatment with muscarinic cholinergic receptor blocker pirenzepine intravenously, and the results compared with saline infusion. Pretreatment with pirenzepine severely blunted the TRH-induced GH rise (p- < 0.03 at 45 minutes, $p < 0.05$ at 30 and 60 minutes). Serum TSH response was not affected by pirenzepine pretreatment. Administration of TRH does not routinely stimulate GH secretion, whereas it often does so in endogenous depression (Jackson, 1984). Acetylcholine acts as an important neurotransmitter in the hypothalamic-pituitary control of GH secretion (Delitala, 1984). Acetylcholine is known to reduce the hypothalamic release of somatostatin (SRIH) through a muscarinic mechanism (Richardson et al., 1980); thus increased cholinergic activity may reduce the hypothalamic inhibition, thus allowing the somatotrophs to respond to TRH. Terry and Crowley (1986) elucidated different hypothalamic centers in the somatostaninergic control of TSH and GH secretion, and from this point of view, pirenzipine might be active on the GH-related SRIH system, but not on that controlling TSH release.

Although several findings supporting a cholinergic mediation of neurochemical and neuroendocrine dysregulation in depression have been reported, the specificity of the findings is still inconclusive. Abnormal DST, for example, has been reported in patients with obsessive-compulsive disorder (Insel et al., 1982), dementia (Spar & Gerner, 1982), nondepressed diet clinic patients following weight loss (Edelstein et al., 1983), and in healthy volunteers following caffeine consumption (Uhde et al., 1985). Several abnormalities may be stress-mediated reactions, which in turn may have a strong cholinergic component.

Cholinergic Basis of Phototherapy in Seasonal Affective Disorder

Seasonal affective disorder (SAD) commonly presents with winter depression (Lewy & Sack, 1987; Lewy et al., 1982; Rosenthal et al., 1984), and is relieved by phototherapy (James et al., 1985; Lewy et al., 1983, 1987; Rosenthal et al., 1984, 1986). Dilsaver and associates have applied the thermoregulation paradigm to assess treatment effects on muscarinic (Dilsaver et al., 1987a–c) and nicotinic (Dilsaver & Davidson, 1989; Dilsaver & Hariharan, 1989; Dilsaver & Majchrzak, 1989) mechanisms. Prolonged treatment with bright light blunts the hypothermic response to the muscarinic cholinergic agonist oxotremorine, blocks the supersensitization of muscarinic mechanisms produced by amitryptyline (a muscarinic antagonist), and counteracts the effects of forced swim stress (a means of producing learned helplessness) on muscarinic mechanism.

The findings are consistent with the cholinergic hypothesis of depression with the possibility that light treatment for depressive disorders subsensitizes the central muscarinic mechanisms. Whether a close relationship between intensity and duration of bright light should shift the thermic changes caused by oxotremorine to the right should be an interesting study. Autoradiographic techniques may also be used to assess the effects of bright light and other treatments on mAchR binding in multiple brain regions, as well as on coupling mechanisms and second messenger systems. The latter include polyphosphoinositide cycle, generation of cyclic GMP and flux of calcium ions, and the messengers that mediate the effects of muscarinic agonists in given tissues.

Clinical Applications: Response to Anticholinergic Agents

Anticholinergic agents have euphorigenic (Ben-Barak & Dudai, 1980; Crawshaw & Mullen, 1984; Kaminer et al., 1982; Pullen et al., 1984; Rubenstein, 1978; Smith, 1980) and antidepressant properties (Goldman & Erickson, 1983; Jimmerson et al., 1982; Kasper et al., 1981; Kupfer et al., 1979). Table 3 outlines cholinomimetic and anticholinergic agents with potential clinical usefulness.

Acetylcholine precursors have potential therapeutic value in mania (Cohen et al., 1982). Arecoline, like physostigmine, exerts inhibitory-anergic-depressant effects when given to affective disorder patients as well as to normals (Nurnberger et al., 1982a,b). Reserpine, and possibly propranolol, have been found to have antimanic properties by their ability to block/or decrease central adrenergic activity as well as increase absolute and relative central and peripheral parasympathetic tone (Janowsky et al., 1972). It is thus possible that a combination of a cholinergic agonist or cholinesterase inhibitor and a centrally acting anti-adrenergic agent such as α-methyl dopa, reserpine, or propranolol

Table 3. Drugs with Central Cholinomimetic or Anticholinergic Properties

Cholinesterase inhibitors	Cholinergic agonists
Oral or parenteral	Arecoline
Physostigmine	Oxotremorine
Acetylcholine precursors	Anticholinergics
Choline	Triexyphendyl
Lecithin	Benztropine
Deanol ?	Scopolamine
	Atropine

would be effective as antimanic treatment by shifting the adrenergic-cholinergic balance toward a cholinergic excess. Along the same lines, coadministration of tetrahydrocannabinol and physostigmine may be effective. El-Yousef et al. (1973) found anergia and depression in a group of normal volunteers who received this combination, which was reversed by atropine. Chronic physostigmine administration may produce tolerance and result in a "hypermanic" rebound as reported by Shopsin et al. (1975). This could potentially, however, be prevented by simultaneous administration of antiadrenergic agents.

Lithium may exhibit its effects in part via a cholinergic mechanism. $LiCO_3$ produces increased Ach turnover (Jope et al., 1979). Similarly, Samples et al. (1977) found an increase in physostigmine lethality in lithium-treated rats reversible with antimuscarinic agents. Judd et al. (1977) observed that in normals lithium exerts effects like central cholinomimetic effects, viz., anergia, being drained, lethargic, weak, depressed, and tired. Contrary to these findings, there are reports of lithium exerting antidepressant effects by reducing central cholinergic tone (Niel et al., 1976; Janowsky et al., 1979, 1982a). Lithium's bidirectional effect in manic-depressive illness may be related to its ability to prevent an up-regulation of muscarinic receptors in depression and the dopamine receptors in mania (Pert et al., 1978; Janowsky et al., 1979). Electroconvulsive therapy (Byrne et al., 1980; Dasheiff et al., 1982) or approximations to it, and calcium channel blockade (El-Fakahany et al., 1983) have been reported to down regulate cholinergic receptors. In contrast to the effects of lithium, ECT and verapamil, tricyclic antidepressants induce rapid cycling (Wehr et al., 1979), up-regulate muscarinic receptors and produce muscarinic system supersensitivity.

Although antidepressant activity is not completely related to the anticholinergic activity (as some newer antidepressants have none or minimal anticholinergic effects), there is some evidence that anticholinergic agents do exert euphorigenic-antidepressant effects. Trihexylphenidyl and biperiden (anticholinergic antiparkinsonian drugs) cause a feeling of euphoria, associated with a sense of well being, increased sociability, and reversal of depressive feelings (Jellenic et al., 1981; Jimerson et al., 1981; Kasper et al., 1981). The depression syndrome related to psychostimulant withdrawal may in part be related to cholinergic excess and catecholamine depletion, and response to scopolamine and atropine may further support this hypothesis.

Although investigations suggest that agents with anticholinergic properties exert antidepressant effects, the research design, diagnostic criteria, and outcome measures require more attention.

Interaction of Lithium, Sodium, and Calcium Ions in Cholinergic Mechanisms and the Pathophysiology of Affective Disorders

Dilsaver (1986c) proposes a comprehensive model to explain the interaction of cholinergic mechanisms and effects of sodium, calcium, and other ions on membrane and receptor events. Lithium-induced increase in sodium-potassium ATPase activity in bipolar subjects in vivo (Glen et al., 1968) and decrease in intraneuronal sodium and calcium concentrations correlates with improvement (Naylor et al., 1976). These changes affect the synthesis and release of acetylcholine. Calcium is involved in the transduction processes initiated by activation of the muscarinic receptor and increases the activity of tyrosine hydroxylase (the rate-limiting enzyme in catecholamine synthesis) by binding and activating the calcium-dependent regulator protein (calmodulin). The calcium-calmodulin complexes enhance cyclic AMP generation. Calcium produces muscarinic receptor down-regulation and activates the regulatory enzymes such as the phosphoylases, which induce conformational changes that render the 1 and 2 receptors subsensitive. Calcium, sodium, and lithium all influence the synthesis and release of acetylcholine, and their functional role may help us understand affective illness from a molecular viewpoint.

Muscarinic Receptor Changes in Suicide

Meyerson et al. (1982) studied the (presumably) frontal cortical muscarinic receptors (^{3}H) QNB in brain tissue of eight victims of violent suicide and found a 47% increase in density and no change in affinity relative to homicide controls. Another study (Perry et al., 1977), which examined parietal cortical muscarinic binding, found no difference between elderly patients with depression and age-matched psychiatrically unaffected controls. Stanley and Mann (1983) and Kaufmann et al. (1984) reported no increase in muscarinic receptor density in cortices of suicide victims. In the latter study, the B_{max} of (^{3}H) QNB binding in fmol of ligand/mg protein was 185 ± 80 vs. 170 ± 66 in frontal cortices of suicide and control groups, respectively. The study may have contributed to a falsely negative outcome (increased the type II error) because of a small sample size. The most noteworthy finding, however, was a somewhat lower receptor density and affinity in the hypothalami from suicides. A major difficulty in suicide-receptor binding studies is the problem of justifiable inference; that is, documentation of muscarinic receptor up-regulation in suicides may not reflect muscarinic receptor up-regulation in affective illness. Kaufmann et al. (1984) acknowledge that their suicidal patients were not affective disorder patients. Other problems include concomitant illnesses and psychoactive drug use and analytic techniques that may obscure findings in specified areas of interest.

The muscarinic receptor is not a single protein, but a complex containing binding site, several (five) agonist binding sites (conformations), and an effector (which may consist of a calcium ionophore, guanylate cyclase, and phosphatidylinositol). Binding to the antagonist and agonist sites may be competitive and mutually exclusive (Birdsall & Hulme, 1976). Increased occupancy by (hence response to) agonists might therefore accompany decreased antagonist binding. Direct agonist binding studies are clearly needed. In addition, as coupling between binding site and effector may vary (Ehlert et al., 1982), alleviations in physiological responsiveness might well occur in absence of demonstrable changes in binding. Measures of cyclic GMP synthesis or phosphatidylinositol turnover might more accurately reflect receptor "sensitivity."

Cholinergic Basis of Sleep Deprivation

Sleep deprivation has been a brief but effective therapeutic modality for depression. Responders to total or partial sleep deprivation are more likely to have neurovegetative signs and symptoms. Nasrallah and Coryel (1982) reported that nonsuppression to dexamethasone predicted positive response to sleep deprivation. A significant drop in the telencephalon acetylcholine content of rats was seen after 96 hours of REM sleep deprivation (Bowers et al., 1966). Trusihiya et al. (1969) found a decrease in the tencephalon Ach levels after 24 hours of total but not partial sleep deprivation. Sagelas and Domino (1969) found no changes in whole brain Ach content in animals deprived of 24, 48, and 96 hours and 6 days of REM sleep. Unlike the above studies, isolated brain regions were not analyzed and may have contributed to lack of a positive finding.

In vivo muscarinic binding studies using (^{3}H) scopolamine as receptor ligand and tracer kinetic modeling coupled with autoradiographic technique allows for simultaneous measurement of receptor binding site density in infinite brain regions. This powerful technology used along with pharmacological perturbation studies before and after sleep deprivation may provide greater insights into the neurobiology of sleep deprivation.

Positron Emission Tomography Scanning

Study of pathophysiology of psychiatric conditions has been limited by inability to measure brain function in vivo. Functions capable of being measured in affective disorders include cerebral blood flow, oxygen and glucose utilization, and receptor binding variables. Studies with great promise are those were subjects act as their own controls. Rapidly cycling bipolar disorder patients may be ideal. Cholinergic system supersensitivity induced by pharmacological manipulation (i.e., by scopolamine, arecoline) and accompanying changes in REM sleep that are characteristic of primary depression can be measured during pretreatment and acute and chronic withdrawal periods using a PET scanner. Studies to develop a suitable muscarinic ligand for in vivo PET studies in man are under way (Frey, 1985a,b).

Conclusion

To summarize, considerable evidence points toward a muscarine cholinergic receptor function dysregulation in primary depression. Interaction of the muscarinic cholinergic system with the nicotine and the monoamine systems may provide a more unifying hypothesis in the systems deregulation in affective disorders. These systems have been studied and theories based on other neurochemicals (dopamine and serotonin) and neuropeptides (e.g., somatostatin) have been developed. Although no single "cause" or even correlate of depression has been conclusively demonstrated, a common theme seems to be emerging. Some findings, however, do not support the Ach hypothesis of affective disorder, namely, the minimal efficacy of anticholinergic drugs, and more recently the efficacy of "second generation" antidepressant drugs with no appreciable anticholinergic effects (for example, trazodone, and fluoxetine and citolapram). Furthermore, the anticholinergic profile of conventional tricyclics does not correlate with antidepressant potency. Nonetheless, the role of acetylcholine in mediating stress-responsive neurochemistry in the pathophysiology of depression may be an important one. Absence of definitive etiological specificity in this regard may be due to other factors including diagnosis, effect of comorbid character pathology, and delineation of possible subtypes. The translation of the various dysregulations seen in depression to clinical phenomena will be an interesting area of future research, particularly in context of developments in molecular neurobiology and neuroscience.

References

Aalto J, Hilakivi Z: Difference in sleep wake patterns of the AA and ANA rat lines developed for high and low alcohol intake. *Alcohol* 1988; 3(1): 77–79.

Avery D, Wilschiotz G, Rafaelsen O: REM latency and temperature in affective disorders before and after treatment. *Biol Psychiat* 1982; 17: 463–470.

Ben-Barak Y, Dudai Y: Scopolamine induces an increase in muscarinic receptor level in rat hippocampus. *Brain Res* 1980; 193: 309–313.

Berger M, Doerr P, Lund R, Bronisch T, Zerssen von D: Neuroendocrinological and neurophysiological studies in major depressive disorders: Are there biological markers for the endogenous subtype? *Biol Psychiat* 17: 1217–1242.

Birdsall NJM, Huhme EC: Biochemical studies on muscarinic acetylcholine receptors. *J Neurochem* 1976; 27: 7.

Blake CA, Norman RL, Scaramuzzi RJ, Sawyer CH, Bolin RR: Psychiatric manifestations of artane toxicity. *J Nerv Ment Dis* 1960; 131: 256–259.

Bowers MB, Hartmann EL, Freedman DY: Sleep deprivation and brain acetylcholine. *Science* 1966; 153: 1416–1417.

Bowers MB, Goodman E, Sim VM: Some behavioral changes in man following anticholinesterase administration. *J Nerv Ment Dis* 1964; 138: 383.

Bradbury MW, Burden BJ, Hillhouse EW, Jones MT: Stimulation electrically and by acetylcholine of the rat hypothalamus in vitro. *J Physiol (Lond)* 1974; 239: 269–283.

Bruni JF, Meites J: Effects of cholinergic drugs on growth hormone release. *Life Sci.* 1978; 23: 1315–1358.

Byrne MC, Gottlieb N, McNamara TO: Amygdala kindling induces muscarinic cholinergic receptor declines in a highly specific distribution within the limbic system. *Symp Neurol* 1980; 69: 85–98.

Carroll BJ, Feinberg M, Greden JF et al: A specific laboratory test for the diagnosis of melancholia. *Arch Gen Psychiat* 1981; 38: 15–22.

Carroll BJ, Greden JF, Hasket R et al: Neurotransmitter studies of neuroendocrine pathology in depression. *Acta Psychiat Scand* 1980; 61; Suppl 280: 183–199.

Carroll BJ, Frazer A, Schless A, Mendels J: Cholinergic reversal of manic symptoms. *Lancet* 1973; i: 427.

Chouvet G, Blois R, Debilly G, Jouvet M: The structure of the occurrence of rapid eye movements in paradox sleep is similar in homozygotic twins. *C R Seances Acad Sci (III)* 1983; 296(22): 1063–1068.

Chouvet G: *Structures d'occurrence des activite's phasiques du sommeil paradoxal chez l'animal et chez l'homme*. Ph.D. diss., Université Claude-Bernard, Lyons, France, 1991.

Cioro W, Marchesi D, DeFerri A, Mosti A, Delsingnore R, Govi AM, Gnudi A, Chiodera P: Pirenzepine inhibits growth hormone, but not thyrotropin response to thyrotropin-releasing hormone in patients with endogenous depression. *Psychoneuroendocrinology* 1987; 12: 313–317.

Coble P, Kupfer DJ, Shaw DH: Distribution of REM latency in depression. *Biol Psychiat* 1981; 16: 453–466.

Cohen BM, Lipinsky JF, Altesmar RF: Lecithin in the treatment of mania. *Am J Psychiat* 1982; 139: 1162–1164.

Cozanitis DA: Galanthamine hydrobromide versus neostigmine. *Anaesthesia* 1974; 29: 163–168.

Crashaw JA, Mullen PE: A study of Benzhexol abuse. *Br J Psychiat* 145: 300–303.

Dasheiff RM, Savage DD, McNamara JO: Seizures down-regulate muscarinic cholinergic receptors in hippocampal formation. *Brain Res* 1982; 235: 327–334.

Daszuta A, Gambarelli G: Early postnatal development of EEG and sleep waking cycle in two inbred mouse strains. *Brain Res.* 1985; 354(1): 39–47.

Davis BM, Davis KL: Cholinergic mechanisms and anterior pituitary secretion. *Biol Psychiat* 1981; 303–310.

Delitala G, Maioli M, Pacifico A, Brianda SA, Palermo M, Manneli M: Cholinergic receptor control mechanisms for L-dopa, apomorphine and clonidine-induced growth hormone secretion in man. *J Clin Endocrinol Metab* 1983; 57: 1145–1149.

Dilsaver SC, Davidson RK: Chronic treatment with amitriptyline produces subsensitivity to the hypothermic effects of yohimbine. *Progr Neuropsychopharmacol Biol Psychiat* 1989; 13: 211–215.

Dilsaver SC, Hariharan M: Nicotinic effects of antidepressants. In: Gershon S, Lerer B (Eds.), *New Directions in Affective Disorders,* Springer, New York, 1989, 109–112.

Dilsaver SC, Majchrzak MJ: Bright artificial light produces subsensitivity to nicotine. *Life Sci* 1989; 42: 225–230.

Dilsaver SC: The pathophysiologies of substance abuse and affective disorders: An integrative model? *J Clin Psychopharmacol* 1987; 7: 1–10.

Dilsaver SC, Davidson RK: Effect of chronic cold room stress on the hypothermic response to oxotremorine. *Biol Psychiat* 1987a; 22: 1051–1052.

Dilsaver SC, Davidson RK: Fluoxetine subsensitizes a nicotinic mechanism involved in the regulation of core temperature. *Life Sci* 1987b; 41: 1165–1169.

Dilsaver SC, Snider RM, Alessi NE: Amitriptyline supersensitizes a central cholinergic mechanism. *Biol Psychiat* 22: 495–507.

Dilsaver SC: Cholinergic mechanisms in affective disorders: Future directions for investigation. *Acta Psychiat Scand* 1986a; 74: 312–334.

Dilsaver SC: Pathophysiology of cholinergic supersensitivity in affective disorders. *Biol Psychiat* 1986b; 21: 813–829.

Dilsaver SC: Cholinergic mechanisms in depression. *Brain Res Rev* 1986c; 11: 285–316.

Dilsaver SC, Greden JF: Antidepressant withdrawal-induced activation (hypomania and mania): Mechanisms and theoretical significance. *Brain Res Rev.* 1984; 7: 29–48.

Doerr P, Berger M: Physostigmine-induced escape from dexamethasone in normal subjects. *Biol Psychiat* 18: 261–268.

Dubé S, Sitaram N: Familial inheritance of REM sleep % and arecoline induction response in association with major affective illness. Abstracts of the Proceedings of the CINP, Munich, 1988.

Edelstein CK, Roy-Byrne P, Fawzy FI et al: Effects of weight loss on the dexamethasone suppression test. *Am J Psychiat* 1983; 140: 338–341.

Edwardson JA, Bennett GW: Modulation of corticotropin-releasing factor release from hypothalamic synaptosomes. *Nature* 1974; 251: 425–427.

Ehlert FJ, Roeske WR, Yamamura HI: Muscarinic cholinergic receptor heterogeneity. *Trends Neurosci* 1982; 5: 336.

El-Fakahany E, Richelson E: Effect of some calcium antagonists on muscarinic-mediated cyclic GMP formation. *J Neurochem* 40: 705–710.

El-Yousef MK, Janowsky DS, Davis JM et al: Induction of severe depression by physostigmine in marijuana intoxicated individuals. *Br J Addiction* 68: 321–325.

Ferrari E, Bossolo PA, Vailati A, Martinelli I, Rea A, Nosari I: Variations ciracadiennes des effects d'une substance vagolytique sur le system ACTH-secretant chez l'homme. *Ann Endocrinol (Paris)* 1977; 308: 203–213.

Frey KA, Ehrenkaufer S, Beaucage S, Agranoff BW: Quantitative in vivo receptor binding I. Theory and application of the muscarinic cholinergic receptor. *J Neurosci* 1985; 5: 421–428.

Frey KA, Ehrenkaufer RLL, Agranoff R: Quantitative in vivo receptor binding II. Autoradiographic imaging of muscarinic cholinergic receptors. *Neuroscience* 1985; 5: 2407–2414.

Friedmann JK: A diallel analysis of the genetic underpinnings of mouse sleep. *Physiol Behav* 12: 169–175.

Friedman MJ, Jaffe J: A central hypothermic response to pilocarpine in the mouse. *J Pharmacol Exp Ther* 1969; 167: 34–44.

Ganong WF: Neurotransmitters in pituitary function: Regulation of ACTH secretion. *Fed Proc* 1980; 39: 2923–2930.

Gershon S, Shaw FH: Psychiatric sequelae of chronic exposure to organophosphorus insecticides. *Lancet* 1961; i: 1371–1374.

Gibbs DN, Plotsky PM, deGreef WJ, Neill JD: Effect of histamine and acetylcholine on hypophyseal stalk plasma dopamine and peripheral prolactin levels. *Life Sci* 1979; 24: 2063–2070.

Gillin JC, Janowsky DS, Risch SC: The cholinergic REM induction test (CRIT): A replication. In: Chase MH, McGinty DJ, O'Connor C (Eds.), *Sleep Research*, Los Angeles: University of California, 1988a, Vol. 17, 123.

Gillin JC, Risch SC, Janowsky DS: The cholinergic REM delay test: A preliminary comparison of depressives and controls. In: Chase MH, McGinty DJ, O'Connor C (Eds.), *Sleep Research* 1988b, Vol. 17, 124.

Gillin JC, Sitaram N, Wehr T et al: Sleep and affective illness. In: Post RM, Ballenger JC (Eds.), *Neurobiology of Mood Disorders*, Baltimore: Williams & Wilkins, 1984, 157–189.

Gillin JC: Sleep studies in affective illness: Diagnostic, therapeutic, and pathophysiological implications. *Psychiat Ann* 1982; 13: 367–384.

Gillin JC, Duncan WC, Pettigrew KD, Frankel BL, Snyder F: Successful separation of depressed, normal, and insomniac subjects by EEG sleep data. *Arch Gen Psychiat* 1979b; 36: 85–90.

Gillin JC, Sitaran N, Duncan WC: Muscarinic supersensitivity: A possible model for the sleep disturbance of primary depression? *Psychiat Res* 7: 17–22.

Glen AIM, Ongley EC, Robinson K: Diminished membrane transport in manic-depressive psychosis and recurrent depression. *Lancet* 1968; 2: 241–242.

Golden RN, Potter WZ: Neurochemical and neuroendocrine dysregulation in affective disorders. *Psychiat Clin N Am* 1986; 9: 313–327.

Goldman ME, Erickson CK: Effects of acute and chronic administration of antidepressant drugs in the central cholinergic nervous system: comparison with anticholinergic drugs. *Neuropharmacology* 1983; 22: 1222–1285.

Grandison L, Meites J: Evidence for adrenergic medication of cholinergic inhibition of prolactin release. *Endocrinology* 1976a; 99: 775–779.

Hedge CA, De Wied D: Corticotropin and vasopressin secretion after hypothalamic implantation of atropine. *Endocrinology* 1971; 88: 1257–1259.

Hillhouse EW, Burden J, Jones MT: The effects of various putative neurotransmitters on the release of corticotropin releasing hormone from the hypothalamus of the rat in vitro. I. The effect of acetylcholine and nor-adrenaline. *Neuroendocrinology* 1975; 17: 1–11.

Hill P, Wyndber EL: Smoking and cardiovascular disease. *Am Heart J* 1974; 87: 491–496.

Hobson JA: The reciprocal interaction model of sleep cycle control: Discussion in the light of Giuseppe Moruzzi's concepts. In: Pompeiano O, Ajmone N (Eds.) *Brain Mechanisms and Perceptual Awareness,* New York: Raven, 1981, 379–404.

Hobson JA, McCarley RW, Wyzinksi PW: Sleep cycle oscillation: Reciprocal discharge by two brainstem neuronal groups. *Science 1975; 189: 55–58.*

Hobsen JA, McCarley RW, Pivik RT, Freedman R: (1974). Selective firing by cat pontine brain stem neurons in desynchronized sleep. *J Neurophysiol* 1974; 38: 497–511.

Hobson JA, McCarley RW: Spontaneous discharge rates of rat cerebellar purkinje cells in sleep and waking. *Electroenceph Clin Neurophysiol* 1972; 457–469.

Insel TR, Kalin NH, Cohen RM, et al: Dexamethasone suppression test in patients with obsessive compulsive disorder. *Psychiat Res* 1982; 6: 153–160.

Jackson MD: Hypothalamic releasing hormones: Mechanisms underlying neuroendocrine dysfunction in affective disorders. In: Brown GM, Koslow SH, Reichlin S (Eds.), *Neuroendocrinology and Psychiatric Disorder,* Raven Press, New York, 255–256.

Jaffe JH, Sharpless SK: Pharmacological denervation supersensitivity in the central nervous system. *Res Pub Assoc Res Nerv Ment Dis* (1968); 46: 226–241.

James SP, Wehr TA, Sack DA, Parry BL, Rosenthal NE: Treatment of seasonal affective disorder with light in the evening. *Br J Psychiat* 1985; 147: 424–485.

Janowsky DS, Risch SC, Huey LY, Kennedy B, Zigler M: Effects of physostigmine on pulse, blood pressure, and serum epinephrine levels. *Am J Psychiat* 1985; 142: 738–740.

Janowsky DS, Risch SC: Cholinomimetic and anticholinergic drugs used to investigate an acetylcholine hypothesis of affective disorders in stress. *Drug Dev Res* 1984; 4: 125–142.

Janowsky DS, Risch SC, Huey L, Judd L, Rousch J: Central physostigmine-induced cardiovascular and behavioral changes: Toward an acetylcholine hypothesis of stress. *Psychopharmacol Bull* 19: 675–682.

Janowsky DS, Risch SC, Gillin JC: Adrenergic-cholinergic balance and the treatment of affective disorders. *Prog. Neuro-Psychopharmacol Biol Psychiat* 1983b; 7: 297–307.

Janowsky DS, Judd LL, Groom G: The influence of lithium on cholinergic function. Proceedings of the 12th Collegium Internationale Neuro-Psychopharmacologicum Congress, Jerusalem, Israel, 1982a.

Janowsky DS, Risch SC, Huey LY, Judd LL, Rausch JL: Physostigmine-induced cardiovascular changes, behavioral and neuroendocrine correlations. Presented at the Annual Meeting of the American College of Neuropsychopharmacology, San Juan, Puerto Rico, 1982b.

Janowsky DS, Abrams AA, Groom GP, Judd LL, Clopton P: Lithium administration antagonizes cholinergic behavioral effects in rodents. *Psychopharmacology* 1979; 63: 147–150.

Janowsky DS, Davis JM, El-Yousef MK, Sekerke HJ: Acetylcholine and depression. *Psychosom Med* 1973a; 35(5): 459.

Janowsky DS, El-Yousef MK, Davis JM, Sererke HS: Parasympathetic suppression of manic symptoms of physostigmine. *Arch Gen Psychiat* 1973b; 38: 542–547.

Janowsky DS, El-Yousef MK, Davis JM, Sekerke HJ: Antagonistic effects of physostigmine and methylphenidate in man. *Am J Psychiat* 1973c; 130: 1370–1376.

Janowsky DS, El-Yousef MK, Davis JM, Sererke HS: A cholinergic adrenergic hypothesis of mania and depression. *Lancet* 1972; 2: 632–635.

Jellinec T, Gardos G, Cole J: Adverse effects of antiparkinsonian drug withdrawal. *Am J Psychiat* 1981; 138: 1567–1571.

Jimmerson DC, Nurnberger JI, Simmons S, Gershon ES: Anticholinergic treatment for depression. (Presented at the 135th Annual Meeting of the American Psychiatric Association, Toronto, Canada, May 15–21). *Syllabus* 1982; 218–219.

Jones D, Kelwala S, Bell J, et al: Cholinergic REM sleep induction response—Correlation with endogenous major depressive subtype. *Psychiat Res* 1985; 14: 99–110.

Jope RS: Effects of lithium treatment in vivo and in vitro on acetylcholine metabolism in rat brain. *J Neurochem* 1979; 33: 487–495.

Jouvet M: Cholinergic mechanisms and sleep. In: Waser PG (Ed.) *Cholinergic Mechanisms,* New York: Raven, 1975, 455–476.

Judd LL, Hubbard B, Janowsky DS, Huey LY, Attewell PA: The effect of lithium carbonate on affect, mood and personality of normal subjects. *Arch Gen Psychiat* 1977; 34: 346–351.

Kaminer Y, Munitz H, Wijsenbelc H: Trihyzyphenidyl (artane) abuse: Euphoriant and anxiolytic. *Br J Psychiat* 1982; 140: 473–474.

Karczmar AG, Longo VG, Sotti De Carolis A: A pharmacological model of paradoxical sleep: The role of cholinergic and monoamine systems. *Physiol Behav* 1976; 5: 175–182.

Kasper S, Moises HW, Beckman H: The anticholinergic biperidin in depressive disorders. *Pharmacopsychiatry* 1981; 14: 195–198.

Kaufmann CA, Gillin JC, Hill B, O'Laughlin T, Phillips I, Kleinman JE, Wyatt RJ: Muscarinic binding in suicides. *Psychiat Res* 1984; 12: 47–55.

Kelsoe JR, Gillin JE, Janowsky DJ, Brown JH, Risch SC, Lunkin B: Failure to conform muscarinic receptors on skin fibroblasts. *N Engl J Med* 1984; 312: 861–862.

Kelwala S: Enhanced pilocarpine miosis in depressed patients. Presented at the 137th Annual Meeting of the American Psychiatric Association in Los Angeles, California, May 5–11, NR25, 1984.

Kennedy S, Thompson R, Stancer HC, Roy A, Persad E: Life events precipitating mania. *Br J Psychiat* 1983; 142: 398–403.

Keneko FM: Neurotransmitters involved in ACTH secretion: Catecholamines. *Ann NY Acad Sci* 1977; 297: 509–517.

Krieg JC, Berger M: REM sleep and cortisol response to the cholinergic challenge with RS 86 in normals and depressives. *Acta Psychiat Scand* 1987; 76: 600–602.

Krieg JC, Berger M: Impact of the cholinergic system on the hypothalamic-pituitary-adrenocortical axis and on REM sleep. In: Lerer B, Gershon S (Eds.), *New Directions in Affective Disorders*, New York: Springer-Verlag, 71–74.

Kumbarachi MN, Nastuk WL: Effects of delta 9-tetrahydrocannabinol on excitable membranes and neuromuscular transmission. *Mol Pharmacol* 1980; 17: 344–345.

Kupfer DJ, Hanin I, Spiker DG, Neil J, Coble P: EEG sleep and tricyclic plasma levels in primary depression. *Commun Psychopharmacol* 1979; 3: 73–80.

Kupfer DJ: REM latency: A psychobiological marker for primary depressive disease. *Biol Psychiat* 1976; 11: 159–174.

Kupfer DJ, Foster FG: Interval between onset of sleep and rapid eye movement as an indicator of depression. *Lancet* 1972; 2: 684–686.

Leckman et al: Panic disorder and major depression. *Arch Gen Psychiat* 1983; 40: 1055–1060.

Lennox RH, Hitzemann RJ, Richelson E, Kelsoe JR: Failure to confirm muscarinic receptors on skin fibroblasts in familial affective disorder. *N Engl J Med* 1984; 311: 225–230.

Lewy AJ, Sack RL: Phase typing and bright light therapy of chronobiologic sleep and mood disorders. In: Halaris A (Ed.), *Chronobiology and Psychiatric Disorders*, New York: Elsevier, 1987, 181–206.

Lewy AJ, Sack RL, Fredrickson RH, et al: The use of bright light in treatment of chronobiologic sleep and mood disorders: The phase-response curve. *Psychopharmacol Bull* 1983; 9: 523–525.

Lewy AJ, Kern HA, Rosenthal NE, Wehr TA: Bright artificial light treatment of a manic-depressive patient with seasonal mood cycle. *Am J Psychiat* 1982; 139: 1496–1498.

Lindemann E, Finesinger JE: The subjective response of psychoneurotic patients to adrenaline and mecolyl. *Psychosom Med* 1940; 2: 231–246.

Lloyd C: Life events and depressive disorder: II. Events as precipitating factors. *Arch Gen Psychiat* 1980; 37: 541–548.

Lomax P, Jenden DJ: Hypothermia following systemic and intracerebral injection of oxotremorine in the rat. *Neuropharmacology* 1966; 5: 353–359.

Makara GB, Stark E: The effects of cholinomimetic drugs and atropine on ACTH release. *Neuroendocrinology* 1976; 21: 31–41.

McCarley RW: REM sleep and depression—common neurobiological control mechanisms. *Am J Psychiat* 1982; 139: 565–570.

Mendelson WB, Jacobs LS, Sitaram N, Wyatt RJ, Gillin JC: Methscopolamine inhibition of sleep-related growth hormone secretion. Evidence for a cholinergic secretory mechanism. *J Clin Invest* 1978; 61: 1683–1686.

Meyerson LR, Wennogle LP, Abel MS, Coupet J, Lippa AS, Rauh CE, Beer B: Human brain receptor alterations in suicide victims. *Pharmacol Biochem Behav* 1982; 17: 159–163.

Modestin JJ, Hunger J, Schwartz RB: Uber die depressogene wirkung von physostigmine. *Arch Psychiatr Nervenkr* 1973a; 218: 67.

Modestin JJ, Schwartz RB, Hunger J: Zur frage der beeinflussung schizophrener symptome physostigmine. *Pharmakopsychiatrie* 1973b; 9: 300–304.

Nassrallah HA, Coryell WA: Dexamethasone non-suppression predicts the antidepressant effects of sleep deprivation. *Psychiat Res* 1982; 6: 61–64.

Naumenko EB: Hypothalamic chemoreactive structures and the regulation of pituitary-adrenal function. Effects of local injections of norepinephrine, carbacol and serotonin into the brain of guinea pigs with intact brains and after mesencephalic transaction. *Brain Res* 1968; 11: 1–10.

Naylor GJ, Dick DAT, Dick EG, et al: Erythrocyte membrane cation carrier in mania. *Psychol Med* 1976; 6: 659–663.

Neil JF, Himmelhoch JM, Licata SM: Emergence of myasthenia gravis during treatment with lithium carbonate. *Arch Gen Psychiat* 1976; 33: 1090–1092.

Nomura Y, Kajiyama H, Segawa T: Possible influence of noradrenalin on B-adrenergic and muscarinic receptors in rat heart: Effects of 6-hydroxydopamine, isoproterenol and dimethylimipramine. In Segawa T (Ed.), *Molecular of Neurotransmitter Receptors*, New York: Raven Press, 1983, 83–90.

Nomura Y, Kajiyama H, Segawa T: Alteration in sensitivity to isoproterenol and acetylcholine in rat heart after repeated administration of isoproterenol. *J Pharmacol Exp Ther* 1982; 220: 441–445.

Nurnberg JI, Jr, Sitaram N, Gershon ES, Gillin JC: *Biol Psychiat* 1983; 18: 1161–1165.

Nurnberg J, Gershon ES, Sitaram N, Gillin JC, Brown G, Ebert M, Gold P, Jimmerson D, Kessler L: Dextroamphetamine and arecoline as pharmacogenetic probes in normals and remitted bipolar patients. *Psychopharmacol Bull* 1982a; 17: 80–82.

Nurnberg JI, Jimerson DC, Simmons S, Tammings C, Nadi NS, Gershon ES: Responses to arecoline in normal twins and "well state" patients with affective disorder. Presented at the Society of Biological Society, 13th Annual Convention, Toronto, 1982b.

Oppenheimer G, Ebstein R, Belmaker R: Effects of lithium on the physostigmine-induced behavioral syndrome and plasma cyclic GMP. *J Psychiat Res.* 1979; 14: 133–138.

Overstreet DH: Genetic animal model of depression with cholinergic supersensitivity. *New Directions Affect Disord* 1989; 14: 67–69.

Overstreet DH: Selective breeding for increased cholinergic function: Development of a new animal model of depression. *Biol Psychiat* 1986a; 21: 49–58.

Overstreet DH, Janowsky DS, Gillin JC, Shiromani P, Sutin E: Stress-induced immobility in rats with cholinergic supersensitivity. *Biol Psychiat* 1986b; 21: 657–664.

Overstreet DH, Booth RA, Dana R, Risch SC, Janowsky DS: Enhanced elevation of corticosterone following arecoline administration to rats selectively bred for increased cholinergic function. *Psychopharmacology* 1986c; 88: 129–130.

Overstreet DH: Selective breeding for increased cholinergic sensitivity: Implications for animal models of depression. *Soc Neurosci Abstr* 1984; 10: 289.

Overstreet DH, Russell RW: Selective breeding for differences in cholinergic function: Sex differences in the genetic regulation of sensitivity to the anticholinesterase, DFP. *Behav Neurol Biol* 1984; 40: 227–238.

Overstreet DH, Russell RW: Selective breeding for sensitivity to DFP: Effects of cholinergic agonists and antagonists. *Psychopharmacology* 1982; 78: 150–154.

Overstreet DH, Russell RW, Helps SC, Messenger M: Selective breeding for sensitivity to the anticholinesterase DFP. *Psychopharmacology* 1979; 65: 15–20.

Overstreet DH, Kozar MP, Lynch GS: Reduced hypothermic effects of cholinomimetic agents following chronic anticholinesterase treatment. *Neuropharmacology* 1973; 12: 1017–1032.

Palkovitz M: Neural pathways involved in ACTH regulation. *Ann NY Acad Sci* 1977; 297: 455–476.

Pert A, Rosenblatt J, Sivit C, Pert CB, Bunney WB: Long term treatment with lithium prevents the development of dopamine receptor supersensitivity. *Science* 1978; 201: 171–173.

Perry EK, Perry RH, Blessed G, Tomlinson BE: Necropsy evidence of central cholinergic deficits in senile dementia. *Lancet I* 1977; I: 189.

Pullen GP, Best NR, Maguire J: Anticholinergic drug abuse. A common problem? *Br Med J* 1984; 2xx: 612–613.

Rasmussen KLR, Reite M: Loss-induced depression in an adult Macaque monkey. *Am J Psychiat* 1982; 139: 681–697.

Richardson SB, Hollander CS, D'Elletto R, Greenleaf PW, Thaw C: Acetylcholine inhibits the release of somatostatin from rat hypothalamus in vitro. *Endocrinology* 1980; 107: 122–129.

Risch SC, Janowsky DS, Parker D, et al: Neuroendocrine abnormalities in affective disorders: Possible cholinergic mechanisms. In Post RM, Ballanger J (Eds.), *The Neurobiology of Mood Disorders*, Baltimore: Williams & Wilkins, 1984.

Risch SC, Janowsky DS, Gillin JC: Muscarinic supersensitivity of.anterior pituitary ACTH and B-endorphin release in major depressive illness. *Peptides* 1983; 4: 789–792.

Risch SC, Kalin NH, Janowsky DS, Cohen RM, Pickar D, Murphy DL: Response to central cholinergic stimulation. *Science* 1983; 77: 222.

Risch SC: B-endorphin hypersecretion in depression: Possible cholinergic mechanisms. *Biol Psychiat* 1982; 17: 1071–1079.

Risch SC, Janowsky DS, Judd LL, Rausch JL, Huey LY, Beckman KA, Cohen RM, Murphy DL: Cholinergically stimulated hypothalamic B-endorphin modulation of anterior pituitary prolactin release in humans. *Peptides* 1982; 3: 319–322.

Risch SC, Cohen PM, Janowsky DS, Kalin N, Sitaram N, Gillin JC, Murphy DL: Physostigmine induction of depressive symptomatology in normal human subjects. *J Psychiat Res* 1981a; 4: 89–94.

Risch SC, Kalin NH, Janowsky DS: Cholinergic challenges in affective illness: Behavioral and neuroendocrine correlates. *J Clin Psychopharmacol* 1981b; 1: 186–192.

Risch SC, Cohen RM, Janowsky DS, Kalin NH, Murphy DL: Plasma B-endorphin and cortisol elevations accompany the mood and behavioral effects of physostigmine stimulation. *Science* 1980; 209: 1545–1546.

Rosenberg RS, Bergmann BM, Jeonson H, Arnason BG, Rechtschaffen A: Strain differences in sleep of rats. *Sleep* 1987; 10(6): 537–541.

Rosenblatt JE, Janowsky DS, Davis JM, et al: The augmentation of physostigmine toxicity in the rat by delta 9-tetrahydrocannabinol. *Res Commun Chem Pathol Pharmacol* 1972; 3: 479–481.

Rosenthal NE, Carpenter CJ, James SP, Parry BL, Rogers SLB, Wehr TA: Seasonal affective disorder in children and adolescents. Amer J Psychiat 1986; 3:356–358.

Rosenthal NE, Sack DA, Gillin JC, Levy JA, Goodwin FK, Davenport Y, Mueller PS: Seasonal affective disorder: A description of the syndrome and preliminary findings with light therapy. *Arch Gen Psychiat* 1984; 41: 72–80.

Rowntree DW, Neven S, Wilson A: The effects of di-isopropylfluorophosphonate in schizophrenia and manic depressive psychosis. *J Neurol Neurosurg Psychiat* 1950; 13: 47–62.

Rubinstein JS: Abuse of antiparkinsonism drugs: Feigning of extraphyramidyl symptoms to obtain trihexphenidyl. *J Am Med Assoc* 1978; 239: 2365–2366.

Sagelas T, Erill S, Domino EF: Differential effects of scopolamine and chlorpromazine on REM and NREM sleep in normal male subjects. *Clin Pharmacol Ther* 1969; 10: 522–529.

Samples J, Janowsky DS, Pechnick R, Judd LL: Lethal effects of physostigmine plus lithium in rats. *Psychopharmacology* 1977; 52: 307–309.

Shiromani PJ, Overstreet D, Levy D, Goodrich CA, Campbell SS, Gillin JC:

Increased REM sleep in rats selectively bred for cholinergic hyperactivity. *Neuropsychopharmacology* 1988; 1: 127–133.

Shopsin B, Janowsky DS, Davis JM, Gershon S: Rebound phenomena in manic patients following physostigmine. *Neuropsychobiology* 1975; 1: 180–187.

Shur E, Checkley S, Delgado I: Failure of mianserin to affect autonomic function in the pupils of depressed patients. *Acta Psychiat Scand* 1983; 67: 50–55.

Sitaram N, Nurnberger J, Gershon E, Gillin JC: Faster cholinergic REM sleep induction in euthymic patients with primary affective illness, *Science* 1980; 208: 200–202.

Sitaram N, Nurnberg J, Gershon ES, Gillin JC: Cholinergic regulation of mood and REM sleep: A potential model and marker for vulnerability to affective disorder. *Am J Psychiat* 1982; 139: 571–576.

Sitaram N, Gillin JC: Development and use of pharmacological probes of the CNS in man: Evidence of cholinergic abnormality in primary affective illness. *Biol Psychiat* 1980; 15: 925–955.

Sitaram N, Nurnberger J, Gershon E, Gillin JD: Faster cholinergic REM inductions in euthymic patients with primary affective illness. *Science* 1980; 208: 200–202.

Sitaram N, Mendelson WB, Wyatt RJ, Gillin JC: The time dependent induction of REM sleep and arousal by physostigmine infusion during normal human sleep. *Brain Res* 1977; 122: 562–567.

Sitaram N, Dubé S, Keshavan M, Davies A, Reynal P. The association of supersensitive cholinergic REM-induction and affective illness within pedigrees. *J Psychiat Res* 1987; 21: 487–497.

Sitaram N, Moore AM, Gillin JC: Scopolamine-induced muscarinic supersensitivity in normal man: Changes in sleep. *Psychiat Res* 1979; 1: 9–16.

Spar JE, Gerner R: Does the dexamethasone suppression test distinguish dementia from depression? *Am J Psychiat* 1982; 139: 238–240.

Spiker DG, Coble P, Cofsky J, Foster FG, Kupfer DJ: EEG sleep and severity of depression. *Biol Psychiat* 1978; 13: 485–488.

Stanley M, Mann JJ: Increased serotonin-2 binding sites in frontal cortex of suicide victims. *Lancet* 1983; I: 214.

Subramanian MG, Gala RR: The influence of cholinergic, adrenergic and serotonergic drugs on the afternoon surge of plasma prolactin in overiestomized, estrogen-treated rats. *Endocrinology* 1976; 98: 842–848.

Taylor P: Cholinergic agonists. In: Gilman AG, Goodman LS, Gilman A (Eds.), *The Pharmacological Basis of Therapeutics*, 6th ed, New York: Macmillan, 1980.

Terry LC, Crowley WR: Regulation of growth hormone and thyrotropin secretion by somatostatin systems in rat brain. *Neuroendocrinology* 1986; 42: 218–225.

Tollefson GD, Senogles SE, Frey WH, et al: A comparison of peripheral and central human muscarinic cholinergic receptor affinities for psychotropic drugs. *Biol Psychiat* 1982;17: 555–567.

Trusihiya K, Toru M, Kobayashi T: Sleep deprivation: Changes of monoamine and acetylcholine. *Life Sci* 1969; 8: 867–873.

Uhde T, Post RM, Bierer L: Caffeine-induced escape from dexamethasone suppression. *Arch Gen Psychiat* 42: 737–738.

Valatx JL, Caspuglio R, Paut L, Bailey DW: Genetic study of paradox sleep in mice. *Waking Sleep* 1980; 4(2): 175–183.

Webb WB, Campbell SS: Sleep characteristic relationship of identical and fraternal twins. *Arch Gen Psychiat* 1983; 40: 1093.

Wehr TA, Goodwin FK: Rapid cycling manic-depressives induced by tricyclic antidepressants. *Arch Gen Psychiat* 1979; 15: 149–154.

Weiss JM, Goodman PA, Losito BG, Corrigan S, Charry JM, Bailey WH: Behavioral depression produced by an uncontrollable stressor: Relationship to norepinephrine, dopamine and serotonin levels in various regions of rat brain. *Brain Res Rev* 1981; 3: 167–205.

Wilkins JN, Carlson HE, van Vunakis H, et al: Nicotine from cigarette smoking increases circulating levels of cortisol, growth hormone, and prolactin in male chronic smokers. *Psychopharmacology* 1982; 70: 305–308.

Yoshimura H, Fujiwara M, Ueki S: Biochemical correlates in mouse-killing behavior of the rat: Brain acetylcholine and acetylcholinesterase after administration of delta 9-tetrahydrocannabinol. *Brain Res* 1974; 81: 567–570.

Chapter 4

GABA AND DEPRESSION

Frederick Petty, Gerald L. Kramer, and William Hendrickse

INTRODUCTION

From a historical perspective, the amino acid neurotransmitters are recent candidates for consideration in theories of psychiatric illness. The biogenic amines, particularly norepinephrine and serotonin, have traditionally played a major role in theories regarding the biochemical basis of depression and of its treatment. This is readily explained by a general focus in neuroscience on biogenic amine neurochemistry and neuropharmacology during the formative decades of biological psychiatry. Also, it reflects the early observation that depletion of the biogenic amines with agents such as reserpine was often accompanied by the development of depressive symptoms. Similarly, the first-generation antidepressants were early reported to enhance biogenic amine activity through inhibition of monoamine oxidase or blockade of synaptic reuptake. Thus the catecholamine theory of depression was first published about 25 years ago.

By comparison, the major amino acid neurotransmitter, γ-amino butyric acid (GABA), was not postulated to play a role in mood disorders until about twelve years ago (Emrich, 1980).

A great deal of evidence has accumulated during the last decade to

Frederick Petty, Gerald L. Kramer, and William Hendrickse • Veterans Affairs Medical Center and Department of Psychiatry, University of Texas Southwestern Medical School, Dallas, Texas 75216.

Biology of Depressive Disorders, Part A: A Systems Perspective, edited by J. John Mann and David J. Kupfer. Plenum Press, New York, 1993.

suggest that GABA probably plays an important role, both in the pathophysiology of mood disorders, particularly depression, and in the mechanism of action of antidepressant drugs, as well as electroconvulsive shock. In fact there are now sufficient data to permit the formulation of a GABA hypothesis of mood disorders (Bartholini, Lloyd, Scatton, Zivkovic, & Morselli, 1985; Lloyd, Morselli, & Bartholini, 1987; Lloyd, Zivkovic, Scatton, Morselli, & Bartholini, 1989).

GABA is produced in the brain by decarboxylation of glutamic acid with the enzyme glutamate decarboxylase (GAD). The metabolism of GABA is accomplished by GABA transaminase (GABA-T) and succinyl semialdehyde dehydrogenase. These metabolic pathways are relatively simple. The enzyme for GABA synthesis is intraneuronal and exclusive to presynaptic nerve terminals, while GABA metabolism may be accomplished in glia as well. GABA is a rather ubiquitous neurotransmitter accounting for about one-third of neurons in the mammalian central nervous system. It is also fairly specific to the central nervous system, although small amounts of GABA are found in peripheral tissue such as ovary and adrenal. In contrast to the biogenic amine neurotransmitters, which can be experimentally depleted, pharmacologic manipulations that significantly reduce the levels of GABA in brain result in severe convulsions and eventual death. It should therefore be no surprise that a number of anticonvulsants act by enhancing GABA transmission. It is possible to increase brain levels of GABA by inhibiting the enzyme of GABA degradation without dramatic effects on physiology or behavior. GABA, itself, does not readily cross the blood–brain barrier in that direction, although there is evidence that it moves readily from brain to blood.

GABA is usually considered an inhibitory neurotransmitter. This leads to some conceptual difficulties regarding its behavioral effects. It should be emphasized that most central nervous system function operates through inhibition and that the biogenic amine neurotransmitters themselves are generally inhibitory. Furthermore, inhibition at a neurophysiologic level should not be translated into inhibition at the level of behavior. Even the simplest behavior requires a complicated interaction of both inhibitory and excitatory neurotransmitters. Finally, it is likely that much of GABA activity in areas of the limbic system and cortex pertinent to depression is involved in local circuits and interneurons, which, as will be noted in the concluding part of this chapter, may actually be facilitatory to other neurotransmitters, particularly norepinephrine. The reader is encouraged to examine the data from a neutral perspective and avoid the simplistic notion that "since GABA is inhibitory it is therefore depressogenic." In point of fact it is neither in the limbic system.

The evidence that GABA plays a role in mood disorders and particularly in depression will be presented, beginning with preclinical findings regarding the interactions of antidepressants with the GABA system, where the "up-regulation" of $GABA_B$ receptors appears to provide a unifying framework for antidepressant activity. Next, knowledge obtained from GABA research on animal models of depression is discussed. The second part of this chapter will examine the clinical evidence relating to GABA and mood disorders, particularly measurement of GABA in brain, CSF, and blood from patients suffering with depression. Use of GABA active drugs is briefly reviewed. Finally, a unifying theory of the neurochemistry of mood disorders is presented in an attempt to integrate the major factual and conceptual aspects of GABA into the biogenic amine hypotheses.

Preclinical Studies

Treatments for Mood Disorders Affect GABA Function

Antidepressant Drugs.The action of tricyclic antidepressant drugs and "second-generation" antidepressants has been most extensively studied with respect to the biogenic amines. However, data have accumulated that generally support an increase in GABA function caused by antidepressants. Using push–pull perfusion, Korf and Venema (1983) demonstrated that desipramine enhances the release of endogenous GABA from rat thalamus. In a similar fashion, Sherman and Petty (1982) showed that treatment with imipramine caused an increase in release of endogenous GABA from tissue slices of hippocampus. In a similar fashion, treatment with a variety of antidepressants including imipramine, nomifensin, desipramine, zimelidine, buproprion, maprotiline, and adinazolam on a chronic basis produced a decreased density of $GABA_A$/benzodiazepine receptors in rat brain (Suzdak & Gianutsos, 1985b; Suzdak & Gianutsos, 1986; Suranyi-Cadotte, Dam, & Quirion, 1985; Barbaccia, Ravizza, & Costa, 1986). This decreased B_{max} is compatible with an increased availability of the agonist, presumably GABA, at the receptor. It is of interest that a parallel decrease in the density of noradrenergic β-receptors was also noted in cortex and hippocampus (Suzdak & Gianutsos, 1985b). In a similar fashion, 23 clinically effective antidepressants and 9 of their metabolites were found to fully or partially reverse the inhibitory action of GABA on TBPS binding

(Squires & Saderup, 1988). The latter finding was interpreted as a blockade of GABA inhibition at the picrotoxin binding site linked to the $GABA_A$/benzodiazepine receptor and was similar to the effect of $GABA_A$ receptor agonists. It is pertinent to note that the antidepressants tested showed a GABA agonist-like effect for a variety of different chemical structures ranging from the tricyclic doxepin to the serotonin reuptake blocker citalopram and the experimental antidepressant methylene blue. However, this GABAergic activity was not found for several other antidepressants tested, including desipramine and imipramine. In summary, there is evidence that antidepressants can directly stimulate the release of endogenous GABA and can have effects similar to those of GABA agonists on $GABA_A$ receptor binding.

Lithium. The chronic administration of lithium to rats in doses that produce blood levels similar to those therapeutic in humans leads to an increase in GABA turnover and to an increase in GABA levels in frontal cortex, as well as a decrease in GABA-T activity (Rubio & Losada, 1986). Chronic administration of lithium was also reported to increase the brain concentration of GABA in hypothalamus and amygdala (Gottesfeld, 1976). Chronic low-dose lithium was also reported to significantly increase the levels of GABA in the striatum, midbrain, and pons-medulla, an effect that persisted two days after discontinuation of lithium (Ahluwalia, Grewaal, & Singhal, 1983). This chronic treatment with lithium led to an increase in the activity of GAD in striatum and pons-medulla. However, studies reporting GABA brain concentrations should be interpreted with caution since total levels include GABA which is present in glia and in the neuronal metabolic pool as well as neurotransmitter GABA. Also, since the GABA metabolizing enzyme GABA-T is oxygen dependent while GAD is not, levels of GABA can rise dramatically postmortem. An increase in plasma GABA by lithium in bipolar patients was also reported (Berrettini, Nurnberger, Hare, Gershon, & Post, 1982). Overall the research supports an effect of lithium increasing or enhancing GABA function.

Electroconvulsive Shock. The data on GABA and electroconvulsive shock are somewhat more complex and difficult to interpret. Sackeim et al. (1983) proposed that the mechanism of action of electroconvulsive shock was through an enhancement of GABA activity, which they also suggested to account for the anticonvulsant effects of ECS. Indeed, following a series of electroconvulsive shocks, GABA concentrations in striatum, hippocampus, frontal cortex, hypothalamus, and olfactory bulbs are noted to increase (Bowdler, Green, Minchin, & Nutt, 1983;

Ferraro, Golden, & Hare, 1990). However, repeated electroconvulsive shock did not change spontaneous release of GABA from tissue slices of cortex, hippocampus, or striatum, but did decrease potassium stimulated GABA release (Green & Vincent, 1987a). These investigators replicated earlier work demonstrating an increased GABA content in striatum and showed the synthesis rate to be inhibited in striatum and hippocampus but not in cortex. Green and Vincent proposed that repeated electroconvulsive shock led to long-term inhibition of GABA synthesis, which would seem contradictory with their finding of increased GABA content. Additional research is needed to clarify these issues, particularly in light of the regional differences found and the recent report of differences in effects on the GABA system depending on electrode placement (Ferraro et al., 1990).

The $GABA_B$ Receptor and the Mechanism of Action of Antidepressant Treatments. The best documented neurochemical effect of chronic antidepressant treatment continues to be "down-regulation" of the postsynaptic noradrenergic β-receptor. Recently Lloyd and co-workers (1985) have demonstrated that chronic treatment with antidepressants also leads to an increase in the density of $GABA_B$ receptors in the frontal cortex of rats. As shown on Table 1 this effect is found with a variety of antidepressant drugs, including norepinephrine, dopamine, and serotonin reuptake inhibitors, monoamine oxidase inhibitors, atypical antidepressants, GABA-mimetics, and electroconvulsive shock. These findings were initially controversial, with at least one laboratory reporting a failure to replicate (Cross & Horton, 1987). However, other laboratories have subsequently replicated the basic findings (Suzdak & Gianutsos, 1986) and have extended them to demonstrating increased $GABA_B$ receptor binding sites in hippocampus (but not frontal cortex) following chronic treatment with lithium and carbamazepine (Motohashi, Ikawa, & Kariya, 1989). Furthermore, an enhancement of the baclofen-induced inhibition of serotonin released from slices of mouse frontal cortex was noted to follow chronic administration of electroconvulsive shock, amitriptyline, desipramine, mianserin, or zimelidine (Gray & Green, 1987). These authors interpreted their findings as demonstrating that chronic antidepressant treatment increased $GABA_B$ receptor function in frontal cortex with respect to its modulation of serotonin release. This is consistent with the reports of increased $GABA_B$ receptor number in this region after antidepressant treatment.

Thus it seems that in addition to their "down-regulation" of post synaptic noradrenergic β-receptors, somatic therapies for mood disorders including most classes of antidepressant drugs, electroconvulsive

Table 1. Effect of Chronic Treatment on Frontal Cortex $GABA_B$ Binding[a]

Compound (mg/kg/d)	Percent of control
A. *Uptake inhibitors*	
Nonspecific	
Amitriptyline (10)	154
Noradrenaline	
Desipramine (5)	151
Maprotiline (10)	135
Viloxazine (10)	188
Dopamine	
Nomifensine (5)	146
Buproprion (5)	154
Serotonin	
Citalopram (10)	172
Fluoxetine (10)	183
Zimelidine (10)	165
B. Monoamine oxidase inhibitor	
Pargyline (20)	142
C. Electroshock	
150mA, 0.5 sec/48h	177
D. GABAmimetics	
Progabide (100)	127
Fengabide (50)	159
Sodium Valproate (100)	120
E. Miscellaneous compounds	
Tazodone (10)	155
Iprindole (5)	129
Mianserine (10)	133
Carbamazepine[b] (50)	139
Lithium[b] (1.5 mg)	129

[a] Compiled from Lloyd et al. (1985) and Motohashi et al., (1989).
[b] Hippocampus. No effect in frontal neocortex

shock, lithium, carbamazepine, progabide, and sodium valproate share a common neurochemical mechanism of action with chronic administration in "up-regulating" the $GABA_B$ receptor. On the surface, "up-regulation" by putative receptor agonists appears to be paradoxical, since classic neuropharmacology usually suggests the opposite, that increased agonist activity leads to decreased numbers of receptors. These $GABA_B$ effects are not yet completely understood, but as pointed out by Lloyd et al. (1985), an increase in the number or density of receptors following chronic treatment with an agonist for that receptor is not unique for the $GABA_B$ receptor since increased dopamine receptor density has also been reported to follow administration of dopamine receptor agonists. Additional research is needed to clarify these issues.

From a clinical viewpoint, these findings suggest that the "upregulation" of the $GABA_B$ receptors is a property shared by antimanic as well as antidepressant agents. Generally speaking, the data regarding the antidepressant efficacy of valproate, lithium, and carbamazepine, at least for the treatment of unipolar depression, is disappointing. However, all three are certainly antimanic agents. Additionally, electroconvulsive shock is an effective treatment for both mania and depression. The effect on $GABA_B$ receptors may reflect a broader spectrum of effects on mood disorders than just antidepressant action.

Animal Model Studies Support a GABA Deficit in Depression

During the last 10 years considerable evidence has accumulated to support the use of animal models in research on the etiology and pathophysiology of mood disorders. Three models in rats have attracted considerable attention and are generally considered to be the most useful, if not the most valid, rodent models of depression in humans. These are behavioral despair, olfactory bulbectomy, and learned helplessness. The interested reader is referred to several excellent recent reviews (Zacharko & Anisman, 1989; Overmier & Hellhammer, 1988).

Behavioral Despair. This model was developed by Porsolt et al. (1978) as a screen for potential antidepressants. It is based on the observation that when a rat or mouse is placed in a cylinder of water from which it cannot escape, it will swim vigorously for a few minutes and then assume a position of immobility with its nose above the surface of the water. On subsequent trials, the animal assumes immobility more rapidly. Antidepressant treatments including tricyclics, atypical antidepressants, electroconvulsive shock, and an "enriched environment," decrease the duration of immobility. That is to say, animals treated with antidepressants continue their vigorous swimming attempts to escape for a longer period of time. Psychomotor stimulants such as an amphetamine are false positives in this test; therefore the complete evaluation includes a test of general open field activity to eliminate nonantidepressant drugs active on the behavioral despair test for reason of increased motor activation only. Otherwise the test has reasonable specificity, with a variety of other psychotropic drugs not active in the model. A reduction of GABA levels in nucleus accumbens, brain stem, and cortex was observed in rats after a pretest session for behavioral immobility (Borsini, Mancinelli, D'Aranno, Evangelista, & Meli, 1987). This reduction in GABA was compensated by administration of aminooxy acetic acid (AOAA), which also reduced immobility time. The GABA deficit was not found in hippocampus, hypothalamus, or olfactory bulb. A series of GABA agonists were tested in this animal model with no effect demonstrated for the $GABA_B$ agonist baclofen. On the other hand $GABA_A$ agonists including AOAA, muscimol, and THIP had an antidepressant-like effect in the model, reducing immobility in a dose-dependent manner (Borsini, Evangelista, & Meli, 1986; Borsini et al., 1987). Also, the GABA-mimetic antidepressant progabide has been shown to have behavioral activity in this model (Lloyd, Morselli, Depoortere, Fournier, Zivkovic, Scatton, Broekkamp, Worms, & Bartholini, 1983).

Olfactory Bulbectomy. This model utilizes rats and involves surgical removal of their olfactory bulbs. After the animals have recovered they demonstrate a pervasive deficit in passive-avoidance learning, which can be reversed with an excellent degree of pharmacologic specificity by antidepressants, and also by treatment with electroconvulsive shock. These treatments are only effective if they are administered on a chronic basis, similar to their therapeutic effect in humans. In bulbectomized rats $GABA_B$ receptor binding in frontal cortex was decreased by about 50%. There were not changes in other brain regions studied including occipital cortex, hippocampus, and amygdala (Lloyd & Pichat, 1985). When the behavioral deficits included by olfactory bulbectomy were reversed in animals by treatment with desipramine for 13 days, there was a correlation between $GABA_B$ receptor binding in frontal cortex and response to desipramine; that is to say, responders showed a higher $GABA_B$ receptor density than nonresponders or controls (Joly, Lloyd, Pichat, & Sanger, 1987). Additional evidence for a GABA deficit in the bulbectomy model is provided by the reversal of behavioral deficits in this model with the $GABA_B$ agonist baclofen (Leonard & Tuite, 1981) and the mixed GABA agonist progabide (Lloyd et al., 1983). Thus in the bulbectomy model, the "depressed" rats have $GABA_B$ receptor density opposite to that caused by antidepressant treatments, and both desipramine and GABA agonists reverse the behavioral deficit as well as the neurochemistry.

Learned Helplessness. When dogs were given inescapable foot shock, about three-fourths demonstrated subsequent inability to perform a simple escape task in a shuttle box (Seligman & Maier, 1967). This stress-induced behavioral depression was called learned helplessness, and a similar behavioral deficit has since been demonstrated in a wide variety of species including rats and humans. The behavioral deficit can be reversed by treating rats with antidepressants including tricyclics, monoamine oxidase inhibitors, atypical antidepressants, and electroconvulsive shock. Analogous to their therapeutic effects in humans, these treatments do not reverse helplessness if administered acutely but only if given over a period of several days. The pharmacologic specificity of the learned helplessness model is good since phenothiazine and other neuroleptics, antihistamines, benzodiazepines, barbiturates, alcohol, stimulants, caffeine, and anticholinergics do not reverse the behavior (Sherman & Petty 1982). Of particular interest is the finding that benzodiazepines, which are thought to exert their anxiolytic activity through the $GABA_A$ receptor, prevent the development of learned helplessness and can do so after only one dose (Sherman, Allers, Petty, & Henn, 1979; Drugan, Ryan, Minor, & Maier, 1984). An anxiogenic β-carboline benzodiazepine receptor ligand can induce helpless behavior in naive animals (Drugan, Maier, Skolnick, Paul, & Brawley, 1985). These findings suggest that an element of anxiety is necessary for the development of depression in this model.

An additional involvement of the GABA system was demonstrated by the finding that animals given controllable shock demonstrated a protection against bicuculline-induced seizures that was not seen in inescapably stressed rats (Drugan, Maier, Skolnick, Paul, & Crawley 1985). Furthermore the development of stress-induced behavioral depression was found to be associated with a decrease in GABA receptor-mediated chloride channel function as shown by benzodiazepine receptor occupancy and chloride ion flux studies (Drugan, Morrow, Weizman, Weizman, Deutsch, Crawley, & Paul, 1989).

Microinjections of GABA into frontal neocortex and hippocampus prevented the development of learned helplessness, while injection of GABA into hippocampus can reverse helpless behavior. Learned helplessness could be induced in naive rats by injection of bicuculline into hippocampus (Sherman & Petty, 1980). Injection of bicuculline into hippocampus, effectively blocking $GABA_A$ receptors, also prevented cortical injection of tricyclic antidepressants from reversing helpless behavior, suggesting that the hippocampal GABA system plays a crucial role in both development and reversal of learned helpless behavior. Additional support for this is provided by the finding that release of endogenous GABA from hippocampal slices of animals subjected to inescapable stress is decreased while chronic administration of imipramine in naive rats increased hippocampal GABA release. In helpless animals receiving chronic imipramine, these effects were additive, with the neurochemical deficits restored in the same time frame as the learned helplessness behavior is reversed (Sherman & Petty, 1982). Neurotransmitter flux through the small neuronal hippocampal GABA pool is also decreased in helpless rats (Petty, Sherman, & Saquitne, 1981). GABA agonists such as muscimol and progabide can reverse the helpless behavior caused by stress (Lloyd et al, 1983; Poncelet, Martin, Danti, Simon, & Soubrie, 1987).

In a complex series of studies Plaznik et al. (1988) concluded that decreased activity of brain norepinephrine, serotonin, dopamine, and GABA all contributed to behavioral suppression after inescapable stress, emphasizing not only the involvement of GABA but the complexity of the neuronal systems involved.

The antidepressant activity of GABA agonists in learned helplessness was not reduced by picrotoxin, a $GABA_A$ receptor antagonist, but was influenced by prazosin, an α-adrenergic blocker as well as by penbutolol, a β-receptor blocker (Poncelet et al., 1987). These and similar findings led these investigators to conclude that the functional activity of GABAergic drugs in reversing animal depression (and perhaps also human depression) involves a final common pathway of noradrenergic rather than GABAergic processes. Our conceptual theoretical model of the neurochemistry of learned helplessness (Sherman & Petty, 1982; Petty, McChesney & Kramer, 1985) (Fig. 1) incorporates this. As detailed below, a unified hypothesis of antidepressant activity can be formulated explaining GABAergic function as a mediation of noradrenergic transmittal.

The combined evidence from several laboratories on GABA and learned helplessness therefore strongly suggests that a GABA deficit is produced by inescapable stress. This deficit is functionally responsible for the subsequent behavioral deficit called learned helplessness. Restoration of this GABA deficit leads directly to the reversal of the behavioral depression.

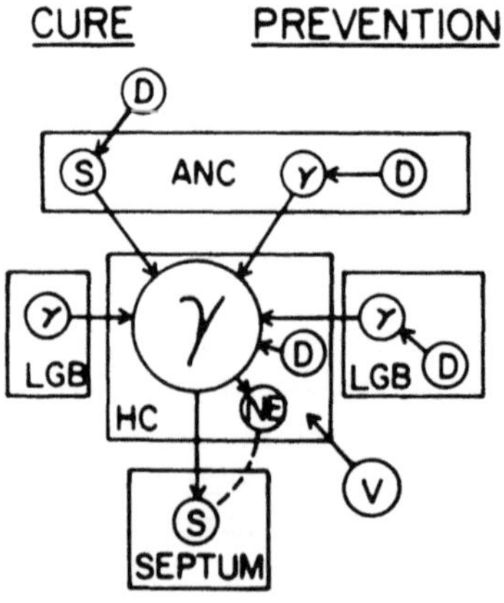

Figure 1. Neurochemistry of learned helplessness. Hypothetical map from data obtained with regional microinjections and tissue release. Abbreviations: FNC, frontal neocortex; HPC, hippocampus; SEP, septum; LGB, lateral geniculate body; S, serotonin; NE, norepinephrine; γ, GABA; V, diazepam; D, desipramine. Directionality of pathways was determined by blocking GABA in hippocampus with bicuculline. Also, microinjection of glutamate into cortex caused helplessness in naive rats (Petty et al., 1985).

In summary, preclinical research supports a GABA hypothesis of mood disorders:

1. Somatic therapy for mood disorder facilitates GABA.
2. Somatic treatments for mood disorder "up-regulate" the $GABA_B$ receptor, but only if given chronically.
3. Animal models of depression show GABA deficits, with GABA agonists serving as antidepressants and antidepressants activating GABA in the models.

Clinical Studies

Clinical studies supporting a role for GABA in mood disorders involve examination of GABA activity in brain, spinal fluid, or blood from individuals suffering from depression, and other mood disorders as well as studies of the clinical effect of GABA agonists on patients with mood disorders.

Brain

Activity of the GABA-synthesizing enzyme glutamate decarboxylase (GAD) was studied in eight regions of postmortem brain by Perry et al. (1977). These investigators compared multi-infarct dementia, dialysis encephalopathy, and senile dementia of the Alzheimer type to depression and to non-mentally ill patients, and found that GAD activity was significantly decreased in frontal cortex, occipital cortex, caudate nucleus, and substantia nigra of the subjects with depression compared to control. This decreased activity was not found in temporal cortex, parietal cortex, hippocampus, or mamillary bodies. In contrast, Korpi et al. (1988) failed to find any significant differences between concentrations of GABA in five brain areas including frontal cortex, and amygdala in postmortem brains of suicide victims compared with normal control subjects. On the other hand, Cheetham et al. (1988) found the density of brain $GABA_A$/benzodiazepine receptor sites to be significantly greater in frontal cortex of depressed suicides compared to controls. This increase in receptor binding could not be attributed to drug effects and could be interpreted as reflecting a chronically lowered GABA activity in frontal cortex of the depressives. The same group also failed to replicate Perry's finding (Cheetham, Cromptom, Katona, Parker, & Horton, 1988) by finding GAD activity to be the same in frontal and temporal cortex of depressed suicides as controls. Cross et al. (1987) have also reported a negative study in which $GABA_B$ receptors in cortex and hippocampus of depressed suicides were not altered compared to control subjects. However, GABA levels from cortex of depressed patients undergoing psychosurgery for intractable psychiatric illness including obsessive-compulsive disorder and depression had a significantly negative correlation with mood (Honig et al., 1988). That is to say, patients with the lowest GABA levels in cortex had the worst depression as measured by ratings scales. This correlation was not seen for other amino acids.

The available studies of GABA function in brain are suggestive, in some cases, of a GABA deficit associated with depression, but also include some studies not demonstrating such a correlation.

Cerebrospinal Fluid

There are six reports in the literature of GABA levels in cerebrospinal fluid of patients with depression and one of patients diagnosed with schizoaffective disorder. All these studies report GABA levels in depressed patients to be lower than those of control (Table 2).

Four of the six studies that measured CSF GABA in depressed patients found it to be significantly lower, with the two negative studies employing either a different aliquot of CSF (Post et al., 1980) or ambiguous diagnostic criteria (Zimmer et al., 1980). Generally CSF GABA levels are decreased to between 55% and 92% of control values. In the largest and best designed study (Gerner et al., 1984) this difference between depression and control was highly significant and patients with schizophrenia or mania were not significantly different from control.

There is considerable evidence to suggest that CSF levels of GABA reflect brain GABA activity (Loscher, 1982; Grove et al., 1982). Certainly the consensus of evidence of CSF studies in depression supports a GABA deficit hypothesis of depression.

The specificity of the findings with GABA in depression in brain and spinal fluid has not been widely studied, but the few reports that have examined GABA function in brain and CSF in schizophrenia have not shown any consistent abnormality (Lloyd & Morselli, 1987).

Blood

Most studies on peripheral GABA function in psychiatric patients have studied plasma levels of GABA. However, the GABA synthesized

Table 2. CSF GABA and Mood Disorders

	GABA levels				
Diagnosis (criteria)	*n*	Percent of control	*p*	Effect size	Reference
Primary depression (RDC)	37	73	.004	1.01	Gerner et al. (1984)
Primary depression (RDC)	17	55	.01	0.78	Gold (1980)
Primary depression (RDC)	24	73	.006	0.74	Gerner and Hare (1981)
Schizoaffective (RDC)	6	59	NS	0.69	Bowers et al. (1980)
Depression (ICD9)	13	69	0.05	0.60	Kasa et al. (1982)
Depression (Unspecified)	6	77	NS	0.20	Zimmer et al. (1981)
Depression (RDC)	16	92	NS	0.18	Post et al. (1980)
Unmedicated euthymic Bipolar disorder (Unspecified)	10	101	NS	0.03	Berretinni et al. (1986)
Li-treated bipolar disorder	20	105	NS	0.12	Berretinni et al. (1983)

enzyme, GAD in plasma, has been shown to be lower in symptomatic patients with both unipolar and bipolar depression (Kaiya, Namba, Yoshida, & Nakamura, 1982). Similarly, activity of the GABA-metabolizing enzyme $GABA_T$ in platelets was found to be low in euthymic bipolar patients off medications (Berrettini, Umberkoman-Wiita, Nurnberger, Vogel, Gershon, & Post, 1980). Except for these two studies, other examinations of peripheral GABA utilized levels of plasma GABA (pGABA) itself.

The precise extent into which pGABA reflects CNS activity cannot be directly or conclusively established in humans. However, a variety of indirect data are available to suggest that pGABA is a meaningful index of brain activity. First, unlike the biogenic amine neurotransmitters for which large concentrations are found in peripheral organs such as the adrenal, 99% of GABA in humans is located inside the central nervous system (Zachman, Tocci, & Nyhan, 1966) and a similar proportion of the GABA synthesized enzyme GAD is located within the CNS. Nevertheless, small amounts of GABA are found in peripheral tissues including mesenteric plexus, pancreas, and kidney (Erdo & Kiss, 1986). The physiologic significance of peripheral GABA is not well known.

One approach to this problem would be simultaneous measurement of CSF and plasma GABA. In animal experiments, the correlation between CSF and pGABA is quite high, $r = 0.67$ in rats (Boheln, Huot, & Palfreyman, 1979) and $r = 0.59$ in dogs (Loscher, 1982). In humans, children with convulsions (Loscher, Rating, & Siemes, 1981) and patients with a variety of neurological diagnoses taking medications (Schmidt & Loscher, 1982) showed no consistent correlation between CSF and plasma levels, nor did a series of neurological cases examining CSF and whole blood GABA (Enna, Ziegler, Lake, Wood, Brooks, & Butler, 1980). However, abnormalities in GABA function occur in a variety of neurological diseases including convulsive disorders, and these conditions may also lead to abnormalities in the blood–brain barrier, further complicating the relationship between blood and CNS GABA. In normal adult controls, free CSF GABA correlates well with free plasma GABA ($r = 0.51, p < 0.01$) (Uhlhaas, Langae, Wappenschmidt, & Olek, 1986). The only CSF/plasma GABA study of direct relevance to this proposal was that of Berrettini et al. (1982), who found a correlation of 0.34 (n = 10, n.s.) between CSF and pGABA in medication-free bipolar patients. A close examination of Berrettini's data shows a trend for a positive linear correlation, which would likely be significant with a larger number of subjects. Research in progress in our laboratory will help to clarify this important issue.

Plasma GABA levels certainly reflect brain GABA activity since pharmacological manipulations changing brain levels of GABA do so to pGABA in similar proportion (Loscher, 1979; Ferkany, Butler, & Enna, 1978a; Ferkany, Smith, Seifert, Caprioli, & Enna, 1978b; Bohlen et al, 1979). Of particular interest is the demonstration that pGABA levels increase after intracerebroventricular administration of ethanolamine-O-sulfate, a specific inhibitor of GABA metabolism, suggesting a central origin for pGABA (Loscher, 1979). This increase in pGABA is not seen after intraperitoneal administration of the drug. Likewise, an increase in pGABA occurred after the increase in CNS GABA when GABA metabolism was inhibited with γ-vinyl GABA (Apud, Racagni, Iuliano, Cocchi, Casanueva, & Muller, 1981; Loscher, 1979; Loscher, 1981). Conversely, decrease in pGABA is seen after administration of a GAD inhibitor (Ferkany, 1978b). Furthermore, it may be that the small amounts of GABA found in peripheral tissues are not derived from glutamate through GAD as is brain GABA, but rather from putrescine (Caron, Cote, & Kremzner, 1988). Therefore a change in activity of brain GAD should be reflected in plasma GABA. Further corroboration for the idea that pGABA is derived from CSF is provided by the finding that pGABA levels are almost identical to those in CSF (Loscher & Schmidt, 1984), suggesting that there is not an active gradient between these two compartments.

Therefore, the weight of the evidence suggests that pGABA is a good reflection of CNS GABA and possibly CSF GABA. It is clear that changes in brain GABA result in similar pGABA changes.

Stability of Plasma GABA, a Putative Biological Marker for Mood Disorders. Plasma GABA in monozygotic twins shows a high correlation and seems to be stable over time (Berrettini et al., 1982). Berrettini and colleagues (1984) in fact suggested that "pGABA could be a biochemical marker for genetic vulnerability to affective illness" and "pGABA should differentiate between well and ill relatives of a proband who has low plasma GAB." Our research to date has generally supported Berrettini's suggestions.

For the last 10 years our group has performed a number of experiments on pGABA in mood disorders. This, along with the pioneering work of Berrettini and colleagues, provides a framework of support for the idea that plasma GABA is a stable biological traitlike marker for depression, and perhaps for other mood disorders.

However, if a particular biological marker is to be considered as a marker for a particular illness or disease state, it is important to examine other possible environmental and constitutional factors that might influence levels of the marker. One is reminded of the earlier studies suggesting urinary MHPG as a marker for mania and depression. Although the parent compound, norepinephrine, continues to play a major conceptual role in our thinking about mood disorders, subsequent studies show that urinary MHPG levels could be profoundly influenced by environmental factors such as exercise and activity.

Factors that do not Influence GABA Levels in Plasma. We have previously shown that pGABA in control subjects was not affected by gender, mild exercise, menstrual cycles, cholinergic challenge, or post prandial or colonic status (Petty, Kramer, & Feldman, 1987). Since CSF levels of GABA in nonhuman primates are known to exhibit circadian fluctuations (Perlow, Enna, O'Brien, Hoffman, & Wyatt, 1979), it was important to determine whether pGABA levels in humans vary during the course of a normal working day. Plasma GABA levels were obtained in 20 normal controls at 8 a.m., 12 noon, and 4 p.m., and there were no significant differences found between the mean values of pGABA at these times of day (Petty, unpublished data).

The peripheral activity of other neurotransmitters such as serotonin has been shown to vary during the course of a year. We therefore determined pGABA levels in a group of normal controls at bimonthly intervals and have found no suggestion of a seasonal or circumannual effect. Furthermore, we have not found any significant correlations between pGABA and height, weight, or age either in normal controls or in patients with mood disorders. These findings are consistent with some (Loscher & Schmidt, 1984) but not all (Ferraro & Hare, 1985) reports relating CSF GABA levels to age in normals. In summary, although a biological marker for psychiatric illness does not have to be independent of environmental influences, it certainly makes it easier to study. This appears to be the case for pGABA, at least in normal controls.

Plasma GABA and Depression. We have repeatedly demonstrated that plasma levels of GABA in patients with primary unipolar depression are significantly lower than in controls. (Petty & Schlesser, 1981; Petty & Sherman, 1984; Petty & Coffman, 1984; Petty & Sherman, 1982; Petty et al., 1987; Petty & Kramer, 1989; Petty, Kramer, Dunnam, & Rush, 1989a; Petty, Kramer, Dunnam, Backman, & Rush, 1989b). In our most recent research (Petty, Kramer, Dunnam, & Rush, 1990a,b) we have examined plasma levels of GABA in 110 patients with primary unipolar depression and 76 controls (Fig. 2). Examination of the frequency distribution of plasma levels of GABA in normal controls reveals a generally normal distribution with a natural cutoff at 100 pMole/ml. About 6% of normal controls are below this level and 94% above it. Comparison with patients with primary unipolar depression, however, reveals 44% of depressives to have low pGABA levels ($p < 0.0001$). Also, the mean level of pGABA in depressives is significantly lower than control (111 ± 27 vs. 126 ± 22, $p < 0.05$).

Plasma GABA is a Traitlike Marker of Depression. There are a variety of criteria that must be fulfilled to demonstrate that a particular biological correlate of a psychiatric illness (or for that matter any disease) is truly a trait marker (Rush, Gullion, & Roffwarg, 1990). Otherwise, it is possible that a biological abnormality seen in a psychiatric illness is a consequence of the state of the patient at a point in time, in this case the state of depression. For example, weight loss is a state which usually reverts to normal with recovery from the depressed state. Abnormal dexamethasone suppression is another example of a statelike marker of depression. Alternately a biologic abnormality might be a consequence of the illness or "scar," much as cirrhosis of the liver can be seen following development of alcoholism.

Several of our recent findings suggest that pGABA in patients with primary unipolar depression has traitlike characteristics. First, there is no correlation between severity of depression as measured by the Hamilton Rating Scale for depression and pGABA; second, there is no correlation between age of onset or duration of depressive illness and pGABA levels. In a 6–12-month follow-up GABA levels did not significantly change in patients with primary unipolar depression while Hamilton Rating Scale scores dropped from 24 to about 12 on the aver-

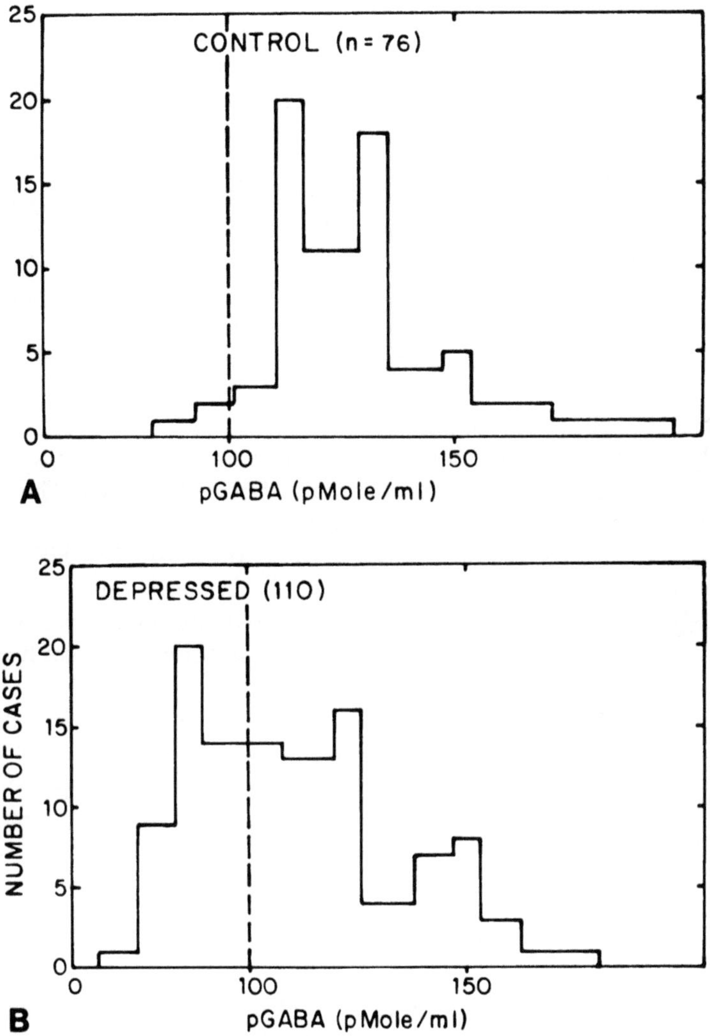

Figure 2. Frequency distribution of plasma GABA in normal controls (A) and in patients with primary unipolar depression (B).

age. To date, our findings in patients with depression may be summarized thus: Close to half patients with primary unipolar depression have abnormally low pGABA; this does not correlate with severity or duration of illness and does not revert to normal on short-term followup. So far, pGABA shows characteristics of being a traitlike marker for depression.

Specificity of pGABA in Depression. Our original research with pGABA in mood disorders (Petty & Schlesser, 1981) showed patients with secondary depression and bipolar depression to have levels of pGABA in the control range, while manic patients had high levels of pGABA. Our most recent series of experiments does not replicate this earlier finding, in that we now find patients with bipolar depression, secondary depression, and mania to all have a distribution of pGABA similar to that of the primary unipolar depressives and all significantly lower than control. The most obvious explanation for this seeming discrepancy involves the assay procedure used to determine pGABA levels. Our earlier work involved a gas chromatographic (GC) procedure, which utilizes high-temperature derivitization with strong acid. These conditions will convert dipeptides and conjugates of GABA into free GABA. Not surprisingly, the levels of pGABA we obtained with the GC technique were 2 to 4 times higher than those of our most recent studies. This recent work utilized a high-performance liquid chromatography (HPLC) assay originally developed by Hare (Hare & Manyam 1980). The HPLC technique is specific for free GABA; therefore, the higher levels of GABA we previously found in bipolar and secondary patients may have reflected higher levels of conjugated GABA in these conditions. Research underway should clarify these issues.

If low pGABA in mood disorders, regardless of polarity (bipolar versus unipolar) or primacy (primary versus secondary), is replicated and confirmed to be a trait marker, low pGABA may represent the manifestation of a neurochemical or genetic defect which is necessary but not sufficient for the development of symptomatology of depression or mania in a subset of patients.

Regarding specificity of low pGABA to mood disorders, our earlier research and a recent report (Moss, Burns, Maddock, & Tarter, 1990) suggest an abnormality in GABA function to occur in alcoholism. This need not detract from the finding in mood disorders since a familial as well as pathologic correlation between alcoholism and mood disorders is well established. Our unpublished research in progress along with previous experiments suggests that normal levels of GABA are found in patients with panic disorder, schizophrenia, anorexia nervosa, and bulimia. To date there are no data on patients with obsessive-compulsive disorder or generalized anxiety. The issue of specificity of low pGABA to mood disorders is therefore not yet adequately determined.

A good bit of work remains to be done in proving low pGABA to be a trait marker for mood disorders. pGABA should be studied in patients in long-term follow up and in children and adolescents with major depression experiencing their first episode of mood disorder. If children and adolescents with depression have low pGABA, this would provide powerful confirmation of its traitlike characteristics and confirm that it is not a "scar" caused by many years of illness.

GABA Agonists in Mood Disorders

Considering GABA's ubiquity as a neurotransmitter and its importance in CNS function it is not surprising that few drugs are available for use in humans that primarily act through a direct action on the GABA system. Cycloserine is the only drug used in humans that is known to decrease brain GABA levels, probably by inhibition of GAD. This drug is used as an antitubercular and has been known for some time to cause severe mental depression as a side effect (Lewis, Calden, Thurston, & Gilson, 1957).

Two antidepressants have been developed in Europe which are GABA-mimetics, that is to say, they have a GABA-like action and are mixed $GABA_A/GABA_B$ receptor agonists. These are progabide and fengabine (Scatton, Lloyd, Zivkovic, Dennis, Claustre, Dedek, Abrilla, Langer, & Bartholini, 1987). Progabide has been shown to be as effective an antidepressant as imipramine for treating patients with major depressive disorder. Fengabine, while slightly less effective than imipramine in major depression, was more effective in patients with minor depression or dysthmia (Musch & Garreau, 1986). Unfortunately, progabide has a significant liver toxicity that may preclude its marketing in the United States. Fengabine, while safer, may also not reach the American market because of its being perhaps somewhat less effective for major depression than standard treatments.

There are several drugs in use that are indirect GABA agonists or facilitate GABA transmission. Most noteworthy among these are the benzodiazepines, and diazepam increases GABA in human CSF (Loscher & Schmidt, 1987). Of the benzodiazepines, alprazolam has been most widely studied in patients with major depressive disorder. Twelve of the 15 studies published show alprazolam to be more effective than placebo or as effective as standard antidepressant drug. Overall, these studies show that patients treated with alprazolam experience an approximate 50% drop from baseline in depressive symptoms and about 20% more improvement than placebo. The clinical response to alprazolam in patients who respond is more rapid than to tricyclic antidepressants, with most of the clinical improvement evident during the first week of treatment. A careful analysis of the clinical response of patients receiving alprazolam for depression shows that the core symptoms of depression improved along with the expected improvement in anxiety and sleep (Petty & Trivedi, 1992).

Lithium increases pGABA and has antidepressant effects particularly in bipolar depression (Berrettini, Nurnberger, Hare, Simmons-Alling, Gershon, & Post, 1983). The use of lithium as an adjunctive treatment to enhance the efficacy of the tricyclics and other antidepressants has become widespread and may relate to lithium's effect on GABA function.

Valproic acid (or sodium valproate) increases brain levels of GABA. Efficacy of valproate acid in the treatment of mania is now well documented (Emrich, Dose, & Von Zerssen, 1985) and a multicenter placebo control comparison with lithium in the treatment of acute mania is underway at this writing (December, 1992). There are few data regarding the effectiveness of valproate as an antidepressant, either alone or in conjunction with tricyclics. Our GABA hypothesis would predict that valproate should be useful in the treatment of depression.

A new anticonvulsant, γ-vinyl GABA (GVG), should be available in the United States soon. GVG is an irreversible inhibitor of $GABA_T$. Nothing is known at present concerning its use in psychiatric patients, but anecdotal reports from its anticonvulsant trials suggest it may have antidepressant and perhaps anxiolytic effects. Development of safe and "cleaner" drugs that affect the GABA system should provide an exciting opportunity for psychopharmacology research in the decade to come.

The GABA Theory of Mood Disorder

A consistent picture emerges of GABA deficits being involved in depression. The major problem in interpreting these findings remains their integration into the existing theories of neurochemical determinants of mood disorders. From a simplistic perspective one might argue that since GABA is usually considered to be an inhibitory neurotransmitter, high levels of inhibition (and therefore high GABA) should be seen in patients with depression. This is not necessarily true, since GABA services a facilitatory function at times and since inhibition of another inhibitory neurotransmitter would also have a generally facilitatory effect. At any rate there is a long and complicated path between the synapse and behavior.

Recent research suggests that a GABA hypothesis for mood disorders in fact complements the existing monamine hypotheses and does not detract from their validity. Interactions between GABA and the monoamines have been well documented (Bartholini, 1984a,b). GABA is reported to inhibit serotonin release from frontal cortex in vitro probably through the $GABA_B$ receptor (Schlicker, Classenm, & Gothert, 1984; Gray & Green, 1987; Green, Minchin, & Vincent, 1987b). Recent work in our laboratory using an in vivo technique with the GABA agonist γ-vinyl GABA, however, showed a facilitation of basal serotonin release in prefrontal cortex (Petty, unpublished data). Our data are consistent with reports that stimulation of GABA receptors in the suprachiasmatic area increases serotonin release (Francois-Bellan, Hery, Faudon, & Hery, 1988) and that the serotonin/GABA axonal interface represents a "privileged" site of interaction between these two neurotransmitters (Bosler, 1989). It is also worth noting that serotonin increases the sensitivity of the $GABA_A$ receptor to GABA (Akasu, 1988).

In spite of its being traditionally considered an "inhibitory" neurotransmitter in cortex, GABA is known to facilitate noradrenergic transmission. (The data regarding GABA's inhibitory effects are mostly from iontophoretic brain stem experiments at the locus coeruleus.) Long-term administration of antidepressant drugs potentiates norepinephrine-induced cyclic AMP accumulation caused by the $GABA_B$ agonist baclofen (Suzdak & Gianutsos, 1985a). This $GABA_B$ receptor potentiation of a norepinephrine-induced cyclic AMP accumulation occurs in rank order of potency of $GABA_B$ binding assays (Karbon, Duman, & Enna, 1984) and is probably postsynaptic. The $GABA_A$ receptor mediated facilitation of norepinephrine is likely to be presynaptic (Suzdak & Gianutsos, 1985b) or may in fact involve another transsynaptic neuronal circuit as yet to be defined (Dennis et al., 1985). A carrier for GABA uptake has been identified in noradrenergic nerve endings in rat cortex and hippocampus but not on serotonin terminals (Bonanno & Raiteri, 1987a). This carrier is a $GABA_A$ receptor located presynaptically on the noradrenergic nerve terminal which regulates the release of norepinephrine (Bonanno & Raiteri, 1987b). GABA-mimetics including progabide and fengabine increase the turnover of norepinephrine in hypothalamus and septum and their chronic administration causes desensitization of the noradrenergically stimulated adenylate cyclase (Zivkovic, Scatton, Dedek, & Bartholini, 1982; Scatton et al., 1987).

Therefore in the cortex, particularly the frontal cortex, the effects of GABA on serotonin terminals are possibly facilitatory in vivo (Strahlendorf, Lee, Netzeband, & Strahlendorf, 1988). The effects of serotonin on norepinephrine are facilitatory (Poncelot et al., 1987) and the effects of GABA on norepinephrine are facilitatory. This might account for why GABA hypofunction in the central nervous system, presumably reflected in CSF and plasma, could accompany some mood disorders. This also explains why therapeutic manipulations that increase brain GABA activity have antidepressant and perhaps antimanic activity. In this context the GABA hypothesis of mood disorder is no longer a competitor or alternative for the well-established monoaminergic theories but rather complements theories of serotonin and norepinephrine function in depression.

These findings are summarized in Fig. 3, which is modified from Lipinski et al. (1988). This unified theory of the neurochemistry and pharmacology of affective illness suggests that the antidepressant and antimanic effects of presently available somatic treatments for mood disorder, including GABAergic drugs, produce their final action through a postsynaptic adrenergic receptor, and this activity is modulated both presynaptically and postsynaptically by GABA. Clinical support for a GABA/norepinephrine interaction was recently provided by the significant correlation noted between CSF GABA and MHPG (the norepinephrine metabolite) in controls (Roy, DeJong, Ferraro, Adinoff, Gold, Rubinow, & Linnoila, 1989).

Summary and Directions for Future Research

Overall the data presented suggest that there is an important involvement of GABA in the pathophysiology of mood disorders and that GABA is involved in the mechanisms of therapeutic action of antidepressants and perhaps antimanic drugs. Future research should first determine why and how the GABA deficit develops. Is this due to decreased synthesis, increased metabolism, increased clearance, or a combination of factors? In this respect studies on the GABA synthesizing

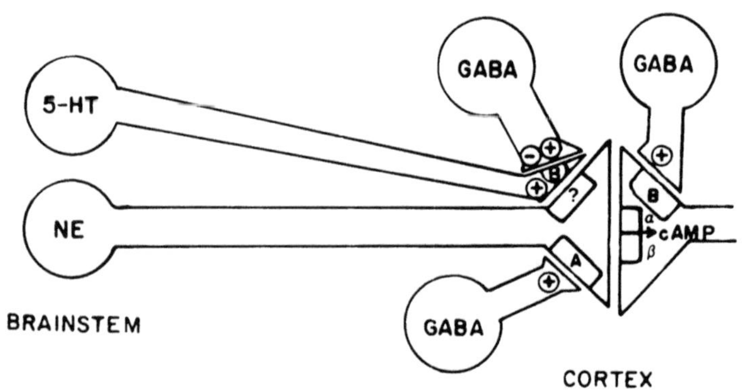

Figure 3. Unified theory of neurochemistry of mood disorders. See text for details.

and degrading enzymes in both controls and patients with mood disorders need to be undertaken. Renal clearance of GABA should also be examined. Do patients with low GABA respond preferentially to GABA agonists such as progabide, benzodiazepines, lithium, or valproate acid? Such studies should be feasible and perhaps provide for more rational treatment of depression. Finally, if low GABA indeed is shown to be a trait marker of mood disorders, familial and genetic studies are necessary to establish the precise mechanism of transmission and familial association.

Acknowledgments

Work was supported by the Department of Veterans Affairs and a Department of Veterans Affairs Research Career Development Award, and also by NIMH grant No. MH37899, NIMH Mental Health Clinical Research grant No. MH41115, NIAAA grant No. AA07234, and the John Schemmerhorn Fund. We thank Dinah Turner for excellent secretarial support and Kenneth Z. Altshuler, M.D. for administrative assistance.

References

Ahluwalia P, Grewaal DS, Singhal RL: Changes in γ-aminobutyric acid and glutamic acid and glutamic acid decarboxylase in discrete regions of rat brain following lithium administration and withdrawal. *Drug Dev Res* 1983; 3: 153–161.

Akasu T: 5-hydroxytryptamine facilitates GABA-induced depolarization in bullfrog primary afferent neurons. *Neurosci Lett* 1988; 92: 270–274.

Apud JA, Racagni G, Iuliano E, Cocchi D, Casanueva F, Muller EE: Role of central nervous system-derived or circulating γ-aminobutyric acid on prolactin secretion in the rat. *Endocrinol* 1981; 108: 1505–1510.

Barbaccia ML, Ravizza L, Costa E: Maprotiline: An antidepressant with an unusual pharmacological profile. *J Pharm Exper Ther* 1986; 236: 307–312.

Bartholini G: Experimental basis for the antidepressant action of the GABA receptor agonist progabide. *Neurosci Lett* 1984a; 47: 351–355.

Bartholini G: Pharmacology of the GABAergic system: Effects of progabide, a GABA receptor agonist. *Psychoneuroendocrinol* 1984b; 9: 135–140.

Bartholini G, Lloyd KG, Scatton B, Zivkovic B, Morselli PL: The GABA hypothesis of depression and antidepressant drug action. *Psychopharmacol Bull* 1985; 21: 385–388.

Berrettini WH, Umberkoman-Wiita B, Nurnberg JI, Jr, Vogel WH, Gershon ES,

Post RM: Platelet GABA-transaminase in affective illness. *Psychiat Res* 1980; 7: 255–260.

Berrettini WH, Nurnberger JI, Jr, Hare T, Gershon ES, Post RM: Plasma and CSF GABA in affective illness. *Brit J Psychiat* 1982; 141: 483–487.

Berrettini WH, Nurnberger JI, Jr, Hare TA, Simmons-Alling S, Gershon ES, Post RM: Reduced plasma and CSF γ-aminobutyric acid in affective illness: Effect of lithium carbonate. *Bio Psychiat* 1983; 18: 184–195.

Berrettini WH, Goldin LR, Nurnberger JI, Jr, Gershon ES: Genetic factors in affective illness. *J Psychiat Res* 1984; 18: 329–350.

Berrettini WH, Nurnberger JI, Jr, Hare TA, Simmons-Alling S, Gershon ES: CSF GABA in euthymic manic-depressive patients and controls. *Biol Psychiat* 1986; 21: 844–846.

Bohlen P, Huot S, Palfreyman MG: The relationship between GABA concentrations in brain and cerebrospinal fluid. *Brain Res* 1979; 167: 297–305.

Bonanno G, Raiteri M: A carrier for GABA uptake exists on noradrenaline nerve endings in selective rat brain areas but not on serotonin terminals. *J Neural Transm* 1987a; 69: 59–70.

Bonanno G, Raiteri M: Release-regulating $GABA_A$ receptors are present on noradrenergic nerve terminals in selective areas of the rat brain. *Synapse* 1987b; 1: 254–257.

Borsini F, Evangelista S, Meli A: Effect of GABAergic drugs in the behavioral despair test in rats. *Eur J Pharmacol* 1986; 121: 265–268.

Borsini F, Mancinelli A, D'Aranno V, Evangelista S, Meli A: On the role of endogenous GABA in the forced swimming test in rats. *Pharmacol Biochem Behav* 1987; 29: 275–279.

Bosler O: Ultrastructural relationships of serotonin and GABA terminals in the rat suprachiasmatic nucleus. Evidence for a close interconnection between the two afferent systems. *J Neurocytol* 1989; 18: 105–113.

Bowdler JM, Green AR, Minchin MCW, Nutt DJ: Regional GABA concentration and [^{3}H]-diazepam binding in rat brain following repeated electroconvulsive shock. *J Neural Transm* 1983; 56: 3–12.

Bowers MB, Jr, Gold BI, Roth RH: CSF GABA in psychotic disorders. *Psychopharmacol* 1980; 70: 279–282.

Caron PC, Code LJ, Kremzner LT: Putrescine, a source of γ-aminobutyric acid in the adrenal gland of the rat. *Biochem J* 1988; 251: 559–562.

Cheetham SC, Crompton MR, Katona CLE, Parker SJ, Horton RW: Brain $GABA_A$/benzodiazepine binding sites and glutamic acid decarboxylase activity in depressed suicide victims. *Brain Res* 1988; 460: 114–123.

Cross JA, Horton RW: Are increases in $GABA_B$ receptors consistent findings following chronic antidepressant administration? *Eur J Pharmacol* 1987; 141: 159–162.

Dennis T, Curet O, Nishikawa T, Scatton B: Further evidence for, and nature of, the facilitatory GABAergic influence on central noradrenergic transmission. *NS Arch Pharmacol* 1985; 331: 225–234.

Drugan RC, Ryan SM, Minor TR, Maier SF: Librium prevents the analgesia and shuttlebox escape deficit typically observed following inescapable shock. *Pharmacol Biochem Behav* 1984; 21: 749–754.

Drugan RC, Maier SF, Skolnick P, Paul SM, Crawley JN: An anxiogenic benzodiazepine receptor ligand induces learned helplessness. *Eur J Pharmacol* 1985; 113: 453–457.

Drugan RC, Morrow AL, Weizman R, Weizman A, Deutsch SI, Crawley JN, Paul SM: Stress-induced behavioral depression in the rat is associated with a decrease in GABA receptor-mediated chloride ion flux and brain benzodiazepine receptor occupancy. *Brain Res* 1989; 487: 45–51.

Emrich HM, Zerssen DV, Kissling W, Moller HJ, Windorfer A: Effect of sodium valproate on mania. *Arch Psychiatr Nervenkr* 1980; 229: 1–16.

Emrich HM, Dose M, Von Zerssen D: The use of sodium valproate, carbamazepine and oxcarbazepine in patients with affective disorders. *J Affect Disord* 1985; 8: 243–250.

Enna SJ, Ziegler MG, Lake CR, Wood JH, Brooks BR, Butler IJ: Cerebrospinal fluid γ-aminobutyric acid. In: Wood JH (Ed.), *Neurobiology of Cerebrospinal Fluid*, New York: Plenum Press, 1980; 189–196.

Erdo SL, Kiss B: Presence of GABA, glutamate decarboxylase, and GABA transaminase in peripheral tissues: A collection of quantitative data. In: Erdo SL, Bowery NG (Eds.), *GABAergic Mechanisms in the Mammalian Periphery*, New York: Raven Press, 1986.

Ferkany JW, Butler IJ, Enna SJ: Effect of drugs on rat brain, cerebrospinal fluid and blood GABA content. *J Neurochem* 1978a; 33: 29–33.

Ferkany JW, Smith LA, Seifert WE, Jr, Caprioli RM, Enna SJ: Measurement of γ-aminobutyric acid (GABA) in blood. *Life Sci* 1978b; 22: 2101–2128.

Ferraro TN, Hare TA: Free and conjugated amino acids in human CSF influence of age and sex. *Brain Res* 1985; 338: 53–60.

Ferraro TN, Golden GT, Hare TA: Repeated electroconvulsive shock selectively alters γ-aminobutyric acid levels in the rat brain: Effect of electrode placement. *Convulsive Ther* 1990; 6(3): 199–208.

Francois-Bellan AM, Hery M, Faldon M, Hery F: Evidence for GABA control of serotonin metabolism in the rat suprachiasmatic area. *Neurochem Int* 1988; 134(4): 455–462.

Gerner RH, Hare TA: CSF GABA in normal subjects and patients with depression, schizophrenia, mania, and anorexia nervosa. *Am J Psychiat* 1981; 138: 1098–1101.

Gerner RH, Fairbanks L, Anderson GM, Young JG, Scheinin M, Linnoila M, Hare TA, Shaywitz BA, Cohen DJ: CSF neurochemistry in depressed, manic, and schizophrenic patients compared with that of normal controls. *Am J Psychiat* 1984; 141: 1533–1540.

Gold BI, Bowers MB, Jr, Roth RH, Sweeney DW: GABA levels in CSF of patients with psychiatric disorders. *Am J Psychiat* 1980; 137: 362–364.

Gottesfeld Z: Effect of lithium and other alkali metals on brain chemistry and behavior. *Psychopharmacologia* 1976; 45: 239–242.

Gray JA, Green AR: $GABA_B$-receptor mediated inhibition of potassium-evoked release of endogenous 5-hydroxytryptamine from mouse frontal cortex. *Br J Pharmacol* 1987; 91: 517–522.

Green AR, Vincent ND: The effect of repeated electroconvulsive shock on GABA synthesis and release in regions of rat brain. *Br J Pharmacol* 1987a; 92: 19–24.

Green AR, Minchin CW, Vincent ND: Inhibition of GABA release from slices prepared from several brain regions of rats at various times following a convulsion. *Br J Pharmacol* 1987b; 92: 13–18.

Grove J, Schecter PJ, Hanke NFJ, de Smet Y, Agid Y, Tell G, Koch-Weser J: Concentration gradients of free and total γ-aminobutyric acid and homocarnosine in human CSF: Comparison of suboccipital and lumbar sampling. *J Neurochem* 1982; 39: 1618–1622.

Hare TA, Manyam NVB: Rapid and sensitive ion-exchange fluorometric measurement of γ-aminobutyric acid in physiological fluids. *Anal Biochem* 1980; 101: 349–355.

Honig A, Bartlett JR, Bouras N, Bridges PK: Amino acid levels in depression: a preliminary investigation. *J Psychiatr Res* 1989; 22(3): 159–164.

Joly D, Lloyd KG, Pichat P, Sanger DJ: Correlation between the behavioral effect of desipramine and $GABA_B$ receptor regulation in the olfactory bulbectomized rat. *Br J Pharmacol* 1987: 125.

Kaiya H, Namba M, Yoshida H, Nakamura S: Plasma glutamate decarboxylase activity in neuropsychiatry. *Psychiat Res* 1982; 6: 335–343.

Karbon EW, Duman RS, Enna SJ: $GABA_B$ receptors and norepinephrine-stimulated cAMP production in rat brain cortex. *Brain Res* 1984; 306: 327–332.

Kasa K, Otsuki S, Yamamoto M, Sato M, Kuroda H, Ogawa N: Cerebrospinal fluid γ-aminobutyric acid and homovanillic acid in depressive disorders. *Biol Psychiatry* 1982; 17: 877–883.

Korf J, Venema K: Desmethylimipramine enhances the release of endogenous GABA and other neurotransmitter amino acids from the rat thalamus. *J Neurochem* 1983; 40: 946–950.

Korpi ER, Kleinman JE, Wyatt RJ: GABA concentrations in forebrain areas of suicide victims. *Biol Psychiat* 1988; 23: 109–114.

Leonard BE, Tuite M: Anatomical, physicological, and behavioral aspects of olfactory bulbectomy in the rat. *Int Rev Neurobiol* 1981; 22: 251–286.

Lewis WC, Calden G, Thurston JR, Gilson WE: Psychiatric and neurological reactions to cycloserine in the treatment of tuberculosis. *Dis Chest* 1957; 32: 172–182.

Lipinski JF, Jr, Cohen BM, Zubenko GS, Waternaux CM: Minireview: Adrenoreceptors and the pharmacology of affective illness: A unifying theory. *Life Sci* 1988; 40: 1947–1963.

Lloyd KG, Pichat P: Decrease in $GABA_B$ binding in the frontal cortex of olfactory bulbectomized rats. *Br J Pharmacol* 1985; 87: 36P.

Lloyd KG, Morselli PL: Psychopharmacology of GABAergic drugs. In: Meltzer HY (Ed.), *Psychopharmacology: The Third Generation of Progress,* New York: Raven Press, 1987, 183–195.

Lloyd KG, Morselli PL, Depoortere H, Fournier V, Zivkovic B, Scatton B,

Broekkamp C, Worms P, Bartholini G: The potential use of GABA agonists in psychiatric disorders: Evidence from studies with progabide in animal models and clinical trials. *Pharmacol Biochem Behav* 1983; 18: 957–966.

Lloyd KG, Thuret F, Pilc A: Upregulation of γ-aminobutyric acid $(GABA)_B$ binding sites in rat frontal cortex: A common action of repeated administration of different classes of antidepressants and electroshock. *J Pharmacol Exp Ther* 1985; 235: 191–199.

Lloyd KG, Morselli PL, Bartholini G: GABA and affective disorders. *Med Biol* 1987; 65: 159–165.

Lloyd KG, Zivkovic B, Scatton B, Morselli PL, Bartholini G: The GABAergic hypothesis of depression. *Prog Neuropsychopharmacol Biol Psychiat* 1989; 13: 341–351.

Loscher W: GABA in plasma and cerebrospinal fluid of different species. Effects of γ-acetylenic GABA, γ-vinyl GABA and sodium valproate. *J Neurochem* 1979; 32: 1587–1591.

Loscher W: Relationship between GABA concentrations in cerebrospinal fluid and seizure excitability. *J Neurochem* 1982; 38: 293–295.

Loscher W, Schmidt D: Monitoring of γ-aminobutyric acid in human cerebrospinal fluid: Downward revision of previous control values. *Ther Drug Monit* 1984; 6: 227–231.

Loscher W, Schmidt D: Diazepam increases γ-aminobutyric acid in human cerebrospinal fluid. *J Neurochem,* 49: 152–157.

Loscher W, Rating D, Siemes H: GABA in cerebrospinal fluid of children with febrile convulsions. *Epilepsia* 1981; 22: 697–702.

Moss HB, Yao JK, Burns M, Maddock J, Tarter RE: Plasma GABA-like activity in response to ethanol challenge in men at high risk for alcoholism. *Biol Psychiatry* 1990; 27: 617–625.

Motohashi N, Ikawa K, Kariya T: $GABA_B$ receptors are up-regulated by chronic treatment with lithium or carbamazepine. GABA hypothesis of affective disorders? *Eur J Pharmacol* 1989; 166: 95–99.

Musch B, Garreau M: An overview of the antidepressant activity of fengabine (SL 79.229-00) in open clinical studies. In: Shagass C, Jasiassen RC, Bridger WH, Wiss KJ, Stoff D, Simpson GM (Eds.), *Biological Psychiatry,* New York: Elsevier, 1986; 920–922.

Overmier JB, Hellhammer DH: The learned helplessness model of human depression. In: Simon P, Soubrie P, Widlocher D (Eds.), *An Inquiry into Schizophrenia and Depression,* New York: Karger, 1988, 177–202.

Perlow MJ, Enna SJ, O'Brien PJ, Hoffman HJ, Wyatt RJ: Cerebrospinal fluid γ-aminobutyric acid: Daily pattern and response to haloperidol. *J Neurochem* 1979; 32: 265–268.

Perry EK, Gibson PH, Blessed G, Perry RH, Tomlinson BE: Neurotransmitter abnormalities in senile dementia. *J Neurol Sci* 1977; 34: 247–265.

Petty F, & Trivedi M: Benzodiazepines in the treatment of depression. *Neuropsychopharmacology* 1992 (submitted).

Petty F, Coffman JA: Plasma GABA: A possible indicator of altered GABA function in psychiatric illness. *Neuropharmacol* 1984; 23: 859–860.

Petty F, Kramer GL: Plasma GABA and mood disorders. Society for Neuroscience Annual Meeting, Phoenix, Arizona, October 29–November 3, 1989.

Petty F, Schlesser MA: Plasma GABA in affective illness. *J Affect Disord* 1981; 3: 339–343.

Petty F, Sherman AD: Plasma GABA: A blood test for bipolar affective disorder trait? *Psychol Psychiat Behav* 1982; 7: 431–439.

Petty F, Sherman AD: Plasma GABA levels in psychiatric illness. *J Affect Disord* 1984; 6: 131–138.

Petty F, Sherman AD, Saquitne J: GABAergic modulation of learned helplessness. *Pharmacol Biochem Behav* 1981; 15: 567–570.

Petty F, McChesney C, Kramer G: Intracortical glutamate injection produces helpless-like behavior in the rat. *Pharmacol Biochem Behav* 1985; 22: 531–533.

Petty F, Kramer G, Feldman M: Is plasma GABA of peripheral origin? *Biol Psychiat* 1987; 22: 725–732.

Petty F, Kramer GL, Dunnam D, Rush AJ: Plasma GABA and mood disorders. American College of Neuropsychopharmacology Annual Meeting, Maui, Hawaii, December 10–15, 1989a.

Petty F, Kramer GL, Dunnam D, Backman K, Rush AJ: Plasma GABA and mood disorders. New Clinical Drug Evaluation Unit Program Annual Meeting, Key Biscayne, Florida, May 30–June 2, 1986.

Petty F, Kramer GL, Dunnam D, Rush AJ: Plasma GABA: Stable trait marker for mood disorders? Society for Biological Psychiatry Annual Meeting, New York, New York, May 9–13, 1990a.

Petty F, Kramer GL, Dunnam D, Rush AJ: Plasma GABA in mood disorders. *Psychopharmacol Bull* 1990b; 26(2): 157–162.

Plaznik A, Tamorska E, Hauptmann M, Bidzinski A, Kostowski W: Brain neurotransmitter systems mediating behavioral deficits produced by inescapable shock treatment in rats. *Brain Res* 1988; 447: 122–132.

Poncelet M, Martin P, Danti S, Simon P, Soubrie P: Noradrenergic rather than GABAergic processes as the common mediation of the antidepressant profile of GABA agonists and imipramine-like drugs in animals. *Pharmacol Biochem Behav* 1987; 28: 321–326.

Porsolt RD, Anton G, Blavet N, Jalfre M: Behavioural despair in rats: A new model sensitive to antidepressant treatments. *Eur J Pharmacol* 1978; 47: 379–391.

Post RM, Ballenger JC, Hare TA, Goodwin FK, Lake CR, Jimerson DC, Bunney WE, Jr: Cerebrospinal fluid GABA in normals and patients with affective disorders. *Brain Res Bull* 1980; 5: 755–759.

Roy A, DeJong J, Ferraro T, Adinoff B, Gold P, Rubinow LD, Linnoila M: CSF GABA and neuropeptides in pathological gamblers and normal controls. *Psychiat Res* 1989; 30: 137–144.

Rubio MC, Losada MEO: GABAergic responses to lithium chloride: Dependence on dose, treatment length and experimental condition. In: Racagni G, Donoso

AO (Eds.), *GABA and Endocrine Function*, New York: Raven Press, 1986, 69–77.

Rush AJ, Gullion CM, Roffwarg HP: Research strategies to identify trait-like biological abnormalities in major depression. Presented at the American College of Neuropsychopharmacology, San Juan, Puerto Rico, December 10–14, 1990.

Sackeim HA, Decina P, Prohovnik I, Malitz S, Resor SR: Anticonvulsant and antidepressant properties of electroconvulsive therapy: A proposed mechanism of action. *Bio Psychiat* 1983; 18: 1301–1310.

Scatton B, Lloyd KG, Zivkovic B, Dennis T, Claustre Y, Dedek J, Arbilla S, Langer SZ, Bartholini G: Fengabine, a novel antidepressant GABAergic agent. II. Effect on cerebral noradrenergic, serotonergic and GABAergic transmission in the rat. *J Pharmacol Exp Ther* 1987; 241: 251–257.

Schlicker E, Classenm K, Gothert M: $GABA_B$ receptor-mediated inhibition of serotonin release in the rat brain. *NS Arch Pharmacol* 326: 99–105.

Schmidt D, Loscher W: Plasma and cerebrospinal fluid γ-aminobutyric acid in neurological disorders. *J Neurol Neurosurg Psychiat* 1982; 45: 931–935.

Seligman, MEP, Maier SF: Failure to escape traumatic shock. *J Exp Psychol* 1967; 74: 1–9.

Sherman AD, Petty F: Neurochemical basis of the action of antidepressants on learned helplessness. *Behav Neural Biol* 1980; 30: 119–134.

Sherman AD, Petty F: Additivity of neurochemical changes produced by learned helplessness and imipramine. *Behav Neural Biol.* 1982; 35: 344–353.

Sherman AD, Allers GL, Petty F, Henn FA: A neuropharmacologically relevant animal model of depression. *Neuropharmacology* 1979; 891–893.

Sherman AD, Sacquitine JL, Petty F: Specificity of the learned helplessness model of depression. *Pharmacol Biochem Behav* 16: 449–454.

Squires RF, Saederup E: Antidepressants and metabolites that block $GABA_A$ receptors coupled to 35S-t-butylbicyclophosphorothionate binding sites in rat brain. *Brain Res* 1988; 441: 15–22.

Strahlendorf JC, Lee M, Netzeband JG, Strahlendorf HK: Pentobarbital augments serotonin-mediated inhibition of cerebellar purkinje cells. *Neuroscience* 1988; 107–115.

Suranyi-Cadotte BE, Dam TV, Quirion R: Antidepressant–anxiolytic interaction: Decreased density of benzodiazepine receptors in rat brain following chronic administration of antidepressants. *Eur J Pharmacol* 1985; 106: 673–675.

Suzdak PD, Gianutsos G: Differential coupling of GABA-A and GABA-B receptors to the noradrenergic system. *J Neural Trans* 1985a; 62: 77–89.

Suzdak PD, Gianutsos G: Parallel changes in the sensitivity of γ-aminobutyric acid and noradrenergic receptors following chronic administration of antidepressant and GABAergic drugs. *Neuropharmacol* 1985b; 24: 217–222.

Suzdak PD, Gianutso G: Effect of chronic imipramine or baclofen on GABA-B binding and cyclic AMP production in cerebral cortex. *Eur J Pharmacol* 1986; 131: 129–133.

Uhlhaas S, Lanage H, Wappenschmidt J, Olek K: Free and conjugated CSF and plasma GABA in Huntington's chorea. *Acta Neurol Scand* 1986; 74: 261–265.

Zacharko RM, Anisman H: Pharmacological, biochemical, and behavioral analyses of depression: Animal models. *In:* Koob GF, Ehlers CL, Kupfer DJ (Eds.), *Animal Models of Depression*. Boston: Birkhauser, 1989, 204–238.

Zachmann M, Tocci P, Nyhan WL: The occurrence of γ-aminobutyric acid in human tissues other than brain. *J Biol Chem* 1966; 241: 1355–1358.

Zimmer R, Teelken AW, Meier KD, Acken-Heil M, Zander KJ: Preliminary studies on CSF γ-aminobutyric acid levels in psychiatric patients before and during treatment with different psychotropic drugs. *Prog Neuro-Psychopharmacol* 1980; 4: 613–620.

Zivkovic B, Scatton B, Dedek J, Bartholini G: GABA influence on noradrenergic and serotonergic transmission: Implications in mood regulation. *In:* Langer SZ, et al. (Eds.), *New Vistas in Depression,* Oxford: Pergamon Press, 1982, 195–201.

Chapter 5

BIOLOGICAL RHYTHMS AND DEPRESSIVE DISORDERS

Timothy H. Monk

Introduction

It is increasingly recognized that, like most other creatures on this planet, Homo Sapiens is essentially a rhythmic being. Cycles with a period of about a day (circadian rhythms), less than a day (ultradian rhythms), and more than a day (infradian rhythms) have all been demonstrated in human psychology and physiology, and undoubtedly comprise an important aspect of depressive disorders. This chapter is concerned with the most dominant of these period domains, the circadian rhythms. Other domains are clearly important; the chapter after this one, for example, will be concerned extensively with the ultradian rhythmicity in REM sleep. For a discussion of the infradian rhythm of the menstrual rhythm the reader is recommended a recent book by Severino and Moline (1989).

Human Circadian Rhythms

Although there has always been a strong folklore about regularity in bedtimes leading to health, wealth, and wisdom, the objective study

---

Timothy H. Monk • Human Chronobiology Program, University of Pittsburgh School of Medicine, Western Psychiatric Institute and Clinic, Pittsburgh, Pennsylvania 15213.

Biology of Depressive Disorders, Part A: A Systems Perspective, edited by J. John Mann and David J. Kupfer. Plenum Press, New York, 1993.

of human circadian (approximately 24-hour) rhythms is a comparatively recent development. Not until the 1950s was the scientific investigation of human circadian rhythms begun in earnest and the terms "circadian" and "chronobiology" coined (Halberg, 1969). In both medicine and physiology, early chronobiologists were continually battling against the prevailing view that homeostasis led to an utterly constant physiological milieu, and that the circadian rhythms observed by the chronobiologist were at best trivial and at worst suspect. Partly because of this atmosphere of disbelief, much of the early work was concerned simply with demonstrating that circadian rhythms did exist and that they were nontrivial in magnitude and endogenous in origin. Only in the last one or two decades has the discipline really started to define the circadian characteristics of the normal healthy individual and to formulate laws governing the behavior of that circadian system (Wever, 1979; Aschoff, 1981).

The first, and most important, attribute of the human circadian system is that it is indeed endogenous and self-sustaining, continuing to generate circadian rhythms in mood, performance, and physiology even when all external time cues are removed and sleep is suspended. This is illustrated nicely in a study by Froberg (1977) in which 15 young healthy volunteers were subjected to a 72-hour vigil. Figure 1 illustrates the circadian variation that continued to be exhibited in both subjective alertness and body temperature, even though there was no cycling in daylight and darkness or sleep and wakefulness for these subjects. This finding has been replicated in many different studies and subject groups.

The second characteristic of the human circadian system is that much of it is resistant to abrupt changes in schedule. Thus, for example, circadian rhythms in body temperature, cortisol excretion, and REM sleep propensity will continue to cycle on an "old" routine even several days after a "new" routine has been adopted. In the area of transmeridianal flight this leads to the "jet-lag syndrome" (Klein, Wegmann, & Hunt, 1972), in rotating work schedules the "shift maladaption syndrome" (Moore-Ede & Richardson, 1985). Both syndromes (which usually present as sleep disorders, fatigue, malaise, and gastrointestinal distress) are the product of a misaligned circadian system and/or a breakdown in the temporal harmony of its component processes. The duration of this disharmony depends upon both the size and the direction of the change in schedule (Aschoff, Hoffman, Polh, & Wever, 1975) but may last several weeks (Weitzman & Kripke, 1981).

A third characteristic of the healthy circadian system is that it has a natural tendency to "run slow." The "natural" period lenghth of human

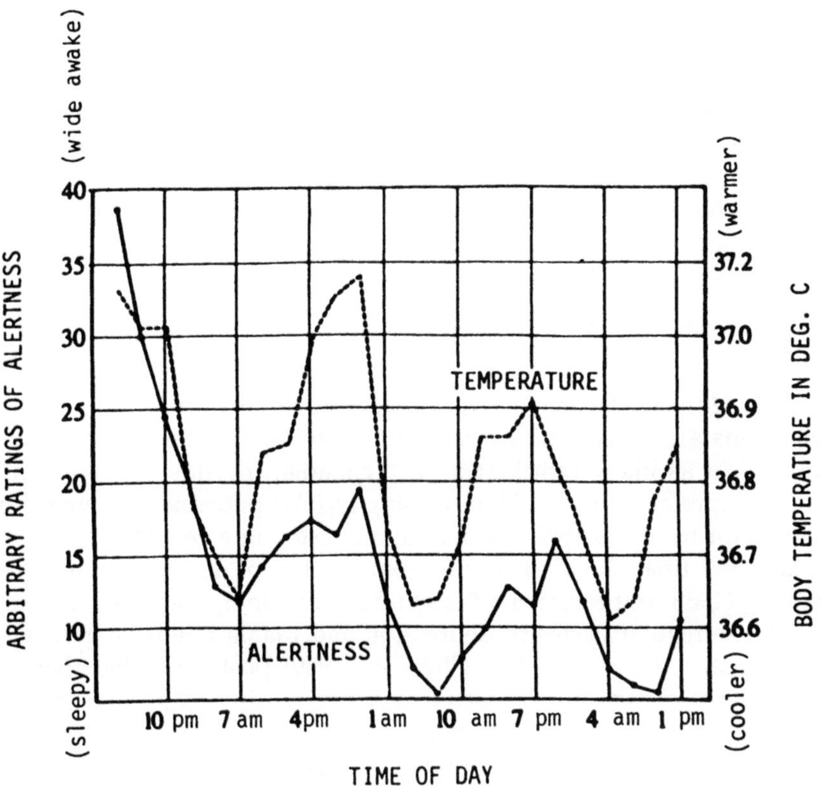

Figure 1. Body temperature and alertness from 15 subjects who experienced 72 hours of constant wakefulness [replotted from Froberg (1977)].

circadian rhythms is determined by isolating people in caves or bunkers where there is no knowledge of either clock time or the daylight/darkness cycle. Subjects are free to go to bed, get up, and take meals whenever they like, and the temporal patterning of these events and of the circadian rhythms (e.g., in body temperature) underlying them can be used to glean a period length ("day" length) that is "natural" to that individual (Wever, 1979). The process is known as "free-running," and the "natural" day length often referred to as "the free-running period" or tau. From study of 147 individuals, Wever determined tau for young people to be approximatey 25 hours with a standard deviation of 0.5 hours around that figure (Wever, 1979). As one gets older, tau appears to shorten toward 24 hours (Weitzman, Moline, et al., 1982; Monk, 1989).

In summary, the healthy human circadian system is an endogenous process leading to circadian rhythms that are resistant to abrupt changes in routine and which tend to run about 1 hour slow relative to the 24 hours of the Earth's rotation and societal time cues. Aspects of interest are (1) the timing (phase) of circadian peaks and troughs in various measures, (2) the amplitude of the peak–trough difference in those measures, (3) the "natural" (free-running) period length of the circadian system, and (4) the rate of phase adjustment of circadian rhythms to an abrupt change in routine.

Circadian Rhythms in Depression

The Desynchronization Hypothesis. Clearly, rhythmically appearing symptoms are those that are most likely to be of interest to the chronobiologist as likely candidates for a circadian rhythm based explanation of depressive illness. Thus, cycling bipolar patients who went from depression to mania (or hypomania) and back in a regular progression yielded the first circadian rhythm explanation of depressive disorders. Halberg (1968) put forward a "desynchronization hypothesis" regarding how these regular cycles of depression and mania might be generated. The hypothesis (discussed in Checkley, 1989) postulates that there are two or more circadian processes which run at different periods due to a lack of entrainment of one or more of them to the 24-hour cycles of the world and society. In any such system a "beating" phenomenon occurs whereby the two cyclic processes move into and out of phase with each other. The cycle length of the symptom changes can then be used to predict the difference in period length between the two circadian processes, as that difference determines the rate at which "beating" occurs (thus, for example, with periods of 22 hours and 26 hours one would expect a change in symptoms every six days).

The theoretical basis for the desynchronization hypothesis is not an implausible one. Free-running studies of normal healthy controls have often revealed "spontaneous internal desynchronization" (Wever, 1975; Weitzman, Czeisler, & Moore-Ede, 1979) whereby the sleep/wake cycle typically runs to a period length that is dramatically longer (or shorter) than the temperature rhythm (e.g., 39.1 hours sleep/wake versus 24.9 hours temperature). Paradoxically, though, Wever (1979) and colleagues have found this phenomenon to be associated with *improvement* in mood, rather than a worsening (the evidence from other groups does not, though, support this). Also, although some regularly cycling bipolars have shown evidence of desynchrony (in the shorter than 24 hours direction) (Kripke, 1983), evidence in favor of the desynchronization hypothesis has largely failed to appear. A free-running study of depressives by Wehr, Sack, Duncan, Mendelson, Rosenthal, Gillin, and Goodwin (1985) notably failed to support the desynchronization hypothesis. As Checkley (1989) concludes, even in depressives who do desynchronize, the circadian periods involved do not have the right numerical values to predict the cycle length of the psychiatric symptoms, and the hypothesis has largely been abandoned.

The Phase Advance Hypothesis. More resilient as a chronobiological explanation for depressive disorders has been the "phase advance hypothesis." This hypothesis has its roots in the findings that many depressives appear to exhibit circadian rhythms that phase earlier in the 24-hour day than those of healthy controls. The most obvious of these is, of course, the early morning awakenings that trouble many depressed patients. Also in the sleep domain, the preponderance of REM sleep early in the night and short REM latencies observed in depression (see following chapter) are all suggestive of an early peaking circadian system, since REM propensity appears to be one of the few sleep architecture parameters that is driven directly by the "deep oscillator" or endogenous circadian pacemaker (Moore-Ede, Sulzman, & Fuller, 1982). Early peaking circadian rhythms are the natural product of a short "natural" (free-running) period continually trying to catch up with the 24-hour day, and the phase advance hypothesis could be restated as the "shortened tau" hypothesis—that the tau of depressives is shorter than 24 hours while that of healthy controls is longer. In the next few paragraphs we shall discuss the empirical evidence for early phasing circadian rhythm and/or a shortened tau in depression.

Arguments concerning the phase advance hypothesis can either be based upon the more soft, circumstantial and/or anecdotal evidence, or upon hard statistically reliable "patient" versus "control" differences. Unfortunately, most of the evidence in favor of the hypothesis comes under the "soft" category, rather than the "hard." As mentioned before, early morning awakenings, and short REM latencies provide circumstantial evidence in favor of the hypothesis. Also added to the list can be "diurnal mood variation" whereby morning mood in depressives is held to be worse than evening mood—a change from the normal pattern and one that bespeaks a dysfunction of the circadian system.

Support for the phase advance hypothesis also comes from the finding that for some patients a phase advance of their sleep wake cycle can produce an improvement in depressive symptoms (Wehr, Wirz-Justice, Goodwin, Duncan, & Gillin, 1979). Indeed an early hope was that chronobiological interventions could become a benign alternative to some of the more radical treatments for depression. Interestingly, phase shifts induced by transmeridianal flight have indeed been shown to precipitate psychiatric disease, with the *direction* of flight predicting the prominence of manic versus depressive symptoms (Jauhar & Weller, 1982).

The other major piece of circumstantial evidence in favor of the phase advance hypothesis is the finding that many of the drugs used to treat affective disorders have the effect of "slowing down" the circadian system (i.e., lengthening tau). Kripke (1983) and Wirz-Justice (1983) document convincing evidence from many different species (including humans) that lithium lengthens tau. MAO inhibitors and tricyclics have also been shown to lengthen tau and thus delay the timing of various circadian rhythms (Wirz-Justice, 1983). In terms of case studies there are some reports of depressives having early phasing circadian rhythms and/or shortened taus. Kripke (1983) cites the example of a rapidly cycling bipolar patient who showed 21.8-hour rhythms in several circadian measures.

When one turns, however, to the "hard" evidence derived from statistical comparison of the circadian rhythms of groups of patients and controls, the evidence for a simple phase advance hypothesis is far from compelling. The acid test is whether free-running depressed patients have shorter temperature taus than free-running controls. Studies of free-running depressives by Wehr, Sack, Duncan, Mendelson, Rosenthal, Gillin, and Goodwin (1985) and Pollak, Alexopoulos, Moline, and Wagner (1989) have markedly failed to find a *consistent* pattern of shorter than 24-hour taus for their patient subjects. Indeed, except for the very occasional shortened tau, the depressives showed free-running period lengths that were well within the range one would expect for their age group and gender.

Even when one weakens the hypothesis to only require a phase advance of 24-hour circadian rhythms in depression, the evidence is, at best, mixed. For example, Von Zerssen, Barthelmes, Dirlich, Doerr, Emrich, Von Lindern, Lund, and Pirke (1985) studied 16 drug-free endogenous depressives and found "little, if any, indication of phase advance" in circadian rhythms of body temperature, urinary potassium, or urinary free cortisol. Ten of the depressives were studied again while in remission and, again, no consistent pattern of advanced phase was found when symptomatic and asymptomatic conditions were compared. Von Zersson et al. did, however, remark upon the reduction in circadian amplitude in depression (see below).

Avery, Wildschiodtz, and Rafaelsen (1982) studied temperature rhythms in nine drug-free depressives (7 unipolar, 2 bipolar) and 12 normal controls. Although there was some evidence for a phase advance in the depressed state compared to the remitted, Avery et al.'s major conclusion was that nocturnal temperature was higher in depression and rhythm amplitude diminished.

A very similar pattern of findings (or rather the lack of them) occurs when cortisol, rather than temperature, is the variable under investigation. Although some subtle phase differences have been found (Jarrett, Coble, & Kupfer, 1983; Sherman & Pfohl, 1985) and there is a well-documented increase in mean daytime level (e.g., Pfohl, Sherman, Schlechte, & Stone, 1985), mostly, the circadian cortisol rhythms in depressives are strikingly similar to those of healthy controls (e.g., Halbreich, Asnis, Shindledecker, Zumoff, & Nathan, 1985; Rubin, Poland, Lesser, Winston, & Blodgett, 1987).

Circadian Dysfunction in General. Given the zeitgeist of the early 1980s when there was so much hope for chronobiological interventions being a possible panacea for depression, many of the studies quoted above concentrated on the lack of a phase advance in their data. This failure to confirm an intuitively very attractive hypothesis rather overshadowed the fact that important circadian dysfunction was indeed being uncovered. A significantly reduced circadian amplitude is clearly indicative of some sort of malfunction of the circadian system.

For example, Beersma, Van Der Hoofdakker, and Van Berkestijn (1983) found no evidence of an endogenous body temperature rhythm in 7 of 15 depressed inpatients deprived of sleep (cf. Fig. 1 of this chapter, where a robust rhythm is seen in healthy controls). Von Zerssen, Dirlich, Doerr, Emrich, Lund, and Ploog (1985) also found a marked diminution of circadian temperature amplitude in a study of 20 medication-free endogenously depressed inpatients. Schulz and Lund (1983) concentrated on body temperature rhythm amplitude differences in 15 symptomatic depressives, 12 remitted depressives, and 10 normal controls. Only when they divided their subjects into those who did or did not have sleep onset REM episodes (SOREMs) did a difference emerge, whence the SOREM group had significantly reduced amplitudes. Thus, an increase in mean level (e.g., Pollak et al., 1989) and decrease in amplitude (Checkley, 1989) appear to be the most robust findings concerning circadian rhythms in depression.

There are two likely suspects to blame for circadian amplitude reductions in depression, namely, either a failure of the endogenous rhythm generation processes, or a failure of the entrainment processes that use zeitgebers (time cues) to keep the circadian system "on-track" to an appropriately aligned 24-hour routine. Entrainment failures may either occur from a breakdown of the biological mechanisms per se (i.e., a failure of "acuity," where the zeitgeber is not adequately "perceived" by the circadian system) or from the absence of appropriate zeitgebers in the life of the patient. The latter is particularly likely when the disorder limits the opportunity to go outdoors or meet people. Entrainment dysfunction is discussed in two later sections referring to Seasonal Affective Disorders (SAD) and Social Zeitgebers.

Process S Deficiency Hypothesis. A recent model of circadian abnormalities in depression has been based upon a currently popular model (the "Process S" model) of the human circadian system put forward by Borbély (1982) and Daan, Beersma, and Borbély (1984). In this general model of the human circadian system there is assumed to be an oscillatory process ("Process C") driving endogenous rhythms in body temperature, cortisol, etc., and a homeostatic (or capacitor-like) process ("Process S") which builds up during wakefulness and is dissipated during slow wave sleep. The timing of sleep onset and offset then depends upon the relationship between these two processes.

The Process S deficiency hypothesis (Borbély & Wirz-Justice, 1982; Borbély, 1988) posits that in the depressive the sleep-dependent component (Process S) does not rise as strongly during wakefulness as it does in the healthy normal. Thus, various features of the depressed patient's sleep/wake cycle characteristics (and the efficacy of sleep deprivation as an antidepressant) can be explained using this quantitative model. Although this model is inherently a circadian one, its major ramifications relate to sleep and sleep architecture, and a detailed examination of the Process S deficiency hypothesis will thus be postponed to the following chapter.

Seasonal Affective Disorder. In the early 1980s researchers at NIMH tapped into a hitherto neglected facet of affective disorders, namely, depression that regularly reappeared on a seasonal basis (Lewy, Kern, Rosenthal, & Wehr, 1982; Rosenthal, Sack, Gillin, Lewy, Goodwin, Davenport, Mueller, Newsome, & Wehr, 1984). The classic presentation of the syndrome is a winter-occurring depression that commences late in the autumn and remits in the spring. Often the syndrome is associated with carbohydrate craving, weight gain, loss of energy, and occasionally hypersomnolence. The background to the syndrome was Lewy, Wehr, Goodwin, Newsome, and Markey's (1980) discovery that the nocturnal secretion of the pineal hormone melatonin could only be suppressed by *daylight* levels of illumination (2500 lux) as opposed to normal artificial levels (typically about 500 lux). The treatment used was thus the intuitive one of using extremely bright levels of artificial illumination in early morning and late afternoon sessions in order to extend the duration of bright light exposure experienced by the patient until it approximated that of a summer day. The approach also related to Kripke's (1981) work regarding photoperiodic mechanisms for depression.

Light treatment worked, and many patients obtained symptomatic relief from the disorder using "light boxes" containing large banks of fluorescent tubes. Considerable media interest was aroused, a catchy title given (SAD—Seasonal Affective Disorder), and an academic society formed to promulgate and control the use of bright lights in the treatment of various disorders (The Society for Light Treatment and Biological Rhythms). The syndrome now has a classification in DSM III-R, confirming its accepted respectability, and numerous "phototherapy clinics" are appearing across the country.

Unfortunately there still remains considerable controversy regarding SAD. The first is whether SAD is a distinct chronobiological disorder as opposed to simply a variant of bipolar depression (Thase, 1989). Although the initial treatment was definitely circadian in origin, further experiments have cast significant doubt on the notion that it is simply a lengthening of the subjective "day" that helps the patients. A summer-based variant of SAD has been described which is clearly outside the circadian photoperiod etiology. Moreover, the effectiveness of treatment for winter depression appears to be independent of the presence of melatonin or its antagonists (Wehr, Jacobsen, Sack, Arendt, Tarmarkin, & Rosenthal, 1986; Rosenthal, Sack, Jacobsen, James, Parry, Arendt, Tamarkin, & Wehr, 1986; Skwerer, Jacobsen, Duncan, Kelly, Sack, Tamarkin, Gaist, Kasper, & Rosenthal, 1988).

Further criticism of SAD and its treatment is that it is essentially a subclinical syndrome (often described as the "winter blues"), which is often, therefore, much more easily treated than would be a true depressive disorder. At the extreme, some would argue that the bright lights deliver only a placebo response. While that criticism is probably overly harsh, the nature of phototherapy as an intervention, and the massive publicity surrounding SAD, has largely precluded investigators being able to even approach appropriate double blind trials that would be required to evaluate any pharmacological treatment (see Rosenthal, Sack, Skwerer, Jacobsen, & Wehr, 1988, for a fuller discussion).

These criticisms aside, it remains the case that certain patients do experience regular cycles of depression during the winter months, and that these patients can obtain symptomatic relief using bright light exposure. Some investigators favor blocks of bright light in the morning hours (Terman, 1988, 1989), others assert that a "phase typing" of the patient should be first performed (e.g., using the timing of the dim light melatonin onset) to determine whether they need to be phase advanced or phase delayed (Lewy, Sack, Singer, White, & Hoban, 1988). Morning lights would be prescribed for the former, evening lights for the latter.

The public acceptance of SAD as a disorder has clearly indicated a need within the population that is being met by light therapy. The extent to which that need relates primarily to psychiatry or to chronobiology has yet to be determined. There is a tremendous amount of work being done on SAD at this time, and a detailed evaluation of all the issues is beyond the scope of the present chapter. Interested readers are referred to the Summer 1988 special issue of the *Journal of Biological Rhythms,* and to the several books on the topic (e.g., Rosenthal and Blehar, 1989) which have recently appeared.

Social Zeitgebers. There has long existed a schism in the field of depressive disorders between those taking a biological approach and those taking a psychosocial one. This difference not only influences the form of treatment given, but also the way the whole disorder is approached. In a recent article, Ehlers, Frank, and Kupfer (1988) present a conceptual model of depression etiology that goes some way toward bridging that gap.

From the biological end, the model concentrates on the circadian dysfunction associated with depression, considering a disentrainment or pathological entrainment (to an aberrant timing) to be a contributory factor toward the occurrence or recurrence of depression. From the psychosocial end, the model considers life events to be agents of change to the social zeitgebers (time cues) normally impinging upon the individual and keeping his or her circadian rhythms appropriately entrained.

The circadian impact of social rhythms is through both the physical zeitgebers of daylight and darkness (experienced, for example, when leaving home for work), and the social zeitgebers inherent in meal and work scheduling and interpersonal contacts. There is strong evidence that both types of zeitgeber are important to the human circadian system (Wever, 1988). Any life event can lead to a loss of social rhythms. Thus, for example, the social rhythm loss consequent upon retirement or bereavement might in some individuals lead to circadian disentrainment and thence to depression.

The social zeitgeber hypothesis is currently under investigation using a longitudinal study of recurrent unipolar depressives treated without medication. An instrument (the Social Rhythm Metric) has been developed to quantify social rhythms (Monk, Flaherty, Frank, Hoskinson, & Kupfer, 1990), and several features of the metric have been found to differ between pharmacologically treated depressives and healthy controls (Monk, Kupfer, Frank, & Ritenour, 1991). More work needs to be done on testing the hypothesis, but the approach appears to be a promising one.

Conclusion

It is probably true to say that all of us working in the area of circadian rhythms in depression are disappointed that the simple phase advance hypothesis did not work out, and that we do not currently have an effective chronobiological treatment for general depression. The fact remains, however, that there *is* a circadian dysfunction associated with depression, and that an understanding of that dysfunction will inevitably lead to a better understanding of depression as a whole.

References

Aschoff J (Ed.): *Handbook of Behavioral Neurobiology,* Vol. 4, New York: Plenum, 1981.

Aschoff J, Hoffman K, Pohl H, Wever RA: Re-entrainment of circadian rhythms after phase-shift of the zeitgeber. *Chronobiologia* 1975; II(1): 23–78.

Avery DH, Wildschiodtz G, Rafaelsen O : Nocturnal temperature in affective disorder. *J Affect Disord* 1982; 4: 61.

Beersma DGM, Hoofdakker RH, Berkestijn HWBM: Circadian rhythms in affective disorders. Body temperature and sleep physiology in endogenous depressives. In: Mendlewicz J, Van Praag HM (Eds.), *Biological Rhythms and Behavior,* New York: Karger, 1983, 114–127.

Borbély AA: A two process model of sleep regulation. *Hum Neurobiol* 1982; 1: 195–204.

Borbély AA: The two-process model of sleep regulation: Implications for sleep depression. In: Kupfer DJ, Monk TH, Barchas JD (Eds.), *Biological Rhythms and Mental Disorders,* New York: The Guilford Press, 1988, 55–81.

Borbély AA, Wirz-Justice A: Sleep, sleep deprivation and depression. *Hum Neurobiol* 1982; 1: 205–210.

Checkley S: The relationship between biological rhythms and the affective disorders. In: Arendt J, Minors DS, Waterhouse JM (Eds.), *Biological Rhythms in Clinical Practice,* London: Wright, 1989, 160–183.

Daan S, Beersma DGM, Borbély AA: Timing human sleep: Recovery process gated by circadian pacemaker. *Am J Physiol* 1984; 246: R161–R178.

Ehlers CL, Frank E, Kupfer DJ: Social zeitgebers and biological rhythms: A unified approach to understanding the etiology of depression. *Arch Gen Psychiatry* 1988; 45: 948–952.

Froberg JE: Twenty-four-hour patterns in human performance, subjective and physiological variables and differences between morning and evening active subjects. *Biol Psychol* 1977; 5: 119–134.

Halberg F: Physiologic considerations underlying rhythmometry with special reference to emotional illness. In: de Ajuriaguerra J (Ed.), *Cycles Biologiques et Psychiatrie,* Paris: Georg, 1968, 73–126.

Halberg F: Chronobiology. *Annu Rev. Physiol* 1969; 31: 675–725.

Halbreich U, Asnis GM, Shindledecker R, Zumoff B, Nathan S: Cortisol secretion in endogenous depression. *Arch Gen Psychiatry* 1985; 42: 909–914.

Jarrett DB, Coble, PA, Kupfer DJ: Reduced cortisol latency in depressive illness. *Arch Gen Psychiat* 1983; 40: 506–511.

Jauhar P, Weller MPI: Psychiatric morbidity and time zone changes: A study of patients from Heathrow Airport. *Br J Psychiat* 1982; 140: 231–235.

Klein KE, Wegmann HM, Hunt BI: Desynchronization as a function of body temperature and performance circadian rhythm as a result of outgoing and homecoming transmeridian flights. *Aerospace Med* 1972; 43: 119–132.

Kripke DF: Photoperiodic mechanisms for depression and its treatment. In: Perris CG, Struwe G, Jansson B (Eds.), *Biological Psychiatry, 1981: Proceedings of the IInd World Congress of Biological Psychiatry,* Amsterdam: Elsevier, 1982, 1249–1252.

Kripke DF: Phase-advance theories for affective illness. In: Wehr TA, Goodwin FK (Eds.), *Circadian Rhythms in Psychiatry,* Pacific Grove, CA: The Boxwood Press, 1983, 41–69.

Lewy AJ, Kern HA, Rosenthal NE, Wehr TA: Bright artificial light treatment of a manic-depressive patient with a seasonal mood cycle. *Am J Psychiat* 1982, 139: 1496–1498.

Lewy AJ, Sack RL, Singer CM, White DM, Hoban TM: Winter depression and phase shift hypothesis for bright light's therapeutic effects: History, theory, and experimental evidence. *J Biol Rhythms* 1988; 3: 121–134.

Lewy AJ, Wehr TA, Goodwin FK, Newsome DA, Markey SP: Light suppresses melatonin secretion in humans. *Science* 1980; 210: 1267–1269.

Monk TH: Sleep disorders in the elderly. *Clin Ger Med* 1989; 5: 331–346.

Monk TH, Flaherty JF, Frank E, Hoskinson K, Kupfer DJ: The social rhythm metric: An instrument to quantify the daily rhythms of life. *J Nerv Ment Dis* 1990; 178: 120–126.

Monk TH, Kupfer DJ, Frank E, Ritenour AM: The social rhythm metric (SRM): Measuring daily social rhythms over a 12 week period. *Psychiat Res.* 1991; 36; 195–207.

Moore-Ede MC, Richardson GS: Medical implications of shift-work. *Annu Rev Med* 1985; 36: 607–617.

Moore-Ede MC, Sulzman FM, Fuller CA: *The Clocks That Time Us,* Boston: Harvard University Press, 1982.

Pfohl B, Sherman B, Schlechte J, Stone R: Pituitary-adrenal axis rhythm disturbances in psychiatric depression. *Arch Gen Psychiat* 1985; 42: 897–903.

Pollak CP, Alexopoulos GS, Moline ML, Wagner DR: Circadian period and phase of body temperature and plasma cortisol rhythms of depressives living in isolation from all temporal cues. *Sleep Res.* 1989; 18: 438.

Rosenthal NE, Blehar M: *Seasonal Affective Disorders and Phototherapy,* New York: The Guilford Press, 1989.

Rosenthal NE, Sack DA, Gillin JC, Lewy AJ, Davenport Y, Mueller PS, Newsome DA, Wehr TA: Seasonal affective disorder: a description of the syndrome and preliminary findings with light therapy. *Arch Gen Psychiat* 1984; 41: 72–80.

Rosenthal NE, Sack DA, Jacobsen FM, James N, Parry B, Arendt J, Tamarkin L, Wehr TA: Melatonin in seasonal affective disorder and phototherapy. *J Neural Trans* 1986; 21: 257–267.

Rosenthal NE, Sack DA, Skwerer RG, Jacobsen FM, Wehr TA: Phototherapy for seasonal affective disorder. *J Biol Rhythms* 1988; 3: 101–120.

Rubin R, Poland R, Lesser I, Winston RA, Blodgett ALN: Neuroendocrine aspects of primary endogenous depression. *Arch Gen Psychiat* 1987; 44: 328–336.

Schulz H, Lund R: Sleep onset REM episodes are associated with circadian parameters of body temperature: A study in depressed patients and normal controls. *Biol Psychiat* 1983; 18: 1421–1426.

Severino SK, Moline ML: *Premenstrual Syndrome: A Clinician's Guide,* New York: Guilford Press, 1989.

Sherman B, Pfohl B: Rhythm-related changes in pituitary-adrenal function in depression. *J Affect Disord* 1985; 9: 55–61.

Skwerer RG, Jacobsen FM, Duncan CC, Kelly KA, Sack DA, Tamarkin L, Gaist PA, Kasper S, Rosenthal NE: Neurobioloy of seasonal affective disorder and phototherapy. *J Biol Rhythms* 1988; 3: 135–154.

Terman M: On the question of mechanism in phototherapy for seasonal affective disorder: Considerations of clinical efficacy and epidemiology. *J Biol Rhythms* 1988; 3: 155–172.

Terman M: Light therapy. In: Kryger MH, Roth T, Dement WC (Eds.) *Principles and Practice of Sleep Medicine,* Philadelphia: W.B. Saunders, 1989, 717–722.

Thase ME: Affective disorder with other forms of recurrent depression. In: Rosenthal N, Blehar M (Eds.), *Seasonal Affective Disorders,* New York: Guilford Press, 1989, 64–78.

Wehr TA, Jacobsen FM, Sack DA, Arendt J, Tamarkin L, Rosenthal NE: Phototherapy of seasonal affective disorder. *Arch Gen Psychiat* 1986; 43: 870–875.

Wehr TA, Sack DA, Duncan WC, Mendelson WB, Rosenthal NE, Gillin JC, Goodwin FK: Sleep and circadian rhythms in affecting patients isolated from external time cues. *Psychiat Res* 1985; 15: 327–339.

Wehr TA, Wirz-Justice A, Goodwin FK, Duncan WC, Gillin JC: Phase advance of the circadian sleep-wake cycle as an antidepressant. *Science* 1979; 206: 710.

Weitzman ED, Czeisler CA, Moore-Ede MC: Sleep–wake, neuroendocrine and body temperature circadian rhythms under entrained and non-entrained (free-running) conditions in man. In: Suda M, Hayaishi O, Nakagawa H

(Eds.), *Biological Rhythms and Their Central Mechanisms,* Amsterdam: Elsevier, 1979, 199–227.

Weitzman ED, Kripke DF: Experimental 12-hour shift of the sleep-wake cycle in man: Effects on sleep and physiologic rhythms. In: Johnson LC, Tepas DI, Colquhoun WP, Colligan MJ (Eds.), *Variations in Work-Sleep Schedules: Effects on Health and Performance, Advances in Sleep Research,* Vol. 7, New York: Spectrum Publications, 1981.

Weitzman ED, Moline ML, Czeisler CA, Zimmerman JC: Chronobiology of aging: Temperature, sleep-wake rhythms and entrainment. *Neurobiol Aging* 1982; 3: 299–309.

Wever RA: The circadian multi-oscillator system of man. *Int J Chronobiol* 1975; 3: 19–55.

Wever RA: *The Circadian System of Man: Results of Experiments Under Temporal Isolation,* New York: Springer-Verlag, 1979.

Wever RA: Order and disorder in human circadian rhythmicity: Possible relations to mental disorders. In: Kupfer DJ, Monk TH, Barchas JD (Eds.), *Biological Rhythms and Mental Disorders,* New York: Guilford Press, 1988, 238–324.

Wirz-Justice A: Antidepressant drugs: Effects on the circadian system. In: Wehr TA, Goodwin FK (Eds.), *Circadian Rhythms in Psychiatry,* Pacific Grove, CA: The Boxwood Press, 1983, 235–264.

Zerssen von D, Barthelmes H, Dirlich G, Doerr P, Emrich HM, Von Lindern L, Lund R, Pirke KM: Circadian rhythms in endogenous depression. *Psychiat Res.* 1985; 16: 51–63.

Zerssen von D, Dirlich G, Doerr P, Emrich HM, Lund R, Ploog D: Are biological rhythms disturbed in depression? *Acta Psychiat Belg* 1985; 85: 624–635.

Chapter 6

SLEEP DISORDERS IN DEPRESSIVE DISORDERS

Daniel J. Buysse and David J. Kupfer

INTRODUCTION

The relationship between sleep and mood holds considerable promise for understanding the neurobiology of depressive disorders. It is axiomatic that depression—whether situational, endogenous, unipolar, neurotic, or melancholic—is accompanied by sleep disturbance. However, the mood–sleep relationship is not a simple bidirectional one. For instance, most healthy subjects report little change in mood (or mild dysphoria) as a consequence of partial or total sleep deprivation, but approximately 60% of depressed patients note transient antidepressant effects after sleep deprivation. These basic observations suggest not only that mood and sleep interact with each other, but that they interact differently in depressed and nondepressed individuals.

The purpose of this chapter is to highlight several persistent questions regarding the study of sleep in depressive disorders, and to address the significance of depressive sleep disturbances. Specifically, we will focus on the following questions: (1) What is being measured by sleep studies? (2) What are the characteristics of sleep in depression and subtypes of depression? (3) Are sleep changes in depression episodic (statelike) or persistent (traitlike)? (4) Are the sleep changes in depres-

Daniel J. Buysse and David J. Kupfer • Department of Psychiatry, University of Pittsburgh School of Medicine, Western Psychiatric Institute and Clinic, Pittsburgh, Pennsylvania 15213.

Biology of Depressive Disorders, Part A: A Systems Perspective, edited by J. John Mann and David J. Kupfer. Plenum Press, New York, 1993.

sion primary, or are they epiphenomena? (5) How sensitive and specific are the sleep findings in depressive disorders? (6) What implications do sleep findings have for the neurobiology of depression?

While we clearly do not have complete answers to these questions, it is also clear that the gaps in our knowledge may prove at least as interesting and important as the answers in hand.

What Is Being Measured by Sleep Studies?

Comparing Rapid Eye Movement (REM) sleep latency values in groups of depressed and nondepressed subjects appears to be straightforward. However, simple quantitative measurements ignore many complexities that characterize sleep and its measurements. Sleep and wakefulness are complex physiological states, which can be *described* but not *defined* by any single physiological measurement; by extension, "sleep studies" measure selected general characteristics of the sleep state, but not sleep itself. The study of sleep in healthy and depressed subjects is further hampered by the fact that, although a number of theories have been suggested, we do not know the function of sleep or its different stages.

A great deal has been learned about the basic neurobiology and neurophysiology of sleep, mostly from animal work using classic neuroscience techniques such as ablation and stimulation studies, single unit recordings, and surface electroencephalography. The generation of wakefulness and Non-Rapid Eye Movement (NREM) sleep relies on activity from widespread CNS sites, ranging from the ponto-medullary junction to the midbrain, basal forebrain, and cortex. REM sleep appears to be more discretely localized to the dorsolateral pontine tegmentum. A number of neurotransmitters (including serotonin, norepinephrine, acetylcholine, GABA, and peptides) are involved in the generation of sleep and the cyclic alternation between NREM and REM (Institute of Medicine, 1990). In general, sleep centers and sleep-related neurotransmitters show a good deal of redundancy and overlap, making it difficult to identify the unique and specific roles for most centers or transmitters.

Standard sleep studies in human subjects include measurement of electroencephalographic activity (from one vertex EEG channel), eye movement, activity, and muscle tension. These three variables are used to define REM sleep and four stages of NREM sleep (with Stages 3 and 4 often being combined as "Slow Wave Sleep" (SWS) or "Delta" sleep). Recordings are scored by assigning a sleep stage to each epoch, usually 30- or 60-second segments, in a recording. Epochs across the night are often summarized in terms of *sleep continuity* measures (which describe overall amounts of sleep, wakefulness, and ratios of sleep to wakefulness); *sleep architecture* (describing amounts and/or percentages of various sleep stages); and *REM measures* (describing estimates of actual eye movement activity and duration of REM periods). Thus, summary measures describe several different types of information, including amounts and percentages of sleep stages, ratios, measures of periodic and cyclic phenomena, and time durations and intensity of EEG and eye movement activities.

Although traditional sleep studies generate an impressive number and range of variables for statistic analyses, many of these are highly intercorrelated, and all are based on fairly crude categorizations of a small number of electrophysiological variables measured repeatedly during a night. Furthermore, not all sleep variables are consistently defined. One specific example is REM latency, which may have significantly different values *in the same patient* depending on the specific definition being used (Knowles et al., 1982; Reynolds et al., 1983b).

Sleep studies summarize physiological events that have characteristic ultradian and circadian rhythmicity. Most summary measures of sleep and many between-group comparisons do not detail this temporal information and instead report total amounts of sleep or ratios for the entire night. In disorders such as depression, which may involve underlying circadian and ultradian rhythm abnormalities, such summaries may neglect important temporal information.

A number of techniques have been developed to provide more detailed information about brain activity during sleep, in both depressed patients and healthy subjects. One such technique, period and amplitude analysis of the sleep EEG signal, provides a quantitative measure of the *number* of EEG waves meeting certain frequency or amplitude criteria (Feinberg et al., 1978; Kupfer et al., 1984a). Period/amplitude analysis is particularly helpful for examining amounts of SWS, or the distribution of SWS across successive NREM periods. It is more sensitive than traditional sleep stage scoring, which does not reflect the slow frequency activity that occurs throughout NREM stages 2–4. A related technique, automated rapid eye movement analysis, uses a computer program to count the actual *number* of eye movements in REM sleep epochs (Kupfer et al., 1984a); results from this type of analysis correlate strongly with visual estimates of the amount of eye movements.

Another more sophisticated technique to examine frequency characteristics of the EEG during sleep is power spectral analysis (Borbély et al., 1984). Rather than categorizing sleep by one of five stages, or counting the number of waves in certain frequency range, spectral analysis measures the comparative *power density* associated with different

EEG frequencies across a specified time interval. This technique has been especially useful in providing an intensity measure of NREM sleep, characterized by a saturating exponential rise with increased duration of wakefulness and an exponential decline during nocturnal sleep (Borbély, 1982).

Examples of summary data using traditional sleep stage scoring, period/amplitude analysis, automated REM analysis, and spectral analysis are shown in Fig. 1 for a healthy young adult subject.

Future efforts in examining the sleep EEG in depression are likely to include measurements of "dimensionality," a characteristic of time-series data (such as EEG) based on nonlinear functions. In basic terms, dimensionality is a means of quantifying the complexity of an EEG signal. Wakefulness and REM, with their mixed-frequency EEG signals, are more complex than the regular, low-frequency signal seen in SWS; wakefulness and REM are also characterized by higher dimensionality (Ehlers et al., 1991).

In summary, the routine measurement of sleep in depression may oversimplify a complex physiological state. In doing so, it risks the

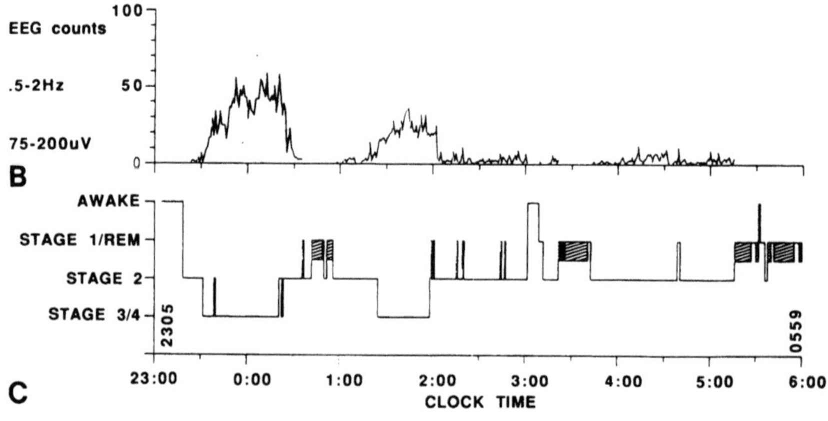

Figure 1. One night of sleep in a healthy control subject (37-year-old female) showing the results of (A) automated REM detection; (B) automated delta analysis; (C) visually scored

possibility of missing important, albeit subtle, changes in the sleep of depressed patients.

What Are the Characteristics of Sleep in Depression and Subtypes of Depression?

Each of the three major categories of traditional sleep measures—sleep continuity, sleep architecture, and REM measures—is affected in depression (Kupfer, 1976; Reynolds & Kupfer, 1987). With regard to

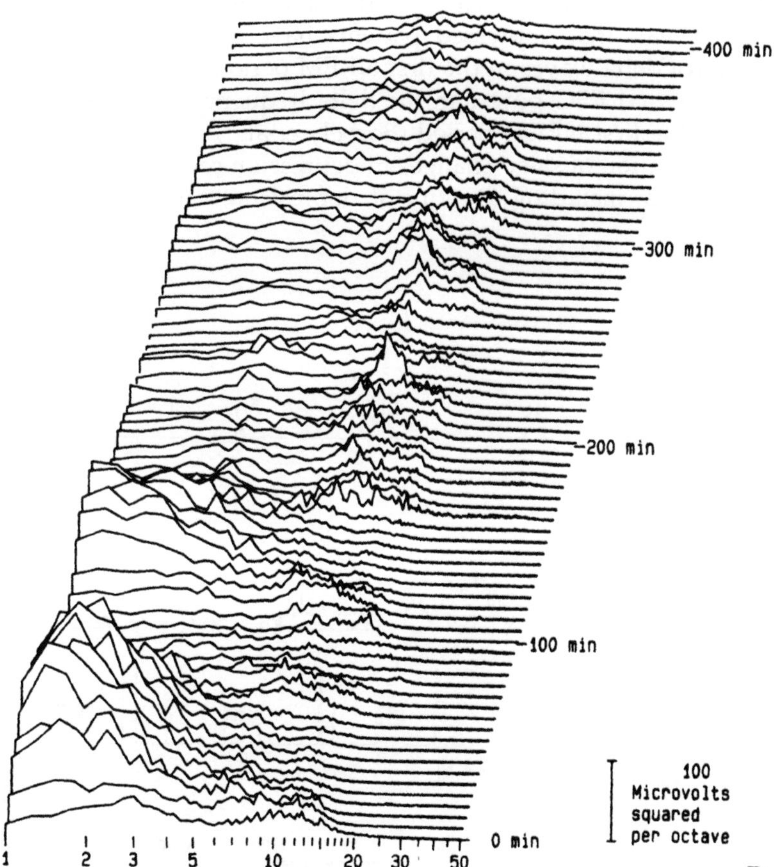

sleep stages; and (D) compressed spectral array. (The compressed spectral arrays were provided by Dr. C.L. Ehlers of the Research Institute of the Scripps Clinic, La Jolla, California.)

sleep continuity measures, depressed patients often have longer sleep onset latencies, more frequent awakenings, shorter sleep duration, decreased "sleep efficiency," and increased early morning wakefulness relative to healthy control subjects. NREM sleep architecture in depression is more variably affected, but often includes increased State 1 sleep and decreased SWS. This last finding is not uniformly reported (e.g., Riemann and Berger, 1989). Finally, REM sleep abnormalities in depression include the hallmark shortening of REM latency (which also represents the duration of the first NREM period), increased phasic REM activity, and a "redistribution" of REM sleep (both in terms of time and phasic activity) toward the early part of the night (Vogel et al., 1980).

Using the more sophisticated methods of period analysis or spectral analysis reveals that depressed patients have a reduced number and frequency of delta waves in the first NREM period (Kupfer et al., 1984b; 1990). Decreased EEG power in the delta frequency range is also most marked in the first NREM period (Borbély et al., 1984).

Interpretation of sleep changes in depression can be complicated by several factors, including severity of depression, age, gender, and de-

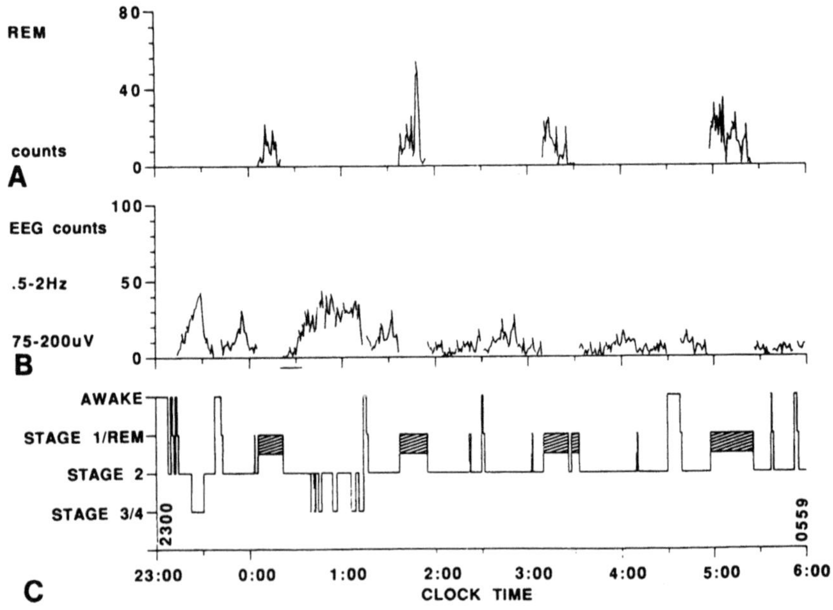

Figure 2. One night of sleep in a depressed outpatient (31-year-old female) showing the results from (A) automated REM detection; (B) automated delta analysis; (C) visually scored sleep stages; and (D) compressed spectral array. Note that the depressed patient has

pressive subtype. The relationship between sleep EEG findings and severity of depression has yielded inconclusive findings. Several reports have suggested an inverse relationship between REM latency and severity of depression, measured by global clinical impressions, global scores on rating scales, or specific items on rating scales (Spiker et al., 1978; Reynolds et al., 1983b; Giles et al., 1986a). However, other studies have failed to find a relationship between severity of depression and REM latency defined either as a continuous or dichotomous variable (Giles et

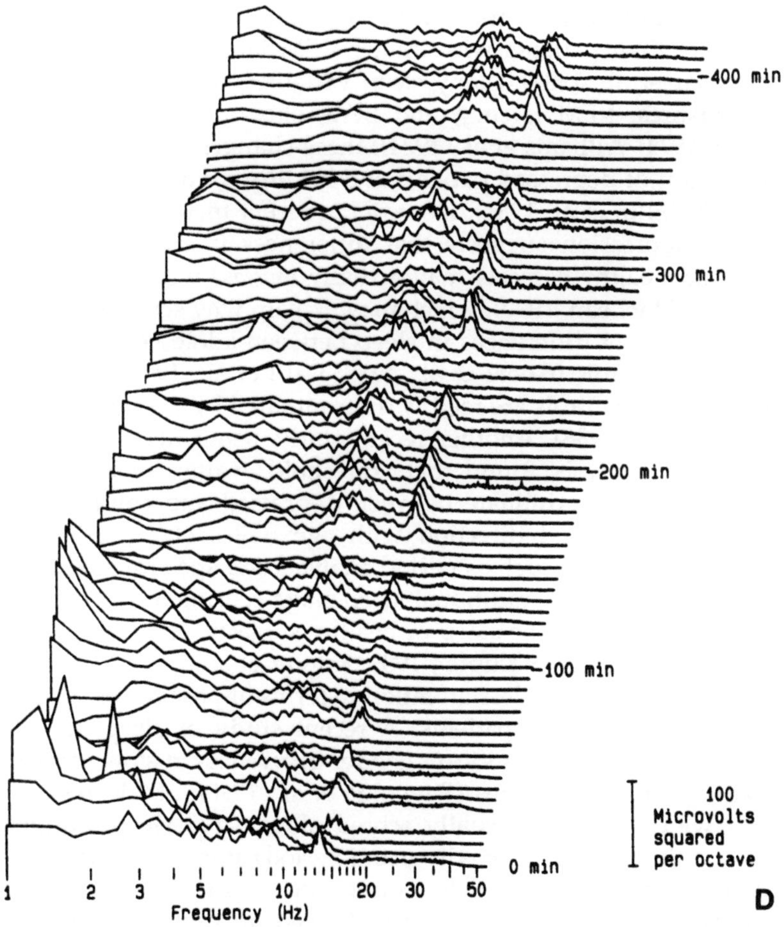

reduced REM latency, less SWS, and an altered temporal profile of REM, compared to the control subject in Fig. 1. (The compressed spectral arrays were provided by Dr. C.L. Ehlers of the Research Institute of the Scripps Clinic, La Jolla, California.)

al., 1987b; Ansseau et al., 1984; Kumar et al., 1987). While Feinberg et al. (1982) found no correlation between reported measures of sleep disturbance (assessed by the Hamilton rating scale) and corollary measures using sleep EEG, a similar study demonstrated significant correlations between reported and recorded measures of sleep continuity (Kupfer & Frank, 1984).

The age of subjects may further confound analysis of sleep–mood associations. For instance, the "bimodal distribution" of REM latencies sometimes reported in depressed patients (Coble et al., 1981; Schulz et al., 1979) may be at least partially explained by age effects. Patients with sleep-onset REM periods are older at the time of study and at age of illness onset than patients with longer REM latencies (Ansseau et al., 1984; Kumar et al., 1987). Many sleep variables, including sleep continuity measures, REM latency, and SWS, show consistency with age (Feinberg, 1974; Williams et al., 1974); these changes are essentially parallel in groups of depressed patients and healthy control subjects (Ulrich et al., 1980; Gillin et al, 1981). However, a recent meta-analysis of sleep studies in depression suggests that differences in sleep between depressed and healthy subjects may actually *increase* with increasing age (Knowles & MacLean, 1990), which may explain why several studies fail to show consistent differences in sleep between adolescent depressives and healthy controls (Goetz et al., 1987; Dahl et al., 1990), while older depressives and controls are readily discriminated by the same measures (Reynolds et al., 1988). Another recent study, using Receiver Operator Characteristic analysis, found that "threshold" values of REM latency which optimally discriminate between depressed and control subjects varied in the expected direction with age (Giles et al., 1990). More sophisticated analyses such as period analysis and spectral analysis may be more sensitive than traditional sleep stage scoring at distinguishing age-related changes in the sleep of healthy subjects, especially in younger age ranges (Ehlers & Kupfer, 1989; Kupfer et al., 1986b; 1989a).

Gender has received less attention in terms of possible effects on the sleep EEG of depressed patients. A recent study of sleep in 302 depressed patients found significant gender differences in slow wave sleep variables across five decades (Reynolds et al., 1990b). Male–female differences were noted in visually scored slow wave sleep, automated counts of delta frequency activity using period/amplitude analysis, and in the temporal pattern of delta counts. In contrast to these findings, sleep continuity and REM measures showed no robust gender-related differences.

Depressive subtypes may also show specific sleep characteristics. Early studies focused on the *primary/secondary* distinction, with the former being characterized by shorter REM latencies and higher amounts of phasic REM activity (Coble et al., 1976; Kupfer et al., 1978; Akiskal et al., 1982). However, a subsequent study comparing primary depressives to patients with depression secondary to other psychiatric disorders found no reliable differences between the groups (Thase et al., 1984). More consistent differences have been found between *endogenous* and *nonendogenous* subtypes, with the former group demonstrating more "severe" deviations from control values (Feinberg et al., 1982; Rush et al., 1982; Mendlewicz et al., 1984; Giles et al., 1986a). Compared to other dichotomous classifications of depression (such as primary/secondary or family history subtypes), the endogenous/nonendogenous distinction yields more consistent differences in sleep EEG measures (Giles et al., 1987b; Kupfer et al., 1982). This distinction also appears robust when factors such as age and severity of illness are partialled out (Feinberg & Carroll, 1984).

The *psychotic subtype* of major depression has also been associated with severe sleep EEG changes, including an increased frequency of sleep-onset REM periods (Kupfer et al., 1980). In contrast to other depressive subtypes, psychotic depressives may also demonstrate decreased REM sleep percentage and decreased phasic REM activity (Thase et al., 1986). Adolescents with psychotic major depression have shorter REM latencies than those with the nonpsychotic subtype (Naylor et al., 1990). Older delusional and nondelusional depressives are difficult to distinguish with traditional sleep EEG measures, but differences are observed with spectral analysis, indicating increased power in high-frequency bandwidths in the psychotic group (Kupfer et al., 1989b).

Patients with *atypical* depression differ from health controls in terms of short REM latency, but they do have the sleep continuity disturbances characteristic of those with endogenous depression (Quitkin et al., 1985). The *anergic/hypersomnic* primarily bipolar depressed patients studied by Thase et al. (1989) were found to have *neither* short REM latency *nor* sleep continuity disturbance. However, it is also clear that these patients do not have the severe daytime sleepiness characteristic of narcoleptic or sleep apneic patients (Nofzinger et al., 1991). Patients with *dysthymia* show attenuated versions of the "typical" depressive sleep stigmata (Akiskal et al., 1984).

Only minor differences between *bipolar* and *unipolar* depressed patients have been reported. For instance, bipolar depressed patients have a tendency for longer sleep time and more fragmented REM sleep periods, but no difference in REM latency or other measures (Giles et al., 1986b; Duncan et al., 1979). Although mania is clearly associated with shorter sleep time, reduced REM latency is also seen (Hudson et al., 1988).

One final distinction that has received some attention is the *inpatient/outpatient* dichotomy. Most sleep studies in depression have focused on inpatient populations, although consistent findings have also been noted in outpatients samples (e.g., Rush et al., 1982; Giles et al., 1987b). Mean REM latencies are generally longer in outpatient than inpatient depressed groups, although few statistical comparisons have been reported. Reynolds et al. (1982) reported that groups of matched inpatient and outpatient depressives did not show significant differences in traditional sleep EEG measures, but the outpatients showed more adaptation effects, as indicated by differences between two recording nights. By contrast, Kupfer et al. (1986a) found REM latencies ≤ 30 minutes in 25% of younger inpatients, but only 16% of younger outpatients, and significantly shorter mean REM latency in the inpatient group. In a study of unmatched patient groups, Buysse et al. (1990) found that outpatient depressives showed very few of the typical sleep EEG changes associated with depression, despite having depression scores comparable with the inpatients. Moreover, the outpatient group was very similar to the control group in terms of REM sleep timing across the night, while the inpatient group was distinctly different.

In summary, consistent sleep findings have been reported in patients with major depression, but these findings vary as a function of age, and probably as a function of severity. Of the various subtypes of depression, the endogenous subtype has been associated with the most consistent sleep changes.

Are the Sleep Changes in Depression Episodic (Statelike) or Persistent (Traitlike)?

Most of the studies discussed above have been cross sectional, studying depressed patients only during an episode of major depression. Therefore, it is reasonable to ask whether the observed changes persist throughout an episode or into clinical remission. Such data can provide important information regarding brain function in depression, and specifically, whether brain function changes with clinical state.

An early study examined sleep EEG findings in medication-free depressed inpatients, studied each day for five consecutive weeks (Coble et al., 1979). Mean REM latency was found to be stable across this interval, and individual values correlated strongly between the first and fifth weeks. Sleep continuity measures also remained stable over the five weeks, while measures of tonic and phasic REM activity showed small but significant decreases with time.

A similar finding, obtained with a very different methodology, has been reported by Kupfer et al. (1988). Nineteen depressed patients were studied during an index episode of depression, and were restudied very early in the course of a recurrence. Several REM sleep measures, including REM latency, amount of REM sleep, and measures of phasic REM activity, were more abnormal during the recurrence, despite a *shorter* duration of illness than in the index episode.

Clinical recovery is associated with a decrease in the frequency of sleep-onset REM periods. Schulz et al. (1979) found that 31% of 90 sleep recordings in 6 depressed patients included SOREMPs; after recovery, 58 medication-free studies in four of these patients showed only 5% SOREMPs.

Other sleep changes in depression may be more persistent. For instance, Hauri et al. (1974) reported that remitted medication-free unipolar depressed patients had longer sleep latencies, greater intermittent wakefulness, more Stage 1 sleep, and less delta sleep compared to healthy controls; no differences in REM latency or amount of REM sleep were noted. Greater intraindividual and intragroup variability also characterized the remitted depressives. These results contrast with findings in a group of 10 remitted bipolar patients reported by Knowles et al. (1986), who did not differ from controls except in the number of arousals.

Rush et al. (1986) performed longitudinal studies in 13 patients during and after an episode of depression. No differences in any polysomnographic measures—including REM latency, phasic REM activity, delta sleep, and sleep continuity measures—were found between illness and recovery, suggesting that sleep EEG measures are persistent, or at least slow to normalize. Similar results were subsequently reported by Giles et al. (1987a), who found stability of categorically defined REM latency from pretreatment baseline through 24 months of drug-free follow-up; in 20/25 patients, REM latency category remained stable or changed only once during the follow-up.

Another recent study, including 8 depressed patients, allowed a drug-free interval of 3 months before patients were restudied (Riemann and Berger, 1989). Sleep latency, intermittent wakefulness, and REM density all showed significant changes toward control values, and REM latency, although not significantly changed, also tended toward longer (more "normal") values. SWS showed no change, but was not found to be reduced, even during the episode. These findings argue that at least some sleep EEG changes, particularly REM sleep changes, are more episodic or statelike.

Conflicting results have been found in studies using cholinergic agonists as a probe for sleep regulation in remitted depressed patients. Sitaram and colleagues (1980) found more rapid induction of REM sleep in remitted depressives than controls using intravenous arecoline. However, Riemann and Berger (1989) found that actively depressed patients differed significantly from remitted depressed patients and healthy controls in the degree of REM latency shortening and REM density increase after administration of the cholinergic agonist RS-86.

Family studies provide another method for addressing the question of episodic versus persistent sleep EEG changes. This strategy assumes that sleep changes are a persistent biological trait, and that this trait has some genetic basis. Sleep characteristics such as REM latency and REM sleep responses to cholinergic challenge are correlated within family members, and most highly correlated in identical twins, supporting a genetic component for at least some sleep measures (Webb & Campbell, 1983; Nurnberger et al., 1983; Linkowski et al., 1989). Giles et al. (1989) found a significant association between categorically defined REM latency in depressed probands and depression in the relative. Family members with the "trait" marker of reduced REM latency also have a threefold increase in relative risk for unipolar depression (Giles et al., 1988). While these data do not prove genetic, as opposed to environmental influences, they do provide an intriguing introduction to this question.

Apparent discrepancies in studies addressing the "episodic versus persistent" question may result in part from methodological differences. Not all of the above studies used a longitudinal within-subject design, and others do not employ a control group. Differences are also apparent in the types of patients studied, with some including bipolar patients, and others exclusively unipolar depressives. The proportion of endogenously depressed patients also varies. No studies have used exclusively nonpharmacological treatment, and the duration of drug-free interval is also variable in the studies, ranging from two weeks to three months. Another potential problem is that most of these studies have included at least some inpatient depressives during their acute episode, while remitted depressives and controls are studied as outpatients. Finally, there is no published information available regarding the stability of period/amplitude measures or spectral analysis during depression and into remission.

Nevertheless, it appears that slow wave sleep changes in depression are largely persistent, while measures of phasic and tonic REM are more variable from night to night, and tend to normalize during or after the depressive episode. REM latency, which depends on NREM as well as REM sleep factors, appears to be one of the persistently abnormal variables except in the case of sleep onset REM. REM latency may best be conceptualized as having two potential sources, one a "traitlike" deficiency of slow wave sleep, and the other a "statelike" augmentation of REM sleep and arousal (Kupfer & Ehlers, 1989).

Are the Sleep Changes in Depression Primary, or Are They Epiphenomena?

Although most psychiatric sleep researchers believe that the sleep EEG changes of depression are "primary," it could also be argued that the typical sleep EEG findings associated with depression are epiphenomena, resulting from nonspecific factors such as weight loss and agitation. Evidence to support either view is sparse.

Feinberg and Carroll (1984) found that weight loss was significantly correlated with the number of arousals and phasic REM activity, but not with REM latency or REM sleep time, in groups of endogenous/nonendogenous and unipolar/bipolar patients. No healthy control group was used in this study, which limits any definitive conclusions regarding the effects of weight loss on sleep *per se*.

Mullen et al. (1986) examined the nonspecific effects of weight loss and sleep disruption on sleep EEG characteristics in a total of 28 healthy subjects. Mean weight loss of 4.3 kg over a period of 18 days had no effect on REM latency values. However, repeated partial sleep deprivation (delaying bedtime by one hour and advancing wake-up time by two hours, for a total sleep duration of five hours) led to a reduction in REM sleep latency from a mean of 78.8 minutes to 55.9 minutes. Most studies of sleep in depression do not specify what hours patients actually slept, or whether these conformed to habitual sleep hours. It is possible that some of the sleep characteristics of depression might represent an artifact based on awakening the patients earlier than their habitual Good Morning times, as modeled in the Mullen et al. study. Early awakening, which is certainly conceivable based on the demands of a structured inpatient unit, could lead to relative REM sleep deprivation and evidence of more "intense" REM sleep early in the night. Such a model provides a less adequate explanation for sleep changes in depressed outpatients or family members of depressed probands, and it does *not* explain the deficiency of slow wave sleep or the spontaneous nature of awakenings in depressed patients. Nevertheless, more attention does need to be focused on the issue of sleep timing and sleep duration during sleep studies in depressed patients.

It is also conceivable that sleep changes, and, in particular REM sleep changes, are related nonspecifically to disturbing events in a person's life and not to an inherent biological disorder. Cartwright (1983) investigated this type of nonspecific effect in 29 women undergoing marital separation or divorce. No subject had a personal or family history of major depression, and they were not specifically selected on the basis of depression. Those women with more emotional distress (specifically, higher scores on the Beck Depression Inventory and more "traditional" identities) had the shortest REM latencies and highest REM densities. At one year follow-up, REM latency had increased, but REM activity did not change; the most depressed women at follow-up again had the shortest REM latencies. The major difficulty in viewing this study in the context of "nonspecific" effects is that retrospective classification indicated a diagnosis of remitted or current major depression in 24 of the 29 subjects at intake. Therefore, one could argue that this study was simply another confirmation of REM sleep changes in depressed patients, albeit patients with an apparently low personal or family history diathesis.

Other arguments in favor of nonspecificity can be proposed on the basis of circadian sleep regulation (discussed in greater detail below). Basically, such arguments would suggest that the sleep characteristics of depression are the result of abnormal timing of sleep. Computer simulations based on data from normal subjects undergoing delayed sleep routines can mimic the short REM latencies and prolonged first REM periods observed in depression (MacLean et al., 1983, 1986).

However, it is clear that most depressed patients do not routinely go to sleep 4–6 hours later than most nondepressed people. Furthermore, correlations between nighttime REM latency and REM latency during daytime naps (Kupfer et al., 1981), or between initial REM latency and REM latency after a mid-night awakening (Schulz & Tetzlaff, 1982) have been reported. In these situations, abnormal bedtimes in themselves cannot account for the findings, suggesting that short REM latency is an intrinsic characteristic of sleep in depression, rather than a nonspecific finding that results from altered bedtimes.

In summary, data supporting the idea of nonspecific illness effects as an explanation for depressive sleep disturbances are weak, and they cannot explain the entire "constellation" of sleep changes noted in depression. Nevertheless, careful studies that accurately mimic the nonspecific effects of depression have been very few in number.

How Sensitive and Specific Are the Sleep Findings in Depressive Disorders?

A number of psychiatric disorders have sleep EEG characteristics that resemble those found in depression (Table 1). These findings raise the question of sensitivity and specificity of sleep EEG findings in depression. We have addressed this question in a recent article (Buysse & Kupfer, 1990) and will focus the current discussion on specific aspects of this topic.

As indicated above, sleep studies are by nature *multivariate* measures, while questions of sensitivity and specificity are most often conceptualized from a *univariate* perspective. With regard to sleep studies in depression, the question has focused most often on the sensitivity and specificity of REM latency at a certain "threshold" value as a diagnostic marker in depression. Greater attention has recently been focused on statistical methods such as Receiver Operating Characteristic (ROC) Analysis, which attempt to optimize the diagnostic information obtained by such measures (e.g., Giles et al., 1990; Somoza & Mossman, 1990). However, other single sleep variables or multivariate analyses can also be used to examine the sensitivity/specificity issue, and may lead to improved separation of groups (e.g., Feinberg et al., 1982; Gillin et al., 1979; Reynolds et al., 1988).

Another very important caution in examining the sensitivity/specificity of sleep studies in depression regards the comparison group(s) chosen. Two basic problems can arise: The comparison diagnosis may *overlap* with depression (as in comorbidity or affective "spectrum" disorders), or patients in the comparison groups may have a personal or family *history* of depression, which could influence sleep studies to the extent that depression-related changes are persistent. The first of these problems has been noted in studies of patients with Borderline Personality Disorder (Reynolds et al., 1985b; Akiskal et al., 1985) and Panic Disorder (Dubé et al., 1986). The second problem has been seen in studies of schizophrenic patients, where short REM latencies may be associated with a positive family history of depression (Keshavan et al., 1990). One final caution is that sensitivity/specificity studies tend to be biased with positive results, since studies showing no differences in mean values between groups do not usually report performance characteristics for the variables under study. Having stated these caveats, our previous review indicated that, compared to healthy control subjects, sleep studies in depression have an average sensitivity of about 70%, and specificity of about 85%. Compared to patients with other psychiatric disorders, the average sensitivity and specificity are both approximately 75% (Buysse & Kupfer, 1990).

The question of sensitivity and specificity has received particular attention in comparing the sleep of patients with schizophrenia to patients with depression. Like depression, schizophrenia has been associated with impaired sleep continuity, diminished SWS, and (in some

Table 1. Sleep Findings in Other Psychiatric Disorders[a]

Disorder	References	Similarities to depression	Comments
Anorexia nervosa/bulimia	Neil et al. (1980) Levy et al. (1988) Hudson et al. (1987) Walsh et al. (1985)	Decreased delta sleep in anorexia Sleep continuity disturbance in anorexia Variable decrease in REM latency in anorexia	Comorbidity of anorexia with depression may explain some findings; bulimics' sleep not different from normal subjects
Obsessive Compulsive Disorder	Insel et al. (1982)	Decreased sleep time Decreased delta sleep Decreased REM latency	OCD patients have higher Stage 1 percent and less "intense" early REM than depressives; REM latency shortest in OCD patients with depressive symptoms
Panic disorder	Dubé et al. (1986) Mellman and Uhde (1989)	Increased sleep latency Decreased sleep efficiency	Sleep architecture, REM latency, REM density normal
Generalized Anxiety Disorder	Reynolds et al. (1983a) Papadimitriou et al. (1988)	Sleep continuity disturbance Decreased delta sleep Prolonged sleep latency	Depressives have shorter REM latency, higher REM percent, altered REM temporal distribution compared with GAD
Borderline Personality Disorder	McNamara et al. (1984) Reynolds et al. (1985b) Akiskal et al. (1985)	Decreased REM latency Prolonged sleep latency Increased early REM	Comorbidity with depression may explain findings in BPD patients

Schizophrenia	Zarcone et al. (1987) Ganguli et al. (1987) Hiatt et al. (1985) Kempenaers et al. (1988)	Decreased REM latency (variable finding) Decreased delta sleep Sleep continuity disturbance Abnormal temporal distribution of delta sleep	Temporal distribution of REM and delta similar in schizophrenics and control subjects; REM activity higher in depressives
Mania	Hudson et al. (1988)	Sleep continuity disturbance Decreased REM latency Increased REM density	Delta sleep preserved in mania
Dementia	Reynolds et al. (1985a, 1988) Prinz et al. (1982) Feinberg et al. (1967)	Sleep continuity disturbance Decreased delta Decreased REM latency (variable finding)	REM latency, REM density, REM percent, indeterminate NREM, sleep maintenance distinguish depressed patients
Alcohol dependence	Adamson and Burdick (1973) Snyder and Karacan (1985) Gillin et al. (1990)	Sleep continuity disturbance Prolonged sleep latency Decreased delta Increased early REM	Patients with alcohol dependence and secondary depression show more typical depressive sleep changes

[a] OCD, obsessive-compulsive disorder; GAD, generalized anxiety disorder; BPD, borderline personalitty disorder. Table reprinted from Buysse and Kupfer (1990) by permission.

reports) decreased REM latency (Jus et al., 1973; Kupfer et al., 1970; Hiatt et al., 1985). Direct comparisons of depressed and schizophrenic patients have yielded contradictory results, with some showing distinct sleep profiles for the two groups (Ganguli et al., 1987), and others showing no differences (Zarcone et al., 1987; Kempenaers et al., 1988). Period/amplitude measures demonstrate the same contradictions (Ganguli et al., 1987; Hiatt et al., 1985; Guazzelli et al., 1985).

While these baseline findings may argue for nonspecificity of sleep changes in depression, the issue cannot be resolved definitively at this time. Several strategies can be employed to clarify this issue: (1) The use of challenge paradigms such as REM sleep deprivation (Zarcone et al., 1975; Gillin et al., 1974; Vogel et al., 1975), or cholinergic challenge (Riemann & Berger, 1990); (2) longitudinal studies (Kupfer et al., 1970, 1988); (3) family history studies to determine the "loading" for affective disorders in the families of schizophrenic patients (Keshavan et al., 1990); and (4) controlling for medication use, duration of illness, age, and other clinical factors.

What Implications Do Sleep Findings Have for the Neurobiology of Depression?

Several theories have been proposed to explain the observed links between sleep physiology and depression. Some postulate causal connections between sleep and depression, while others suggest underlying physiological disturbances that lead to both depression and sleep disturbances. Most of these theories are not mutually exclusive, and no single explanation has been sufficient to unite all of the observations regarding sleep and depression.

Neurochemical theories regarding sleep and depression have focused on the cholinergic system and its dynamic interactions with monoaminergic systems. As initially proposed by Janowsky and colleagues (1972), the "cholinergic supersensitivity" theory states that depression is associated with a relative increase in cholinergic sensitivity, and mania with decreased sensitivity. As applied to sleep in depression, the theory postulates that enhanced REM sleep measures (including short REM latency) are another manifestation of cholinergic supersensitivity. The relationship between cholinergic function and sleep is supported by the observation of reciprocal interactions between cholinergic and aminergic neuron systems during the course of REM–NREM cycles in cats (McCarley, 1982); cholinergic systems promote REM sleep, while noradrenergic and serotonergic neurons function as a "REM-off" system.

In support of heightened cholinergic sensitivity in depression, the administration of cholinergic agonists such as arecoline leads to a more rapid appearance of REM sleep in depressed patients than in controls (Sitaram et al., 1982; Berger et al., 1983). Induction of muscarinic supersensitivity by the administration of scopolamine was also found to lead to shortened REM latency (Sitaram et al., 1979). As previously noted, studies also demonstrated the "traitlike" nature of cholinergic sensitivity in depressed patients (Sitaram et al., 1980), as well as specificity of the finding for depression, when compared to panic disorder (Dubé et al., 1985) or eating disorders (Lauer et al., 1988).

The cholinergic supersensitivity theory is appealing because of its simplicity. It has also led to the development of mathematical models of the REN–NREM cycle in healthy subjects and depressed patients (McCarley & Massaquoi, 1986). The major problem with the theory has been difficulty in replicating results. For instance, Berger and colleagues (1983) did not find a differential effect of arecoline in depressed patients, in part because of the awakenings induced by arecoline. Using the oral cholinomimetic RS-86, the same group has failed to find persistence of REM sleep cholinergic sensitivity into recovery (Riemann & Berger, 1989), and has also found heightened sensitivity in schizophrenic patients (Riemann et al., 1990).

Information on neurochemical mechanisms of sleep in depression has also been explored through studies of antidepressant drug effects on sleep. While most effective antidepressant drugs suppress REM sleep and prolong REM sleep latency, specific agents differ in their effects on sleep continuity and SWS. For instance, drugs such as desipramine and fluoxetine appear to increase wakefulness during the night, and have little effect on slow wave sleep (Nicholson & Pascoe, 1988; Shipley et al., 1985), while drugs such as amitriptyline, trazodone, and doxepin improve sleep continuity and may increase slow wave sleep (Roth et al., 1982; Shipley et al., 1984; Montgomery et al., 1983). The major difficulty with drug studies of sleep in depression is that none of these agents is very specific. In addition to their effects on serotonin and/or norepinephrine reuptake, they affect cholinergic and histaminergic systems. Since a drug's neurochemical effects are difficult to specify in vivo, reasoning backward from a drug's effects on sleep to neurochemical mechanisms of depression can be difficult.

Several *circadian rhythm* theories have also been proposed to explain the sleep characteristics of major depression (see also Chapter 5 by Monk, in this volume). The "phase advance" theory postulates that depression is associated with an advance of certain circadian rhythms, including the REM sleep circadian rhythm, relative to other circadian rhythms, such as the sleep–wake–activity rhythm (Wehr & Goodwin, 1983). Supporting evidence comes from several sources: longitudinal studies of bipolar patients, who showed "advanced" REM sleep when switching into depression, studies of healthy subjects in temporal isolation or on 90-minute sleep/wake cycles, who demonstrate short REM latencies when sleep commences near the nadir of the circadian temperature rhythm; and clinical studies, which demonstrate an antidepressant effect of advancing the major sleep period to an earlier clock time (Sack et al., 1985). A related theory suggests a "critical phase" in the circadian temperature rhythm which usually occurs just after waking (Wehr & Wirz-Justice, 1982). In the case of depression, sleep overlaps the critical phase, and depressive symptoms and sleep disturbance result.

Evidence against the phase advance theory has ultimately proven to be more convincing. For instance, measures of circadian temperature rhythms in depression have not demonstrated phase advances (Avery et al., 1986; van den Hoofdakker & Beersma, 1985; Schulz & Lund, 1985), and when circadian phase is estimated by Dim-Light Melatonin Onset (DLMO), patients with a seasonal pattern of depression demonstrate a *delayed* circadian rhythm (Lewy et al., 1988). Studies by Reynolds et al. (1985b) and Buysse et al. (1990) demonstrate that, while short REM latencies may lead to an advance of REM sleep relative to sleep onset in unipolar depressive, there is no advance of REM sleep relative to 24-hour clock time. Finally, studies in healthy control subjects and with computer simulations have shown that changes in sleep–wake timing on the order of 6 hours are needed to produce "depressive" sleep patterns (MacLean et al., 1986).

The "two-process" model of sleep regulation (Borbély, 1982) has also served as the basis for circadian rhythm theories of sleep abnormalities in depression. The model proposes that sleep is regulated by two processes: a homeostatic element (Process S), which is modeled by the exponential decay of EEG power in the delta bandwidth over the night, and which increases as a saturating exponential function with the duration of prior wakefulness; and a circadian function (Process C), which follows the course of the circadian core body temperature rhythm. Borbély and Wirz-Justice (1982) suggested that depression may be associated with a deficient build-up of Process S, which could be associated with reduced slow wave sleep, and indirectly, with reduced REM latency. Many studies have in fact demonstrated deficient slow wave sleep in depression, as well as augmented SWS measures—and clinical improvements—following one night of total sleep deprivation. Moreover, spectral analysis of sleep in depressed patients shows decreased delta power density for the entire night and for the first NREM period

(Borbély et al., 1984). Kupfer et al. (1984b) demonstrated that the number of delta EEG waves in the first NREM period were correlated with the length of the first NREM period (i.e., REM latency), further supporting the theory. In simulations with the two-process model, deficiency in Process S or increased "noise" in the circadian wake threshold can each model aspects of depressive sleep disturbance, including early appearance of REM sleep, sleep continuity problems, and short sleep duration (Beersma et al., 1985).

The two-process model, and particularly the S-deficiency hypothesis, has been less helpful in modeling the REM changes of depression. Furthermore, some studies have failed to demonstrate a decrease in the *level* of EEG delta power, suggesting instead an altered time course (Van den Hoofdakker & Beersma, 1988).

One other circadian view of sleep in depression suggests that depression is associated with decreased *amplitude* of circadian rhythms, such as the body temperature rhythm. Several studies have shown either a higher nocturnal temperature minimum (Avery et al., 1986) or a blunted 24-hour rhythm amplitude in depression (von Zerssen et al., 1985; Lund et al., 1983). If the circadian rhythm in REM sleep propensity were also "flattened, " this might explain the early occurrence of REM sleep and the absence of increasing REM period durations across the night in depressives. Studies with larger numbers of patients and a healthy control sample are needed.

Several findings suggest even more direct connections between sleep and mood in depression. In longitudinal studies of 10–32 days' duration, Hauri and Hawkins (1971) observed that daily fluctuations in depression self-ratings were significantly associated with nightly fluctuations in phasic REM activity; specifically, a high percentage of REM sleep with phasic activity was associated with *lower* depression ratings the following day. Vogel and colleagues have also made important observations linking REM sleep and depression. The most significant findings have been as follows: Repeated specific REM stage deprivation by awakening results in clinical improvement in endogenous (but not reactive) depressives (Vogel et al., 1975); clinical improvement during REM deprivation is significantly associated with an increase in REM during recovery sleep, also known as increased "REM pressure" (Vogel et al., 1977); and clinical improvement is specifically associated with an increase of REM sleep late in the night, which "corrects" the abnormal REM temporal distribution in depressives (Vogel et al., 1980). Subsequent studies have confirmed that REM sleep deprivation leads to differential REM sleep rebound in depressed and control or demented patients (Reynolds et al., 1990a).

Vogel and colleagues proposed that these findings could be explained by decreased REM sleep generation, in combination with decreased REM sleep inhibition (Vogel, et al., 1980). Computer simulations have suggested that decreased REM sleep inhibition alone may be sufficient to explain many of the baseline sleep abnormalities see in depression (Beersma et al., 1984).

Vogel's original observations regarding REM sleep deprivation and depression have also been extended to include antidepressant drug effects. Specifically, it has been proposed that drugs producing large, persistent REM sleep suppression during administration and REM rebound upon withdrawal are antidepressant (Vogel et al., 1990). This suggests that REM sleep deprivation *per se*—whether by repeated awakenings or drugs—has inherent antidepressant effects. It is uncertain whether REM sleep deprivation itself causes changes in the brainstem neurochemical milieu, or whether REM sleep merely *reflects* these changes.

Observations with total and partial sleep deprivation have also contributed to the notion that sleep and mood are directly related in depression. Studies of one night of total sleep deprivation and partial sleep deprivation (with wakefulness in the second half of the night) indicate that approximately 60%–75% of endogenous depressed patients have a short-lived antidepressant response to these maneuvers (Gillin, 1983; Wu & Bunney, 1990). A more "classic" baseline sleep profile, typical endogenous symptoms, diurnal variation in symptoms, lower baseline HVA levels, and normal endocrine responses to dexamethasone and TRH have all been reported to predict a positive antidepressant response to sleep deprivation (Duncan et al., 1980; Reinink et al., 1990; Gerner et al., 1979; Riemann & Berger, 1990). The occurrence of even brief naps following sleep deprivation may lead to a recurrence of depressive symptoms. While early observations suggested that REM sleep in naps may be particularly deleterious (Wiegand et al., 1987), subsequent observations did not find a differential impact of REM versus NREM sleep in naps (Riemann et al., 1989). In fact, at least one study has shown that some patients may *improve* symptomatically when they have a brief nap after sleep deprivation (Gillin et al., 1989).

Several theories have been proposed to explain the antidepressant effect of total or partial sleep deprivation. For instance, the "critical phase" hypothesis is supported by sleep deprivation results, and the fact that sleep deprivation in the second (but not the first) half of the night is antidepressant suggests a role for REM sleep deprivation as a mechanism (Sack et al., 1988). It is also possible that sleep deprivation may be antidepressant because of effects on core body temperature, or the temperature set point (Wehr, 1990). One final hypothesis is that sleep is associated with the release of a depressogenic process or substance which is sequestered or metabolized during wakefulness (Wu & Bunney, 1990). Although no such specific substance or process has been identified, neither has one been sought systematically.

In summary, many theories have been set forth to explain the etiology of sleep abnormalities in depression, or the association between specific sleep patterns and mood. No single theory is entirely satisfactory at this point. The use of increasingly sophisticated methodologies, such as longitudinal studies, neurochemical and behavioral probes, family studies, and alternate EEG analyses, may help to test existing theories, or to generate new hypotheses.

Conclusion

The study of sleep has contributed useful information to our understanding of depressive illness. Sleep studies have been used to examine diagnosis, family history, state and trait abnormalities, and various neurochemical and circadian theories of the pathogenesis of depression.

Over the past 20 years, a number of methodological and analytical refinements have also developed. These include the use of structured diagnoses in patients undergoing sleep studies; consideration of comorbid diagnoses, family psychiatric history, and phase of illness; the use of longitudinal within-subject study designs, the use of pharmacological, endocrine, and sleep deprivation "probes" of sleep regulation; more complex EEG analyses including period/amplitude analysis, automated REM detection, and spectral analysis; and the development of computer simulations and mathematical models to explore underlying sleep mechanisms in healthy and depressed subjects.

Future investigations of sleep in depression will undoubtedly include further improvements of these techniques. However, further work addressing very basic questions is also warranted. In particular, the issue of the specificity of sleep changes in depression requires a more definitive answer. Related factors such as inpatient/outpatient status, the timing of sleep in depressed patients, and the effects of social and physical factors on sleep and circadian rhythms demand specific attention.

Acknowledgments

Supported in part by the National Institute of Mental Health under grants Nos. MH-24652 and MH-30915.

References

Adamson J, Burdick JA: Sleep of dry alcoholics, *Arch Gen Psychiat* 1973; 28: 146–148.

Akiskal HS, Lemmi H, Dickson H, King D, Yerevanian B, Van Valkenburg C: Chronic depressions: Part 2. Sleep EEG differentiation of primary dysthymic disorders from anxious depressions, *J Affect Disord* 1984; 6: 287–295.

Akiskal HS, Lemmi H, Yerevanian B, King D, Belluomini J: The utility of the REM latency test in psychiatric diagnosis: A study of 81 depressed outpatients, *Psychiat Res* 1982; 7: 101–110.

Akiskal HS, Yerevanian BI, Davis GC, King D, Lemmi H: The nosologic status of borderline personality: Clinical and polysomnographic study, *Am J Psychiat* 1985; 142: 192–198.

Ansseau M, Kupfer DJ, Reynolds CF, McEachran AB: REM latency distribution in major depression: Clinical characteristics associated with sleep onset REM periods, *Biol Psychiat* 1984; 19: 1651–1656.

Avery DH, Wildschiodtz G, Smallwood G, Martin D, Rafaelsen OJ: REM latency and core temperature relationships in primary depression, *Acta Psychiat Scand* 1986; 74: 269–280.

Beersma DGM, Daan S, van den Hoofdakker RH: Distribution of REM latencies and other sleep phenomena in depression as explained by a single ultradian rhythm disturbance, *Sleep* 1984; 7: 126–136.

Beersma DGM, Daan S, van den Hoofdakker RH: The timing of sleep in depression: Theoretical considerations, *Psychiat Res* 1985; 16: 253–262.

Berger M, Lund R, Bronisch T, von Zerssen D: REM latency in neurotic and endogenous depression and the cholinergic REM induction test, *Psychiat Res* 1983; 10: 113–123.

Borbély AA: A two process model of sleep regulation, *Human Neurobiol* 1982; 1: 195–204.

Borbély AA, Tobler I, Loepfe M, Kupfer DJ, Ulrich RF, Grochocinski V, Doman J, Matthews G: All-night spectral analysis of the sleep EEG in untreated depressives and normal controls, *Psychiat Res* 1984; 12: 27–33.

Borbély AA, Wirz-Justice A: Sleep, sleep deprivation and depression: A hypothesis derived from a model of sleep regulation, *Human Neurobiol* 1982; 1: 205–210.

Buysse DJ, Jarrett DB, Miewald JM, Kupfer DJ, Greenhouse JB: Minute-by-minute analysis of REM sleep timing in major depression, *Biol Psychiat* 1990; 28: 911–925.

Buysse DJ, Kupfer DJ: Diagnostic and research applications of electroencephalographic sleep studies in depression: Conceptual and methodological issues, *J Nerv Ment Dis* 1990; 178: 405–414.

Cartwright RD: Rapid eye movement sleep characteristics during and after mood-disturbing events, *Arch Gen Psychiat* 1983; 40: 197–201.

Coble P, Foster FG, Kupfer DJ: Electroencephalographic sleep diagnosis of primary depression, *Arch Gen Psychiat* 1976; 33: 1124–1127.

Coble PA, Kupfer DJ, Shaw DH: Distribution of REM latency in depression, *Biol Psychiat* 1981; 16: 453–466.

Coble PA, Kupfer DJ, Spiker DG, Neil JF, McPartland RJ: EEG sleep in primary depression: A longitudinal placebo study, *J Affect Disord* 1979; 1: 131–138.

Dahl RE, Puig-Antich J, Ryan ND, Nelson B, Dachille S, Cunningham SL, Trubnick L, Klepper TP: EEG sleep in adolescents with major depression: The role of suicidality and inpatient status, *J Affect Disord* 1990; 19: 63–75.

Dubé S, Jones DA, Bell J, Davies A, Ross E, Sitaram N: Interface of panic and depression: Clinical and sleep EEG correlates, *Psychiat Res* 1986; 19: 119–133.

Dubé S, Kumar N, Ettedgui E, Pohl R, Jones D, Sitaram N: Cholinergic REM induction response: Separation of anxiety and depression, *Biol Psychiat* 1985; 20: 408–418.

Duncan WC Jr, Gillin JC, Post RM, Gerner RH, Wehr TA: Relationship between EEG sleep patterns and clinical improvement in depressed patients treated with sleep deprivation, *Biol Psychiat* 1980; 15: 879–889.

Duncan WC, Pettigrew KD, Gillin JC: REM architecture changes in bipolar and unipolar depression, *Am J Psychiat* 1979; 136: 1424–1427.

Ehlers CL, Havstad JW, Garfinkel A, Kupfer DJ: Nonlinear analysis of EEG sleep states, *Neuropsychopharmacology* 1991; 5: 167–176.

Ehlers CL, Kupfer DJ: Effects of age on delta and REM sleep parameters, *Electroencephalogr Clin Neurophysiol* 1989; 72: 118–125.

Feinberg I: Changes in sleep cycle pattern with age, *J Psychiat Res* 1974; 10: 283–306.

Feinberg I, Koresko RL, Heller N: EEG sleep patterns as a function of normal and pathological aging in man, *J Psychiat Res* 1967; 5: 107–144.

Feinberg I, March JD, Fein G, Floyd TC, Walker JM, Price L: Period and amplitude analysis of 0.5-3 c/sec activity in NREM sleep of young adults, *Electroencephalogr Clin Neurophysiol* 1978; 44: 202–213.

Feinberg M, Carroll BJ: Biological "markers" for endogenous depression: Effect of age, severity of illness, weight loss, and polarity, *Arch Gen Psychiat* 1984; 41: 1080–1085.

Feinberg M, Gillin JC, Carroll BJ, Greden JP, Zis AP: EEG studies of sleep in the diagnosis of depression, *Biol Psychiat* 1982; 17: 305–316.

Ganguli R, Reynolds CF, Kupfer DJ: Electroencephalographic sleep in young, never medicated schizophrenics, *Arch Gen Psychiat* 1987; 44: 36–44.

Gerner RH, Post RM, Gillin JC, Bunney WE Jr: Biological and behavioral effects of one night's sleep deprivation in depressed patients and normals, *J Psychiat Res* 1979; 15: 21–40.

Giles DE, Biggs MM, Rush AJ, Roffwarg HP: Risk factors in families of unipolar depression. I. Psychiatric illness and reduced REM latency, *J Affect Disord* 1988; 14: 51–59.

Giles DE, Jarrett RB, Roffwarg HP, Rush AJ: Reduced rapid eye movement latency: A predictor of recurrence in depression, *Neuropsychopharmacology* 1987a; 1: 33–39.

Giles DE, Kupfer DJ, Roffwarg HP, Rush AJ, Biggs MM, Etzel BA: Polysomnographic parameters in first-degree relatives of unipolar probands, *Psychiat Res* 1989; 27: 127–136.

Giles DE, Roffwarg HP, Rush AJ, Guzick DS: Age-adjusted threshold values for reduced REM latency in unipolar depression using ROC analysis, *Biol Psychiat* 1990; 27: 841–853.

Giles DE, Roffwarg HP, Schlesser MA, Rush AJ: Which endogenous depressive symptoms relate to REM latency reduction?, *Biol Psychiat* 1986a; 21: 473–482.

Giles DE, Rush AJ, Roffwarg HP: Sleep parameters in bipolar I, bipolar II, and unipolar depressions, *Biol Psychiat* 1986b; 21: 1340–1343.

Giles DE, Schlesser MA, Rush AJ, Orsulak PJ, Fulton CL, Roffwarg HP: Polysomnographic findings and dexamethasone nonsuppression in unipolar depression: A replication and extension, *Biol Psychiat* 1987b; 22: 872–882.

Gillin JC: The sleep therapies of depression, *Prog Neurophsychopharmacol Biol Psychiat* 1983; 7: 351–364.

Gillin JC, Buchsbaum MS, Jacobs LS, Fram DH, Williams RB, Vaughan TB, Mellon E, Snyder F, Wyatt RJ: Partial REM sleep deprivation, schizophrenia and field articulation, *Arch Gen Psychiat* 1974; 30: 653–662.

Gillin JC, Duncan WC, Murphy DL, Post RM, Wehr TA, Goodwin FK, Wyatt RJ, Bunney WE: Age-related changes in sleep in depressed and normal subjects, *Psychiat Res* 1981; 4: 73–78.

Gillin JC, Duncan W, Pettigrew KD, Frankel BL, Snyder F: Successful separation of depressed, normal, and insomniac subjects by EEG sleep data, *Arch Gen Psychiat* 1979; 36: 85–90.

Gillin JC, Kripke DF, Janowsky DS, Risch SC: Effects of brief naps on mood and sleep in sleep-deprived depressed patients, *Psychiat Res* 1989; 27: 253–265.

Gillin JC, Smith TL, Irwin M, Kripke DF, Brown S, Schuckit M: Short REM latency in primarily alcoholic patients with secondary depression, *Am J Psychiat* 1990; 147: 106–109.

Goetz RR, Puig-Antich J, Ryan N, Rabinovich H, Ambrosini PJ, Nelson B, Krawiec V: Electroencephalographic sleep of adolescents with major depression and normal controls, *Arch Gen Psychiat* 1987; 44: 61–68.

Guazzelli M, Maggini C, Landini G, Feinberg I: Similarity of non-REM abnormalities in schizophrenia and depression, *Arch Gen Psychiat* 1985; 42: 834–835.

Hauri P, Chernik D, Hawkins D, Mendels J: Sleep of depressed patients in remission, *Arch Gen Psychiat* 1974; 31: 386–391.

Hauri P, Hawkins DR: Phasic REM, depression, and the relationship between sleeping and waking, *Arch Gen Psychiat* 1971; 25: 56–63.

Hiatt JF, Floyd TC, Katz PH, Feinberg I: Further evidence of abnormal non-rapid-eye-movement sleep in schizophrenia, *Arch Gen Psychiat* 1985; 42: 797–802.

Hudson JI, Lipinski JF, Frankenburg FR, Grochocinski VJ, Kupfer DJ: Electroencephalographic sleep in mania, *Arch Gen Psychiat* 1988; 45: 267–273.

Hudson JI, Pope HG, Jonas JM, Stakes JW, Grochocinski V, Lipinski JR, Kupfer DJ: Sleep EEG in bulimia, *Biol Psychiat* 1987; 22: 820–828.

Insel TR, Gillin JC, Moore A, Mendelson WB, Lowenstein RJ, Murphy DL: The sleep of patients with obsessive-compulsive disorder, *Arch Gen Psychiat* 1982; 39: 1372–1377.

Institute of Medicine Division of Health Sciences Policy, *Basic Sleep Research (A Research Briefing)* Washington, DC: National Academy Press, 1990, 1–12.

Janowsky DS, El-Yousef MK, Davis JM, Sekerke HJ: A cholinergic-adrenergic hypothesis of mania and depression, *Lancet* 1972; 2: 632–635.

Jus K, Bouchard M, Jus AK, Villeneuve A, Lachance R: Sleep EEG studies in untreated, long-term schizophrenic patients, *Arch Gen Psychiat* 1973; 29: 386–390.

Kempenaers C, Kerkhofs M, Linkowski P, Mendlewicz J: Sleep EEG variables in young schizophrenic and depressive patients, *Biol Psychiat* 1988; 24: 833–838.

Keshavan MS, Reynolds CF, Kupfer DJ: Electroencephalographic sleep in schizophrenia: A critical review, *Comprehensive Psychiat* 1990; 30: 34–47.

Knowles JB, Cairns J, MacLean AW, Delva N, Prowse A, Waldron J, Letemendia FJ: The sleep of remitted bipolar depressives: Comparison with sex and age-matched controls, *Can J Psychiat* 1986; 31: 295–298.

Knowles JB, MacLean AW: Age-related changes in sleep in depressed and healthy subjects: a meta-analysis, *Neuropsychopharmacology* 1990; 3: 251–259.

Knowles JB, MacLean AW, Cairns J: Definitions of REM latency: Some comparisons with particular reference to depression, *Biol Psychiat* 1982; 17: 993–1002.

Kumar A, Shipley JE, Eiser AS, Feinberg M, Flegel P, Grunhaus L, Haskett RF: Clinical correlates of sleep onset REM periods in depression, *Biol Psychiat* 1987; 22: 1473–1477.

Kupfer DJ: REM latency: A psychobiologic marker for primary depressive disease, *Biol Psychiat* 1976; 11: 159–174.

Kupfer DJ, Broudy D, Coble, PA, Spiker DG: EEG sleep and affective psychosis, *J Affect Disord* 1980; 2: 17–25.

Kupfer DJ, Ehlers CL: Two roads to rapid eye movement latency, *Arch Gen Psychiat* 1989; 46: 945–948.

Kupfer DJ, Foster G, Coble PA, McPartland RJ, Ulrich RF: The application of EEG sleep for the differential diagnosis of affective disorders, *Am J Psychiat* 1978; 135: 69–74.

Kupfer DJ, Frank E: The relationship of EEG sleep to vital depression, *J Affect Disord* 1984; 7: 249–263.

Kupfer DJ, Frank E, Ehlers CL: EEG sleep in young depressives: First and second night effects, *Biol Psychiat* 1989a; 25: 87–97.

Kupfer DJ, Frank E, Grochocinski VJ, Gregor M, McEachran AB: Electroencephalographic sleep profiles in recurrent depression: A longitudinal investigation, *Arch Gen Psychiat* 1988; 45: 678–681.

Kupfer DJ, Frank E, McEachran AB, Grochocinski VJ: Delta sleep ratio: A biological correlate of early recurrence in unipolar affective disorder, *Arch Gen Psychiat* 1990; 47: 1100–1105.

Kupfer DJ, Gillin JC, Coble PA, Spiker DG, Shaw D, Holzer B: REM sleep, naps, and depression, *Psychiat Res* 1981; 5: 195–203.

Kupfer DJ, Reynolds CF, Ehlers CL: Comparison of EEG sleep measures among depressive subtypes and controls in older individuals, *Psychiat Res* 1989b; 27: 13–21.

Kupfer DJ, Reynolds CF, Grochocinski VJ, Ulrich RF, McEachran A: Aspects of short REM latency in affective states: A revisit, *Psychiat Res* 1986a; 17: 49–59.

Kupfer DJ, Reynolds CF, Ulrich RF, Grochocinski VJ: Comparison of automated REM and slow-wave sleep analysis in young and middle-aged depressed subjects, *Biol Psychiat* 1986b; 21: 189–200.

Kupfer DJ, Targ E, Stack J: Electroencephalographic sleep in unipolar depressive subtypes: Support for a biological and familial classification, *J Nerv Ment Dis* 1982; 170: 494–498.

Kupfer DJ, Ulrich RF, Coble PA, Jarrett DB, Grochocinski V, Doman J, Matthews G, Borbély AA: Application of automated REM and slow wave sleep analysis: I. Normal and depressed subjects, *Psychiat Res* 1984a; 13: 325–334.

Kupfer DJ, Ulrich RF, Coble PA, Jarrett DB, Grochocinski V, Doman J, Matthews G, Borbély, AA: Application of automated REM and slow wave sleep analysis: II. testing the assumptions of the two-process model of sleep regulation in normal and depressed subjects, *Psychiat Res* 1984b; 13: 335–343.

Kupfer DJ, Wyatt RJ, Scott J, Snyder F: Sleep disturbance in acute schizophrenic patients, *Am J Psychiat* 1970; 126: 1213–1223.

Lauer C, Zulley J, Krieg J-C, Riemann D, Berger M: EEG sleep and the cholinergic REM induction test in anorexic and bulimic patients, *Psychiat Res* 1988; 26: 171–181.

Levy AB, Dixon KN, Schmidt H: Sleep architecture in anorexia nervosa and bulimia, *Biol Psychiat* 1988; 23: 99–101.

Lewy AJ, Sack RL, Singer CM, White DM, Hoban TM: Winter depression and the phase-shift hypothesis for bright light's therapeutic effects: History, theory, and experimental evidence, *J Biol Rhythms* 1988; 3: 121–134.

Linkowski P, Kerkhofs M, Hauspie R, Susanne C, Mendlewicz J: EEG sleep patterns in man: A twin study, *Electroencephalogr Clin Neurophysiol* 1989; 73: 279–284.

Lund R, Kammerloher A, Dirlich G: Body temperature in endogenously depressed patients during depression and remission. In: Wehr TA, Goodwin FK (Eds.), *Circadian Rhythms in Psychiatry*, Pacific Grove, CA: The Boxwood Press, 1983, 77–88.

MacLean AW, Cairns J, Knowles JB: REM latency and depression: Computer simulations based on the results of phase delay of sleep in normal subjects, *Psychiat Res* 1983; 9: 69–79.

MacLean AW, Knowles JB, Vetere C: REM sleep and depression: Further use of computer simulation to test the phase-advance hypothesis, *Psychiat Res* 1986; 19: 25–36.

McCarley RW: REM sleep and depression: Common neurobiological control mechanisms, *Am J Psychiat* 1982; 139: 565–570.

McCarley RW, Massaquoi SG: A limit cycle mathematical model of the REM sleep oscillator system, *Am J Physiol* 1986; 251: R1011–R1029.

McNamara E, Reynolds CF, Soloff PH, Mathias R, Rossi A, Spiker D, Coble PA, Kupfer DJ: Electroencephalographic sleep evaluation of depression in borderline patients, *Am J Psychiat* 1984; 141: 182–186.

Mellman TA, Uhde TW: Electroencephalographic sleep in panic disorder: A focus on sleep-related panic attacks, *Arch Gen Psychiat* 1989; 46: 178–184.

Mendlewicz J, Kerkhofs M, Hoffmann G, Linkowski P: Dexamethasone suppression test and REM sleep in patients with major depressive disorder, *Br J Psychiat* 1984; 145: 383–388.

Montgomery I, Oswald I, Morgan K, Adam K: Trazodone enhances sleep in subjective quality but not in objective duration, *Br J Clin Pharmacol* 1983; 16: 139–144.

Mullen PE, Linsell CR, Parker D: Influence of sleep disruption and calorie restriction on biological markers for depression, *Lancet* 1986; 2: 1051–1055.

Naylor MW, Shain BN, Shipley JE: REM latency in psychotically depressed adolescents, *Biol Psychiat* 1990; 28: 161–164.

Neil J, Merikangas JR, Foster RF, Spiker DG, Kupfer DJ: Waking and all-night sleep EEGs in anorexia nervosa, *Electroencephalogr Clin Neurophysiol* 1980; 1: 9–15.

Nicholson AN, Pascoe PA: Studies on the modulation of the sleep-wakefulness continuum in man by fluoxetine, a 5-HT uptake inhibitor, *Neuropharmacology* 1988; 27: 597–602.

Nofzinger EA, Thase ME, Reynolds CF, Himmelhoch JM, Mallinger A, Houck P, Kupfer DJ: Hypersomnia in bipolar depression: Part 1: A comparison with narcolepsy using the multiple sleep latency test, *Am J Psychiat* 1991; 148: 1177–1181.

Nurnberger J Jr, Sitaram N, Gershon ES, Gillin JC: A twin study of cholinergic REM induction, *Biol Psychiat* 1983; 18: 1161–1165.

Papadimitriou GN, Linkowski P, Kerkhofs M, Kempenaers C, Mendlewicz J: Sleep EEG recordings in generalized anxiety disorder with significant depression, *J Affect Disord* 1988; 15: 113–118.

Prinz PN, Peskind ER, Vitaliano PP, Raskind MA, Eisdorfer C, Zembeuznikov W, Gerber CJ: Changes in the sleep and waking EEGs of nondemented and demented elderly subjects, *J Am Ger Soc* 1982; 30: 86–93.

Quitkin FM, Rabkin JG, Stewart JW, McGrath PJ, Harrison W, Davies M, Goetz R, Puig-Antich J: Sleep of atypical depressives, *J Affect Disord* 1985; 8: 61–67.

Reinink E, Bouhuys N, Wirz-Justice A, van den Hoofdakker R: Prediction of the antidepressant response to total sleep deprivation by diurnal variation of mood, *Psychiat Res* 1990; 32: 113–124.

Reynolds CF, Buysse DJ, Kupfer DJ, Hoch CC, Houck PR, Mattzie J, George CJ: Rapid eye movement sleep deprivation as a probe in elderly subjects, *Arch Gen Psychiat* 1990a; 47: 1128–1138.

Reynolds CF, Kupfer DJ: State-of-the-art review—sleep research in affective illness: state of the art circa 1987, *Sleep* 1987; 10: 199–215.

Reynolds CF, Kupfer DJ, Houck PR, Hoch CC, Stack JA, Berman SR, Zimmer B:

Reliable discrimination of elderly depressed and demented patients by electroencephalographic sleep data, *Arch Gen Psychiat* 1988; 45: 258–264.

Reynolds CF, Kupfer DJ, Taska LS, Hoch CC, Spiker DG, Sewitch DE, Zimmer B, Marin RS, Nelson JP, Martin D, Morycz R: EEG sleep in elderly depressed, demented, and healthy subjects, *Biol Psychiat* 1985a; 20: 431–442.

Reynolds CF, Kupfer DJ, Thase ME, Frank E, Jarrett DB, Coble PA, Hoch CC, Buysse DJ, Simons AD, Houck PR: Sleep, gender, and depression: An analysis of gender effects on the electroencephalographic sleep of 302 depressed outpatients, *Biol Psychiat* 1990b; 28: 673–684.

Reynolds CF, Newton TF, Shaw DH, Coble PA, Kupfer DJ: Electroencephalographic sleep findings in depressed outpatients, *Psychiat Res* 1982; 6: 65–75.

Reynolds CF, Shaw DF, Newton TF, Coble PA, Kupfer DJ: EEG sleep in outpatients with generalized anxiety: A preliminary comparison with depressed outpatients, *Psychiat Res* 1983a; 8: 81–89.

Reynolds CF, Soloff PH, Kupfer DJ, Taska LS, Restifo K, Coble PA, McNamara ME: Depression in borderline patients: A prospective EEG sleep study, *Psychiat Res* 1985b; 14: 1–15.

Reynolds CF, Taska LS, Jarrett DB, Coble PA, Kupfer DJ: REM latency in depression: Is there one best definition?, *Biol Psychiat* 1983b; 18: 849–863.

Riemann D, Berger M: EEG sleep in depression and in remission and the REM sleep response to the cholinergic agonist RS 86, *Neuropsychopharmacology* 1989; 2: 145–152.

Riemann D, Berger M: The effects of total sleep deprivation and subsequent treatment with clomipramine on depressive symptoms and sleep electroencephalography in patients with a major depressive disorder, *Acta Psychiat Scand* 1990; 81: 24–31.

Riemann D, Gann H, Hohagen F, Olbrich R, Fleckenstein P, Berger M: REM sleep in depression, anxiety disorders and schizophrenia. The influence of cholinergic stimulation with RS 86. In: Horne JA (Ed.), *Sleep '90*, Proceedings of the 10th European Congress on Sleep Research, Strasbourg, May 1990, Bochum: Pontenagel Press, 1990, 235–237.

Riemann D, Wiegand M, Zulley J, Lauer Ch, Schreiber W, Berger M: The effect of the occurrence of REM sleep during morning naps on mood after sleep deprivation in patients with major depressive disorder. In: Horne J (Ed.), *Sleep '88*, New York: Gustav Fischer Verlag, 1989, 230–232.

Roth T, Zorick F, Wittig R, McLenaghan A, Roehrs T: The effects of doxepin HC1 on sleep and depression, *J Clin Psychiat* 1982; 43: 366–368.

Rush AJ, Erman MK, Giles DE, Schlesser MA, Carpenter G, Vasavada N, Roffwarg HP: Polysomnographic findings in recently drug-free and clinically remitted depressed patients, *Arch Gen Psychiat* 1986; 43: 878–884.

Rush AJ, Giles DE, Roffwarg HP, Parker CR: Sleep EEG dexamethasone suppression test findings in outpatients with unipolar major depressive disorders, *Biol Psychiat* 1982; 17: 327–341.

Sack DA, Duncan W, Rosenthal NE, Mendelson WE, Wehr TA: The timing and duration of sleep in partial sleep deprivation therapy of depression, *Acta Psychiat Scand* 1988; 77: 219–224.

Sack DA, Nurnberger J, Rosenthal NE, Ashburn E, Wehr TA: Potentiation of antidepressant medications by phase advance of the sleep-wake cycle, *Am J Psychiat* 1985; 142: 606–608.

Schulz H, Lund R: On the origin of early REM episodes in the sleep of depressed patients: A comparison of three hypotheses, *Psychiat Res* 1985; 16: 65–77.

Schulz H, Lund R, Cording C, Dirlich G: Bimodal distribution of REM sleep latencies in depression, *Biol Psychiat* 1979; 14: 595–600.

Schulz H, Tetzlaff W: Distribution of REM latencies after sleep interruption in depressive patients and control subjects, *Biol Psychiat* 1982; 17: 1367–1376.

Shipley JE, Kupfer DJ, Dealy RS, Griffin SJ, Coble PA, McEachran AB, Grochocinski VJ: Differential effects of amitriptyline and of zimelidine on the sleep electroencephalogram of depressed patients, *Clin Pharmacol Therapeut* 1984; 36: 251–259.

Shipley JE, Kupfer DJ, Griffin SJ, Dealy RS, Coble PA, McEachran AB, Grochocinski VJ, Ulrich R, Perel JM: Comparison of effects of desipramine and amitriptyline on EEG sleep of depressed patients, *Psychopharmacology* 1985; 85: 14–22.

Sitaram N, Moore AM, Gillin JC: Scopolamine-induced muscarinic supersensitivity in normal man: Changes in sleep, *Psychiat Res* 1979; 1: 9–16.

Sitaram N, Nurnberger JI, Gershon ES, Gillin JC: Cholinergic regulation of mood and REM sleep: Potential model and marker of vulnerability to affective disorders, *Am J Psychiat* 1982; 139: 571–576.

Snyder S, Karacan I: Sleep patterns of sober chronic alcoholics, *Neuropsychobiology* 1985; 13: 97–100.

Somoza E, Mossman D: Optimizing REM latency as a diagnostic test for depression using receiver operating characteristic analysis and information theory, *Biol Psychiat* 1990; 27: 990–1006.

Spiker DG, Coble P, Cofsky J, Foster FG, Kupfer DJ: EEG sleep and severity of depression, *Biol Psychiat* 1978; 13: 485–488.

Thase ME, Himmelhoch JM, Mallinger AG, Jarrett DB, Kupfer DJ: Sleep EEG and DST findings in anergic bipolar depression, *Am J Psychiat* 1989; 146: 329–333.

Thase ME, Kupfer DJ, Spiker DG: Electroencephalographic sleep in secondary depression: A revisit, *Biol Psychiat* 1984; 19: 805–815.

Thase ME, Kupfer DJ, Ulrich RF: Electroencephalographic sleep in psychotic depression: A valid subtype?, *Arch Gen Psychiat* 1986; 43: 886–893.

Ulrich RF, Shaw DH, Kupfer DJ: Effects of aging on EEG sleep in depression, *Sleep* 1980; 3: 31–40.

Van den Hoofdakker RH, Beersma DGM: On the explanation of short REM latencies in depression, *Psychiat Res* 1985; 16: 155–163.

Van den Hoofdakker RH, Beersma DGM: On the contribution of sleep wake physiology to the explanation and treatment of depression, *Acta Psychiat Scand (Suppl.)* 1988; 341: 53–71.

Vogel GW, Buffenstein A, Minter K, Hennessey A: Drug effects on REM sleep and on endogenous depression, *Neurosci Biobehav Rev* 1990; 14: 49–63.

Vogel GW, McAbee R, Barker K, Thurmond A: Endogenous depression improvement and REM pressure, *Arch Gen Psychiat* 1977; 34: 96–97.

Vogel GW, Thurmond A, Gibbons P, Sloan K, Boyd M, Walker M: REM sleep reduction effects on depression syndromes, *Arch Gen Psychiat* 1975; 32: 765–777.

Vogel GW, Vogel F, McAbee RS, Thurmond AJ: Improvement of depression by REM sleep deprivation, *Arch Gen Psychiat* 1980; 37: 247–253.

von Zerssen D, Barthelmes H, Dirlich G, Doerr P, Emrich HM, von Lindern L, Lund R, Pirke KM: Circadian rhythms in endogenous depression, *Psychiat Res* 1985; 16: 51–63.

Walsh BT, Goetz R, Roose SP, Fingeroth S, Glassman AH: EEG-monitored sleep in anorexia nervosa and bulimia, *Biol Psychiat* 1985; 20: 947–956.

Webb WB, Campbell SC: Relationships in sleep characteristics of identical and fraternal twins, *Arch Gen Psychiat* 1983; 40: 1093–1095.

Wehr TA: Effects of wakefulness and sleep on depression and mania. In Montplaisir J, Godbout R (Eds.), *Sleep and Biological Rhythms: Basic Mechanisms and Applications to Psychiatry,* New York: Oxford University Press, 1990, 42–86.

Wehr TA, Goodwin FK: Biological rhythms in manic-depressive illness. In: Wehr TA, Goodwin FK (Eds.), *Circadian Rhythms in Psychiatry,* Pacific Grove, CA: The Boxwood Press, 1983, 129–184.

Wehr TA, Wirz-Justice A: Circadian rhythm mechanism in affective illness and in antidepressant drug action, *Pharmacopsychiatry* 1982; 15: 31–39.

Wiegand M, Berger M, Jurgen Z, Lauer C, von Zerssen D, The influence of daytime naps on the therapeutic effect of sleep deprivation, *Biol Psychiat* 1987; 22: 389–392.

Williams RL, Karacan I, Hursch CJ: Changes in sleep variables across age groups. In: *Electroencephalography (EEG) of Human Sleep: Clinical Applications,* New York: John Wiley & Sons, 1974, 69–95.

Wu JC, Bunney WE: The biological basis of an antidepressant response to sleep deprivation and relapse: Review and hypothesis, *Am J Psychiat* 1990; 147: 14–21.

Zarcone V, Azumi K, Dement W, Gulevich G, Kraemer H, Pivik T: REM phase deprivation and schizophrenia II, *Arch Gen Psychiat* 1975; 32: 1431–1436.

Zarcone VP, Benson KL, Berger PA: Abnormal rapid eye movement latencies in schizophrenia, *Arch Gen Psychiat* 1987; 44: 45–48.

Chapter 7

THYROID AXIS AND DEPRESSIVE DISORDERS

Andrew Winokur

Introduction

Interest has been directed to the relationship between depressive disorders and thyroid axis function for many years. In the past 20 years, numerous studies have examined various indices of thyroid axis function in patients with depressive disorders and effects of thyroid axis hormones on mood and behavior. A third line of interest has involved effects of various psychotropic drugs used in the treatment of affective disorders on thyroid axis function. The major focus of this chapter will be on studies examining the interaction between the thyroid axis and depression. However, we start with a brief overview of reasons for interest in a role of the thyroid axis in depression.

Background to Studies of Thyroid Axis Function and Depression

Recognition of a critical role for thyroid hormones in brain development dates back over one hundred years to the observation of cretinism, associated with severe mental retardation, in cases of untreated neonatal

Andrew Winokur • Department of Psychiatry, University of Pennsylvania, Philadelphia, Pennsylvania 19104.

Biology of Depressive Disorders, Part A: A Systems Perspective, edited by J. John Mann and David J. Kupfer. Plenum Press, New York, 1993.

hypothyroidism. Subsequently, numerous laboratory studies, as well as clinical observations, have documented profound effects of altered thyroid states on brain development and function. For example, hypothyroidism has been reported to be associated with hypoplasia of the cerebral cortex, reduced development of several characteristic types of brain cells, decreases in RNA content and reduced protein synthesis, a reduction in the activity of many of the key brain enzymes associated with metabolism, diminished parameters of EEG function, deficits in the growth and differentiation of neurons, and abnormalities in the formation of myelin (reviewed in Ford & Cramer, 1977). More recently, numerous studies have reported effects of thyroid hormone manipulations on various parameters of CNS neurotransmitter function. Thus, affects of thyroid manipulations have been observed on benzodiazepine receptors (Gavish, Weizman, Okun, & Youdim, 1986), on serotonin and substance P content (Savard, Mérand, Di Paolo, & Dupont, 1983), opioid receptors (Bhargava, Ramaro, & Gulati, 1989), and γ-amino butyric acid (GABA) uptake (Mason, Walker, & Prange, 1987). In a review article published more than a quarter of a century ago, Waldstein (1966) described data accumulated over the previous fifty years indicating extensive interactions between thyroid hormones and catecholamines in the periphery. More recently, numerous studies had demonstrated effects of thyroid hormone treatments on CNS catecholamine activity. For example, Schildkraut, Winokur, Draskozy, and Hensle (1971) reported that addition of thyroxine (T_4) to chronic administration of imipramine in rats offset the slowing of norepinephrine turnover produced by the tricyclic antidepressant drug. Frazer, Pandey, Mendels, Neely, Kane, and Hess (1974) reported that addition of tri-iodothyronine (T_3) to chronic imipramine treatment resulted in a reduction in the amount of (^{3}H)-cyclic AMP produced. Schmidt and Schultz (1985) reported that chronic treatment with T_3 or T_4 resulted in a decreased number of beta adrenergic receptors in cerebral cortex and a reduction in norephinephrine-stimulated cyclic AMP formation. In contrast to the previously cited findings, Mason, Bondy, Nemeroff, Walker, and Prange (1987) reported that administration of T_3 or T_4 resulted in an up-regulation of cortical beta adrenergic receptors. The reasons for the discrepancies in findings across studies are not clear, but they may reflect methodological differences. Nevertheless, the results of these and many other studies illustrate that peripheral thyroid hormones can produce pronounced effects on various CNS neurotransmitter systems.

A direct effect for T_4 in the CNS has been suggested by Sterling (1975), who hypothesized that it may act as neurotransmitter. Dratman and colleagues have conducted a series of studies examining a possible neurotransmitter role for thyroid hormones in the CNS. In a review of the role of thyroid hormones in the CNS, Dratman, Crutchfield, and Gordon (1984) pointed out that the relatively small concentrations of thyroid hormones in brain are very tightly regulated in the face of severe degrees of hypothyroidism or hyperthyroidism, suggesting a key role for thyroid hormones in brain function. Autoradiographic techniques demonstrated a selective uptake of radioactively labeled T_3 in discrete regions of the CNS such as the dentate gyrus of the hippocampus. Moreover, following intravenous administration of radioactively labeled T_3, selective uptake of iodothyronine in synaptosomes, the nerve terminal containing subfraction of brain tissue, was observed (Dratman, Crutchfield, Axelrod, Colburn, & Thoa, 1976), consistent with a neurotransmitter function for this hormone. Finally, this group demonstrated effects of direct administration of T_3 in the CNS (Goldman, Dratman, Crutchfield, Jennings, Maruniak, & Gibbons, 1985). In hypothyroid rats, 1.5 nmol T_3/100 gm body weight administered by intraventricular injection produced a significant increase in heart rate, whereas intravenous administration of the same dose of T_3 did not alter heart rate. In summary, the tight regulation of brain thyroid hormone concentrations, the discrete regional and subcellular localization, and the observation of physiological responses following intraventricular administration all support a significant role for thyroid hormones in the CNS, possibly including a neurotransmitter function.

Prange, Lipton, and co-workers conducted an extensive series of studies examining interrelationships between thyroid hormones and antidepressant compounds (Breese, Prange, & Lipton, 1974). Early studies demonstrated that hyperthyroidism enhanced and hypothyroidism diminished imipramine toxicity in rats. These findings prompted the investigators to suggest that thyroid hormones might enhance the antidepressant effects of the tricyclic compounds. Clinical studies exploring this hypothesis are described later in this chapter.

Behavioral effects associated with interactions of thyroid hormones and tricyclic antidepressant compounds were examined in various animal models (Breese, Cooper, Prange, Lotti, & Lipton, 1974). Administration of T_3 in combination with imipramine produced a significant increase in rectal temperature compared to imipramine alone. In the DOPA potentiation test of Everett (1966), a putative screen for antidepressant activity, T_3 potentiated the effects of imipramine in producing behavioral activation. Interestingly, T_3 administered by itself produced equivalent effects to those observed with imipramine.

Behavioral effects related to administration of a thyroid axis hormone have also been cited for the hypothalamic releasing hormone thyrotropin releasing hormone (TRH). TRH was the first of the hypothamalic releasing hormones to be isolated and characterized. Soon after its characterization, TRH was demonstrated to be distributed widely throughout the central nervous system (Winokur and Utiger, 1974). Moreover, TRH was shown to be highly concentrated in the synaptosomal fraction (Winokur, Davies, & Utiger, 1977) and to exert direct effects on neuronal activity (Yarbrough, 1979). Moreover, TRH receptors have been demonstrated to be distributed widely throughout the CNS, with highly discrete patterns of localization (Manaker, Winokur, Rostene, & Rainbow, 1985). These and other findings have prompted a number of investigators to speculate that TRH may function as a neurotransmitter in the CNS in addition to its traditional hypothalamic hypophysitropic function (Winokur & Utiger, 1972). Plotnikoff, Prange, Breese, Anderson and Wilson (1972) reported that TRH was active in the DOPA potentiation test of Everett (1966), demonstrating similar potency to the effects produced by the prototypical tricyclic antidepressant compound imipramine. Strikingly, TRH was equally active when administered to mice that had previously been hypopysectomized. Results from this study suggested that TRH might exert mood-elevating effects. Moreover, the finding that TRH maintained activity in animals with a lesioned pituitary–thyroid axis suggested that this tripeptide exerted direct effects on the CNS. Subsequently, numerous studies have demonstrated that TRH administration is associated with a number of physiological and behavioral responses (Yarbrough, 1979). For example, microinjection of TRH, either intraventricularly or directly into discrete brain stem regions, produces a number of autonomic responses including alteration in heart rate, blood pressure, respiration, and gastrointestinal activity (see Winokur, 1990 for review). Among the numerous behavioral effects reported for TRH, one of the most frequently observed findings is that administration of the tripeptide to animals that had previously been sedated with barbiturates, ethanol, or other CNS depressant compounds produces a pronounced reduction in sedation and in hypothermia (Breese et al., 1974). In studies carried out in the hibernating ground squirrel, Stanton, Winokur and Beckman (1980) observed that microinjection of small amounts of TRH into the dorsal hippocampus resulted in full behavioral arousal from hibernation. Moreover, these investigators reported that effects produced by TRH in the ground squirrel were state dependent. Thus, in euthermic, awake ground squirrels, TRH produced effects on brain temperature, EEG activity, and metabolism that were opposite in direction to the effects produced in the hibernating ground squirrel and in the euthermic, sleeping ground squirrel (Stanton, Winokur, & Beckman, 1981). Thus, these investigators speculated that TRH may play a role in the regulation of states of arousal within the central nervous system.

In summary, numerous studies have demonstrated effects of hormones of the hypothalamic-pituitary-thyroid (HPT) axis on brain function, including alterations of neurotransmitter and receptor activity, and indications of direct neurotransmitter function. Moreover, a number of behavioral effects have been reported following administration of these thyroid axis hormones. Of particular relevance to the present chapter are the reports cited above of activity both of peripheral thyroid hormones and of TRH in animal behavioral models for antidepressant compounds.

Thyroid Axis Hormones in Depressive Disorders

Many studies have examined elements of thyroid axis function in depressed patients. The notion that monitoring thyroid hormone levels may be relevant to psychiatric illness was reinforced by reports from several groups indicating that a substantial percentage (from 5% to 33%) of recently hospitalized acute psychiatric patients demonstrated hyperthyroxinemia (Spratt, Port, Moller, McDougall, Bazer, & McLaughlin 1982; Cohen & Swigar 1979; Borst, Eli, & Burman 1983; and Levy, Jensen, Laus, Agle, & Engel, 1981), though not all studies have identified such thyroid abnormalities (Kramlinger, Gharib, Swanson, & Maruta, 1984). Hyperthyroxinemia has been reported across diagnostic categories, and does not appear to be specific for depression. Moreover, when patients were followed longitudinally, thyroid levels were found to return to the normal range within one to two weeks.

Various indices of peripheral thyroid function have been examined in groups of depressed patients and matched control subjects. Among the thyroid parameters assessed have been T_4, T_3, T_3 uptake, reverse T_3 and antithyroid antibodies. While occasional differences in thyroid hormone parameters have been reported, these findings have not been consistently replicated. Whybrow and Bauer (1988) have provided a detailed summary of studies examining iodothyronines in affective disorders. The authors drew attention to one finding that emerged across studies, namely, a tendency for increased total and/or free T_4 without associated alterations in its metabolites (i.e., T_3, or rT_3). They also pointed out that some studies incorporating a longitudinal perspective have identified a fall in total or free T_4 in conjunction with treatment, apparently indicating that this finding is of clinical significance.

A particularly high incidence of thyroid abnormalities has been associated with patients exhibiting rapid cycling bipolar disorder. Cowdry, Wehr, Zis, and Goodwin (1983) reported that 50% of rapid-cycling bipolar patients but none of a non-rapid-cycling bipolar group were found to demonstrate overt hypothyroidism. Bauer, Whybrow, and Winokur (1990) reported an incidence of 23% for grade I hypothyroidism among a group of 30 rapid-cycling patients. Additionally, in this study, 27% of the rapid-cycling patients had grade II hypothyroidism and another 10% were identified as having grade III abnormalities. Thus, in this cohort, a total of 60% of the rapid-cycling patients manifest at least some degree of hypothyroidism. It should be noted that lithium exposure, the treatment of choice for bipolar disorders, will often confound assessment of thyroid function. However, these investigators pointed out that some of their patients with hypothyroid results did not have a prior history of lithium exposure. Some studies have failed to observe a difference in rates of hypothyroidism between rapid-cycling and non-rapid-cycling bipolar patients (Sack, James, Rosenthal, & Wehr, 1988; Bartalena, Pellegrini, Meschi, Antonangeli, Bogazzi, & Dell'Nosso, 1990).

Studies of TSH in Depression

TSH secretion from the pituitary thyrotrope has been shown to be regulated by a diverse range of influences from peripheral (e.g., iodothyronines, glucocorticoids) and central (e.g., TRH, somatostatin) sources (reviewed in Morley, 1981). A circadian pattern for TSH secretion has been reported by Brabant, Prank, Ranft, Schuermeyer, Wagner, Hauser, Kummer, Feistner, Hesch, & von zur Muhlen (1990), with the peak of TSH secretion occurring at night. Sleep deprivation was found to accentuate the nocturnal increase in TSH secretion, whereas sleep dampened this nocturnal surge.

Early studies of TSH secretion patterns in depressed patients produced mixed results. Weeke and Weeke (1978, 1980) observed decreased patterns of TSH secretion at night in depressed patients. In contrast Kijne, Aggernaes, and Fog-Moller (1982), failed to observe a significant difference in the profile of TSH secretion between depressed patients and controls. In a study examining the 24-hour patterns of TSH secretion using a 30-minute sampling regimen in 40 depressed patients and 40 carefully matched control subjects, Rubin, Poland, Lesser, and Martin (1987) reported a significant reduction in the nocturnal pattern of TSH secretion. In this study, T_3 secretion was also diminished in the depressed group. In their discussion of relevant literature on TSH secretion in depression, the authors pointed out that studies demonstrating a significantly reduced TSH secretion profile in depression generally employed multiple sampling techniques over several hours. This technique, as opposed to assessment of single samples, provides an index of thyroid axis physiological regulation.

In some studies, sleep deprivation has been employed as an experimental manipulation to augment the nocturnal TSH surge in depressed patients. Sack et al. (1988) reported that rapid cyclers exhibited a diminished nocturnal rise in TSH and a lack of nocturnal increase in TSH in response to sleep deprivation. Kasper, Sack, Wehr, Kick, Voll, and Vieira (1988) examined the TSH response to total sleep deprivation in a group of depressed patients. In this study, sleep deprivation was associated with a significant increase in TSH levels. The TSH response to sleep deprivation did not differentiate responders from nonresponders to sleep deprivation. In contrast, Baumgartner, Gräf, Kürten, Meinhold, and Scholz (1990) did observe a significant correlation between the magnitude of the TSH response to sleep deprivation and the clinical response to this procedure. Reasons for the divergent findings between these two studies are not clear.

In summary, assessment of TSH secretion patterns has revealed significant differences between depressed patients and controls in many cases. Differences were particularly pronounced during the period of the nocturnal surge, with depressed patients generally demonstrating a diminished TSH secretion profile.

The TRH Stimulation Test in Depression

Numerous studies have examined the TSH response to TRH in depressed patients. The initial reports of TRH administration to depressed patients included the observation that a subset of patients demonstrated diminished or "blunted" TSH responses (Prange, Wilson, Lara, Alltop, & Breese, 1972; Kastin, Ehrensing, Schalch, & Anderson, 1972). This finding has subsequently been reported in a large number of studies involving literally thousands of patients (see: Loosen, 1985; Loosen & Prange, 1982), and it represents one of the most extensively replicated biological markers of depression. Another abnormality of TSH response to TRH administration in depressed patients has been an "exaggerated" response reported in 8% of a large series of depressed patients (Gold, Pottash, & Extein, 1981). These investigators suggested that the exaggerated TSH response to TRH in this subgroup was indicative of clinical or "subclinical" hypothyroidism. They further described that in uncontrolled clinical observations, some depressed patients with an exaggerated TSH to TRH showed a beneficial response to thyroid hormone replacement therapy, either alone or as adjunct to antidepressant treatment. However, controlled clinical trials are needed to evaluate this suggestion.

While the blunted TSH response to TRH has been established for a number of years, the mechanism underlying this response abnormality has not been elucidated. It should be noted, however, that numerous other psychiatric disorders, such as schizophrenia, borderline personality disorder, panic disorder, obsessive-compulsive disorder, and alcoholism, have been reported to demonstrate relatively high incidences of blunted TSH responses (Winokur, 1990). Because of this problem of specificity, the TRH stimulaton test does not appear to offer much potential value as a diagnostic test for depression. A recent suggestion for improving the diagnostic sensitivity of the TRH stimulaton test in depression has been to compare the difference between morning (8 a.m.) and evening (11 p.m.) TRH trials (Duval, Macher, & Mokrani, 1990).

Two other proposals have been offered with regard to the diagnostic utility of the TRH stimulation test in depression. Langer, Koinig, Hatzinger, Schonbeck, Resch, Aschauer, Keshovan, and Sieghart (1986) reported that observation of a change in the TSH response from blunted to normal predicts a positive response to antidepressant therapy. From a slightly different perspective, Kirkegaard (1981) proposed that the change in the TSH response to TRH from baseline to several weeks of antidepressant treatment is of value in predicting long-term clinical outcome—i.e., remission or relapse. While these suggestions are of interest and potential clinical utility, other studies have failed to confirm these observations, and further work is needed to clarify the clinical applicability of the TRH stimulation test in depressed patients.

The Role of Thyroid Axis Hormones in the Treatment of Affective Disorders

Over the past quarter century, several reports have described therapeutic effects of thyroid axis hormones in the treatment of affective disorders. Most studies to date have focused on the role of iodothyronines in potentiating effects of tricyclic compounds in the treatment of major depression and bipolar affective disorder, as well as on the putative mood-elevating effects of TRH in depressed patients. The rationale for proposing therapeutic applications for thyroid axis hormones derives both from earlier clinical reports on the therapeutic efficacy of thyroid hormones in the treatment of psychiatric disorders, such as the classic work of Gjessing (1974) in the treatment of periodic catatonia, and from preclinical studies such as those described earlier in this chapter in which thyroid hormones were reported to potentiate effects of antidepressant compounds (Breese et al., 1974).

In an important study advancing work in this area, Prange, Wilson, Rabon, and Lipton (1969) administered T_3 (25 μg) or placebo to 20 depressed patients who were being treated with imipramine (150 mg/day). In this study, addition of T_3 to imipramine treatment produced a more rapid clinical response in depressed women. Interestingly, significant differences were not observed in the rate of response for male depressed patients treated with imipramine plus T_3 as compared to the imipramine plus placebo male group. Interpretation of this finding is complicated, however, by the fact that the male depressed patients, as a group, appeared to respond more rapidly to imipramine treatment than did female patients. Thus, it would have been difficult to demonstrate, in the male subjects, an advantage for T_3 supplementation in accelerating the rate of response to imipramine. The issue of whether there is a differential rate of response to tricyclic therapy by sex has received relatively limited attention in the literature. The efficacy of T_3 in accelerating response to treatment with imipramine or to amitriptyline has been replicated in some, but not all subsequent studies (reviewed in Bauer & Whybrow, 1988).

Another aspect of the therapeutic effects of iodothyronines in potentiating effects of antidepressant therapy has been the suggestion that tricyclic nonresponders may be converted to responders by the addition of T_3 supplementation. An important study supporting this possibility involved assessment of 12 depressed patients previously identified as being nonresponders to imipramine or amitriptyline therapy alone for adequate periods of time (Goodwin, Prange, Post, Muscettola, & Lipton, 1982). On a double-blind basis, T_3 (25 or 50 μg) was added to tricyclic therapy. Nine of the 12 patients were reported to show significant improvement in the three-week period following addition of T_3, with the improvement in 8 of these patients being rated as pronounced. It should be noted, however, that the addition of T_3 to tricyclic therapy has not been reported to reliably convert nonresponders to responders in all cases. Thus, Gitlin, Weiner, Fairbanks, Hershman, and Friedfeld (1987) added T_3 (25 μg) or placebo pills to 16 depressed patients who had failed to show significant improvement to therapy with imipramine for 4 weeks. After 2 weeks of treatment on one regimen, patients were crossed to the alternative treatment regimen. No significant effect of T_3 supplementation on response to imipramine treatment was observed in this study.

The limited number of well-controlled studies that have been conducted examining the utility of T_3 as a means of reversing nonresponsiveness to tricyclic therapy makes it difficult to evaluate the robustness of this phenomenon.

A recent application of thyroid hormone administration to the treatment of affective disorders has involved the use of high-dose levothyroxine in conjunction with mood-stabilizing agents in patients with rapid-cycling bipolar disorder. This patient group has been reported to show a high rate of nonresponsiveness to treatment with traditional mood-stabilizing agents such as lithium carbonate (Dunner & Fieve, 1974). Bauer and Whybrow (1990) recently reported results obtained from an open trial of high-dose levothyroxine added to ongoing psychotropic medication (primarily lithium carbonate, carbamazepine, or antidepressant drugs) in a group of 11 patients with rapid-cycling disorder. The majority of patients in this open trial showed significant improvement in depression and in mania ratings during high-dose levothyroxine therapy. These promising preliminary findings are of particular interest in light of the fact that this condition often represents a particularly insidious form of affective illness which is highly resistant to standard therapies. Clearly, further studies involving prospective, well-controlled clinical protocols are needed to evaluate the efficacy of high-dose levothyroxine therapy in patients with rapid-cycling bipolar illness.

Another hormone of the thyroid axis, TRH, has been investigated in terms of efficacy in depression. As noted earlier in this chapter, TRH had been reported to be active in the DOPA potentiation test, an animal model screen for compounds with putative antidepressant properties (Plotnikoff et al., 1972). The efficacy of TRH in this animal model provided a rationale for clinical trials in depressed patients. Prange et al. (1972) and Kastin et al. (1972) administered TRH (500 μg, I.V.) or placebo to groups of depressed patients. In both studies, administration of TRH produced rapid though transient improvement in symptoms of depression. Understandably, these reports attracted considerable interest, particularly in light of the strikingly rapid onset of antidepressant effects. However, subsequent clinical trials, which involved highly variable designs, produced very inconsistent results. Thus, at the present time, there is little basis for advocating the use of TRH in the treatment of depressed patients. However, further trials utilizing TRH analogues with longer half-lives and more pronounced CNS effects in depressed patients are needed to assess the therapeutic potential of TRH in a more meaningful manner.

Effects of Psychotropic Drugs on Thyroid Function

The last topic to be discussed in this chapter on the relationship of the thyroid axis and affective disorder involves consideration of effects of psychotropic drugs used in the treatment of affective disorders on thyroid axis function. Lithium carbonate has been recognized to effect thyroid function for many years. Schou, Amidsen, and Olsen (1968) reported that 12 of 330 bipolar patients treated with lithium developed clinically palpable goiters. With discontinuation of lithium therapy, the goiters were noted to resolve in all cases. More extensive investigations of lithium-related thyroid abnormalities revealed not only instances of goiter, but also hypothyroidism and in a few instances overt thyrotoxicosis. Cho, Bone, Dunner, Colt, and Fieve (1979) reported that a high percentage of patients receiving lithium who developed hypothyroidism were rapid-cycling bipolars. In a more recent study, bipolar patients who had been treated with lithium for 1–5 years were observed to have a 44% incidence of goiter, while patients on lithium for more than 10 years had an incidence of 50% (Perrild, Hegedüs, Baastrup, Kayser, and Kastberg, 1990). The incidence of goiter in non-lithium-treated bipolars in this study was 16%. Interestingly, another mood-stabilizing compound, carbamazepine, has been reported to decrease thyroid function in several studies (Herman, Obarzanek, Mikalauskas, Post, & Jimerson, 1991). Whether the antithyroid effects of such chemically different mood-stabilizing compounds as lithium and carbamazepine are related to their clinical efficacy remain to be established.

Several studies have examined effects of tricyclic antidepressants on thyroid axis function. In several instances, treatment with tricyclic antidepressants for several weeks was associated with a significant reduction in indices of thyroid axis function (e.g., Baumgartner, Gräf, Kurken, & Meinhold, 1988; Brady & Anton, 1989). As noted earlier, Bauer and Whybrow (1988) speculated that a reduction in thyroid hormone indices such as T_4 during tricyclic treatment may be associated with response to treatment.

Summary

As reviewed in this chapter, a considerable body of work suggests interactions between thyroid axis function and affective disorders. While assessment of thyroid axis function does not appear to be of specific value in the diagnosis of affective disorders, several findings, including the diminished nocturnal rise in TSH levels and the blunted TSH to TRH, appear to be implicated in the pathophysiological mechanisms of affective disorders. Moreover, recent work on the high incidence of thyroid axis pathology in patients with rapid-cycling bipolar disorder offer further evidence. As noted above, treatment strategies involving T_3 potentiation of tricyclic therapy or high-dose levothyroxine treatment of rapid-cycling bipolar disorder have shown some promise, but are in need of further evaluation. Whether TRH-like analogues offer any promise in the treatment of depression awaits future investigation. Finally, it should be noted that additional progress in preclinical studies investigating the role of thyroid axis hormones on CNS function may provide important new perspectives on the relationship between the thyroid axis and affective disorders.

References

Bartalena L, Pellegrini L, Meschi M, Antonangeli L, Bogazzi F, Dell 'Osso L, Pinchera A, Placidi GF: Evaluation of thyroid function in patients with rapid-cycling and non-rapid-cycling bipolar disorder. *Psychiat Res* 1990; 34: 13–37.

Bauer MS, Whybrow PC: Thyroid hormones and the central nervous system in affective illness: Interactions that may have clinical significance. *Integr Psychiat* 1988; 6: 75–100.

Bauer MS, Whybrow PC: Rapid cycling bipolar effective disorder. II. Treatment of refractory rapid cycling with high-dose levothyroxine: A preliminary study. *Arch Gen Psychiat* 1990; 47: 435–440.

Bauer MS, Whybrow PC, Winokur A: Rapid cycling bipolar affective disorder I. Association with grade I hypothyroidism. *Arch Gen Psychiat* 1990; 47: 427–432.

Baumgartner A, Gräf KJ, Kürten I, Meinhold H: The hypothalamic–pituitary–thyroid axis in psychiatric patients and healthy subjects. Part II. Repeated measurements of thyroxine, free thyroxine, triiodothyronine in patients with major depressive disorder and schizophrenia and healthy subjects. *Psychiat Res.* 1988; 24: 283–305.

Baumgartner A, Gräf KJ, Kürten I, Meinhold H, Scholz P: Neuroendocrinological investigations during sleep deprivation in depression. I. Early morning levels of thyroxine, TH cortisol, prolactin, LH. FSH, cortisol, prolactin, LH, FSH, estradiol and testosterone. *Biol Psychiat* 1990; 28: 556–558.

Bhargava HN, Ramarao P, Gulati A: Changes in multiple opioid receptors in regions of the brain in rats treated chronically with thyroxine. *Neuropharmacology* 1989; 28: 955–960.

Borst GC, Eil C, Burman KD: Euthyroid hyperthyroxinemia. *Ann Internal Med* 1983; 98: 366–378.

Brabant G, Prank K, Rauft U, Schuermeyer Th, Wagner TOF, Hauser H, Hesch

RD, Von Zur Muhlen A: Physiological regulating of circadian and pulsatile thyrotropin secretion in normal thyrotropin secretion in normal man and woman. *J Clin Endocrinol Metab* 1990; 70: 403–409.

Brady KJ, Anton RF: The thyroid axis and desipramine treatment in depression. *Biol Psychiat* 1989; 25: 703–709.

Breese GR, Cooper BR, Prange AJ, Cott JM, Lipton MA: Interactions of thyrotropin-releasing hormone with centrally action drugs. In: Prange AJ (Ed.), *The Thyroid Axis, Drugs and Behavior,* New York: Raven Press, 1974, 115–128.

Breese GR, Prange AJ, Lipton MA: Pharmacological studies and thyroid-imipramine interactions in animals. In: Prange AJ (Ed.), *The Thyroid Axis, Drugs and Behavior,* New York: Raven Press, 1974, 29–48.

Cho JT, Bone S, Dunner DL, Colt E: The effect of lithium treatment on thyroid function in patients with primary affective disorder. *Am J Psychiat* 1979; 136: 115–116.

Cohen KL, Swigar ME: Thyroid function screening in psychiatric patients. *J Am Med Assoc* 1979; 242: 254–257.

Cowdry RW, Wehr TA, Zis AO, Goodwin FK: Thyroid abnormalities associated with rapid-cycling bipolar illness. *Arch Gen Psychiat* 1983; 40: 414–420.

Dratman MB, Crutchfield FL, Axelrod J, Collburn RW, Thoa N: Localization of triiodothyronine in nerve endings fractions of rat brain. *Proc Nat Acad Sci* 1976; 3: 941–944.

Dratman MB, Crutchfield FL, Gordon JT: Thyroid hormones and adrenergic neurotransmitter. In: *Catecholamines, Neuropharmacology and Central Nervous System: Theoretical Aspects,* New York: Alan R. Liss, 1984, 425–436.

Dunner DL, Fieve RR: Clinical factors in lithium prophylaxis failure. *Arch Gen Psychiat* 1974; 30: 229–233.

Duval F, Macher JP, Mokrani MC: Difference between evening and morning thyrotropin responses to protirelin in major depressive episode. *Arch Gen Psychiat* 1990; 47: 443–448.

Everett GM: The DOPA response potentiation test and its use in screening for antidepressant drugs. *Excerpta Med Int Congr Serv No. 122,* 1966.

Ford DH, Cramer EB: Developing nervous system in relation to thyroid hormones. In: Grave GD (Ed.), *Thyroid Hormones and Brain Development,* New York: Raven Press, 1977, 1–17.

Frazer A, Pandey G, Mendels J, Neeley S, Kane M, Hess ME: The effect of tri-iodothyrone in combination with imipramine on (^{3}H)-cyclic AMP production in slices of rat cerebral cortex. *Neuropharmacology* 1974; 13: 1131–1140.

Gavish M, Weisman A, Akun F, Youdim MBH: Modulatory effects of thyroxine treatment on central and peripheral benzodiazepine receptors in the rat. *J Neurochem* 1986; 47: 1106–1110.

Gitlin MJ, Weiner H, Fairbanks L, Hershman JM, Friedfeld N: Failure of T3 to potentiate tricyclic antidepressant response. *J Affect Disord* 1987; 13: 267–272.

Gjessing L: A review of periodic catatonia. *Biol Psychiat* 1974; 8: 23–45.

Gold MS, Pottash AL, Extein IRL: Hypothyroidism and depression. Evidence from complete thyroid function evaluation. *J Am Med Assoc* 1981; 245: 1919–1922.

Goldman M, Dratman MB, Cruthchfield FL, Jennings AS, Maruniak JA, Gibbons R: Intrathecal triiodothyronine administration causes greater heart rate stimulation in hypothyroid rats than intravenously delivered hormone. *J Clin Invest* 1985; 76: 1622–1625.

Goodwin FK, Prange AJ, Jr, Post RM, Muscettola G, Lipton MA: Potentiation of antidepressant effects by L-triiodothyronine in tricyclic nonresponders. *Am J Psychiat* 1982; 139: 34–38.

Herman R, Obarzanek E, Mikalavskas KM, Post RM, Jimerson DC: The effects of carbamazepine on resting metabolic rate and thyroid function in depressed patients. *Biol Psychiat* 1991; 29: 779–788.

Kardon FC, Winokur A, Utiger RD: Thyrotropin releasing hormone (TRH) in rat spinal cord. *Brain Res* 1977; 122: 578–581.

Kasper S, Sack DA, Wehr TA, Kick H, Voll G, Vieira H: Nocturnal TSH and prolactin, secretion during sleep deprivation and prediction of antidepressant response in patients with major depression. *Biol Psychiat* 1988; 24: 631–691.

Kastin AJ, Ehrensing RH, Schalch DS, Anderson MS: Improvement in mental depression with decreased thyrotropin response after administration of thyrotropin-releasing hormone. *Lancet* 1972; ii: 740–742.

Kijore B, Aggernaes H, Fog-Miller F, Anderson HH, Nissen J, Kirkegaard C, Bjdrum N: Circadian variation of serum thyrotropin in endogenous depression. *Psychiat Res* 1982; 6: 277–282.

Kirkegaard C: The thyrotropin response to thyrotropin-releasing hormone in endogenous depression. *Psychoneuroendocrinology* 1981; 6: 189–212.

Kramlinger KG, Gharib H, Swanson DW, Maruth T: Normal serum thyroxine values in patients with acute psychiatric illness. *Am J Med* 1984; 76: 799–801.

Langer G, Koinig G, Hatzinger R, Schonbeck G, Resch F, Aschauer H, Keshavan MS, Sieghart W: Response of thyrotropin to thyrotropin-releasing hormone as predictor of treatment outcome. *Arch Gen Psychiat* 1969; 43: 861–868.

Levy RP, Jensen JB, Laus VG, Ogle DP, Engle IM: Serum thyroid hormone abnormalities in psychiatric disease. *Metabolism* 1981; 30: 1060–1064.

Loosen PT: The TRH-induced TSH response in psychiatric patients: A possible neuro-endocrine marker. *Psychoneuroendocrinology* 1985; 10: 237–260.

Loosen PT, Prange AJ: The serum thyrotropin (TSH) response to thyrotropin-releasing hormone (TRH) in depression: A review. *Am J Psychiat* 1982; 139: 405–416.

Manaker S, Winokur A, Rostene WH, Rainbow TC: Autoradiographic localization of thyrotropin releasing hormone (TRH) receptors in the rat CNS. *J Nueurosci* 1985; 35: 167–174.

Mason GA, Bondy SC, Nemeroff CB, Walker CH, Prange AJ: The effects of thyroid state on beta-adrenergic and serotonergic receptors in rat brain. *Psychoneuroendocrinology* 1987; 12: 261–270.

Mason GA, Walker CH, Prange AJ: Modulation of gamma-aminobutyric acid uptake of brain synaptosomes by thyroid hormones. *Neuropsychopharmacology* 1987; 1: 63–70.

Morley JE: Neuroendocrine control of thyrotropin secretion. *Endocrine Rev* 1981; 2: 396–435.

Perrild H, Hegedüs L, Baastrup PC, Kayser L, Kastberg S: Thyroid function and ultrasonically determined thyroid size in patients receiving long-term lithium treatment. *Am J Psychiat* 1990; 147: 1518–1521.

Plotnitcoff NP, Prange AJ, Breese GR, Anderson MS, Wilson IC: Thyrotropin releasing hormone: Enhancement of DOPA activity by a hypothalamic hormone. *Science* 1972; 178: 417–418.

Prange AJ, Wilson IC, Lara PP, Alltop LB, Breese GR: Effects of thyrotropin-releasing hormone (TRH) in depression. *Lancet* 1972; ii: 999–1002.

Prange AJ, Jr, Wilson IC, Rabon AM, Lipton MA: Enhancement of imipramine antidepressant activity by thyroid hormone. *Am J Psychiat* 1969; 126: 457–469.

Rubin RI, Poland RE, Lesser IM, Martin DJ: Neuroendocrine aspects of primary endogenous depression IV. Pituitary-thyroid axis activity in patients and matched control subjects. *Psychoneuroendocrinology* 1987; 12: 333–347.

Sack DA, James SP, Rosenthal NE, Wehr TA: Deficient nocturnal surge of TSH secretion during sleep and sleep deprivation in rapid-cycling bipolar illness. *Psychiat Res* 1988; 23: 179–191.

Savard P, Mérand Y, Di Paolo T, Dupont A: Effects of thyroid state on serotonin, 5-hydroxyindoleacetic acid and substance P contents in discrete brain nuclei of adult rats. *Neuroscience* 1983; 10: 1399–1404.

Schildkraut JJ, Winokur A, Draskozy PR, Hensle JH: Changes in norepinephrine turnover in rat brain during chronic administration of imipramine and protriptyline: A possible explanation for the delay in onset of clinical antidepressant effects. *Am J Psychiat* 1971; 127: 1032–1039.

Schmidt BH, Schultz JE: Chronic thyroxine treatment of rats down-regulates the noradrenergic cyclic AMP generating system in cerebral cortex. *J Pharmacol Exp Therapeut* 1985; 233: 466–472.

Schou M, Amidson A, Jesen SE, Olsen T: Occurrence of goiter during lithium treatment. *Br Med J* 1968; 3: 710–713.

Spratt DI, Pont A, Miller MB, McDougall IR, Bayer MF, McLaughlin WT, *Am J Med* 1982; 73: 41–48.

Stanton TL, Winokur A, Beckman AL: Reversal of natural CNS depression by TRH action in the hippocampus. *Brain Res* 1980; 137: 470–475.

Stanton TL, Winokur A, Beckman AL: TRH effects in the CNS: dependence on arousal site. *Science* 1981; 214: 678–681.

Sterling K: Thyroid hormone action at the cell level. *New Eng J Med* 1975; 300: 117–123, 173–177.

Waldstein S: Thyroid-catecholamine interrelations. *Ann Rev Med* 1966; 17: 123–132.

Weeke A, Weeke J: Disturbed circadian variation of serum thyrotropin in patients with endogenous depression. *Acta Psychiat Scand* 1978; 57: 281–289.

Weeke A, Weeke J: The 24-hour pattern of serum TSH in patients with endogenous depression. *Acta Psychiat Scand* 1980; 62: 69–74.

Whybrow PC, Bauer MS: Effects of peripheral thyroid hormones on the central nervous system: Relevance to disorders of mood. *Curr Top Neuroendocrinol* 1988; 8: 309–326.

Winokur A: The relevance of thyrotropin-releasing hormone to psychiatric disorders. In: Nemeroff CB (Ed.), *Neuropeptides and Psychiatric Disorders*, Washington, DC: American Psychiatric Press, 1990, 13–281.

Winokur A, Davies R, Utiger RD: Subcellular distribution of thyrotropin-releasing hormone (TRH) in rat brain and hypothalamus. *Brain Res* 1977; 120: 423–434.

Winokur A, Utiger RD: Thyrotropin-releasing hormone: Regional distribution in rat brain. *Science* 1974; 185: 265–267.

Yarbrough GG: On the neuropharmacology of thyrotropin-releasing hormone (TRH). *Prog Neurobiol* 1979; 12: 291–312.

Chapter 8

THE HPA AXIS AND DEPRESSIVE DISORDERS

Roger F. Haskett

INTRODUCTION

During the past two decades, various abnormalities of hypothalamic-pituitary-adrenocortical (HPA) axis function have been described in patients with depressive illness. Abnormalities were initially observed at the most peripheral level, i.e., urinary excretion of cortisol metabolites, but the development of reliable assays for the direct measurement of circulating hormones provided the impetus for an extensive progression of studies. Although the focus of investigation has undergone period reevaluation, the phenomenon of HPA axis dysregulation continues to intrigue investigators of the biology of depression and the "stress response."

The heuristic implications of an association between depression and neuroendocrine dysregulation were quickly recognized. Many depressive symptoms, such as disturbances in appetite, libido, sleep, autonomic activity, appear to reflect disturbances in hypothalamic and limbic functioning, and, as the hypothalamus was known to have a crucial role in neuroendocrine regulation, it was hypothesized that the alterations in peripheral hormonal secretion were additional manifestations of central neurotransmitter dysfunction associated with the pathogenesis of depression (Sachar, Hellman, Roffward, Halpern, Fukushima,

Roger F. Haskett • Department of Psychiatry, University of Pittsburgh School of Medicine, Western Psychiatric Institute and Clinic, Pittsburgh, Pennsylvania 15213-2593.

Biology of Depressive Disorders, Part A: A Systems Perspective, edited by J. John Mann and David J. Kupfer. Plenum Press, New York, 1993.

& Gallagher, 1973). This rationale for the "window into the brain" approach to the study of the biology of depression was based upon the assumption that the hypersecretion of the peripheral hormones was a direct result of increased stimulatory drive in the limbic system, which then cascaded down the HPA axis, essentially unchanged from the hypothalamus to the adrenal cortex. Various neurotransmitter systems, which were hypothesized to be dysfunctional in depressed patients, e.g., catecholamines, serotonin, and acetylcholine, were noted to have important roles in the regulation of HPA functioning, at least in animal studies. It was generally proposed that clarification of the central mechanisms underlying the neuroendocrine dysregulation would reveal important information about the pathogenesis of depression.

Reports of basal hypercortisolemia in depressed patients suggested adrenal hyperfunction and subsequent studies characterized the peripheral abnormality in more detail. This included evidence of an increase in 24-hour mean plasma cortisol levels, a shortened nocturnal period of quiescence, an earlier nadir of secretion, and decreased relative amplitudes in the circadian rhythm (Halbreich, Asnis, Shindledecker, Zumoff, & Nathan, 1985; Linkowski, Mendlewicz, Leclercq, Brasseur, Hubain, Goldstein, et al. 1985; Pfohl, Sherman, Schlechte, & Stone, 1985; Sachar et al., 1973). Recognizing that the HPA axis was also activated by exposure to various stressors, several investigators utilized the dexamethasone suppression test (DST) in an attempt to improve discrimination between depression-related changes in HPA function and non-specific stress-induced increases in adrenocortical secretion. The failure to suppress cortisol after a single dose of dexamethasone, a positive or abnormal DST, was then widely used to identify the presence of HPA axis dysregulation in patients with depressive disorders (Carroll, 1982; Carroll, Curtis, & Mendels, 1976a; Carroll, Curtis, & Mendels, 1976b; Schlesser, Winokur, & Sherman, 1980; Stokes, Pick, Stoll, et al. 1975).

Subsequent investigation of HPA dysregulation in depression has followed two main themes, with the emphasis switching from one to the other. The first attempted to clarify the state or trait characteristics of the observed neuroendocrine changes and to assess their clinical applicability in patients with depressive disorders. Many studies examined the relationship between HPA dysfunction and specific diagnostic subtypes, such as endogenous or melancholic depressions; others sought evidence of links between individual clinical features and the neuroendocrine disturbance, while still other studies assessed the utility of HPA axis abnormalities in the prediction or monitoring of treatment outcome.

Another approach followed a reevaluation of the "window into the brain" and began to address the complexity of the regulatory mechanisms within this closed-loop endocrine system. Investigators attempted to dissect out and examine secretory functions at multiple levels of the HPA axis in an attempt to determine whether peripheral hypersecretion is due to increased drive, impaired feedback, or both. Numerous studies of basal secretion patterns, negative feedback effects, and responses to stimulatory agents, have characterized abnormalities at each level of the axis. These have revealed clear evidence of heterogeneity in the mechanisms underlying the neuroendocrine disturbance and provided new questions about the relationship between these abnormalities and the nosology and pathophysiology of depression.

HPA Dysregulation as a Biological Marker

State and Trait Properties

The state or trait properties of HPA axis dysregulation in depressive disorders have important implications for its role in studies of the pathogenesis of depression and for its utility in clinical practice. A state relationship can have significance for both longitudinal and cross-sectional studies. In the former instance, a state marker that appears concurrently with the clinical syndrome and reverts to normal with its remission has obvious utility in monitoring the course of an illness. In cross-sectional assessments, a correlation between the magnitude of the functional disturbance of the marker and severity, or some other key dimension of the syndrome, may provide information that assists in clinical diagnosis or the refinement of existing nosology.

Many different indicators of HPA axis function have been identified, but the DST remains the most extensively studied marker of this endocrine dysregulation in depressed patients (see Arana, Baldessarini, & Ornsteen, 1985; The APA Task Force on Laboratory Tests in Psychiatry, 1987, for reviews). Overall, in patients with Major Depression, the frequency of a positive DST result is approximately 45%; varying between 50% and 70% for inpatients and between 20% and 40% for outpatients. This observed neuroendocrine heterogeneity in patients with depression is incompletely understood, although several sources of variance have been identified. Certain patient characteristics, such as increased age or severity of depression, are associated with more frequent abnormal DST results, but this test can also be influenced by several metabolic and pharmacological factors that have no specific relationship with depressive disorders.

Longitudinal Studies

Initial interest in the DST's utility as a diagnostic test favored a cross-sectional design for studies, but several investigators examined the longitudinal course of HPA axis dysregulation in depressed patients (Albala & Greden, 1980; Holsboer, Liebl, & Hofschuster, 1982; Bowie & Beaini, 1985). In most cases, it was noted that an abnormal DST result became negative ("normalized") with remission of the depressive syndrome and reverted to abnormal during relapse (Greden, Gardner, King, et al. 1983), suggesting that an abnormal DST was a "state" marker of the depressive syndrome. Additional evidence that HPA dysregulation was a state marker of some central neuronal disturbance came from studies which demonstrated that, despite apparent clinical recovery, persistent or posttreatment DST abnormalities predicted slow recovery or a poor outcome, such as relapse or suicide (Goldberg, 1980; Greden, Albala, Haskett, James, Goodman, Steiner & Carroll, 1980; Bowie, Beaini, & Bowie, 1987; Schweitzer, Maguire, Gee, Tiller, Biddle, & Davies, 1987).

Despite the obvious neuroendocrine and clinical heterogeneity of depression, it was generally presumed that, provided the nonspecific sources of variance were adequately controlled, an abnormal DST reflected HPA dysregulation resulting from limbic/hypothalamic dysfunction associated with depression and that "normalization" of the DST indicated that HPA axis regulatory mechanisms had returned to their premorbid state. It cannot be assumed, however, that all depressed patients with a normal DST result have normal HPA function, since the DST, as commonly performed, evaluates *adrenocortical* secretion in the presence of negative feedback effects of *dexamethasone*. Similarly, the DST's ability to monitor within-subject changes in HPA axis function is also limited, since a proportion of depressed patients may have persistent HPA dysregulation that is undetected by this probe.

Studies of the ACTH and βE response to dexamethasone in depression assessed whether these pituitary peptides also failed to suppress after exogenous glucocorticoid. Although the results showed some disagreement, levels of ACTH and βE did not normally suppress after dexamethasone in a significant proportion of depressed patients (Matthews, Akil, Greden, Charney, Weinberg, Rosenbaum, et al., 1986; Akil *et al.*, 1992; Young *et al.*, 1992). Some depressed patients clearly had demonstrable HPA dysregulation that was not reflected by peripheral cortisol levels. A longitudinal study of postdexamethasone plasma levels of βE in patients, before and during antidepressant treatment, revealed that a persistent or intermittently abnormal βE response during treatment, in the presence of a normal cortisol response to dexamethasone, was associated with a slower response to tricyclics and a more rapid relapse after medication was discontinued (Haskett, Young, & Greden, 1988). This suggests that, while the most peripheral aspects of the system, circulating corticosteroids, appear normal, specific components of the system may be persistently altered. In addition, normalization of HPA axis function at several levels may be necessary before full remission of the syndrome can be expected to occur or be sustained. Further examination of the state relationship between HPA axis dysregulation and depression should not rely exclusively on a single test, such as the DST. Future studies in depression will probably need additional paradigms, which incorporate other positive or negative stimuli and assess secretory responses at multiple levels of this neuroendocrine axis.

Several studies examined the relationship between HPA dysregulation and its ability to predict treatment response. It was initially suggested that the presence of HPA abnormality might be able to distinguish those depressions requiring somatic treatments, although it was acknowledged that a normal DST could not contribute to that question. Many reports were unable to support that proposition, probably because of the generally favorable response rate for patients treated with antidepressants treatment, irrespective of the HPA axis functioning. Similarly, suggestions that HPA abnormalities might assist with the choice of an effective antidepressant were not supported by subsequent studies. One treatment response finding has emerged, however, although it is more evident for patients being treated as inpatients than outpatients. The presence of HPA axis dysregulation generally predicts a poor response to placebo and to specific psychotherapies (Peselow, Stanley, Filippi, Barouche, Goodnick, & Fieve, 1989). Similarly, placebo responders were found to have a low frequency of HPA axis dysregulation, similar to that found in healthy controls.

When patients with a bipolar disorder were studied in various phases of their illness, HPA axis function was generally abnormal in the depressed phase compared to the manic phase, or to measures in healthy controls (Greden DeVigne, Albala, Tarika, Buttenheim, Eiser, et al., 1982). In addition, patients with recurrent depression who had an abnormal DST during one episode were usually found to have similar hormonal changes during subsequent episodes. These findings suggested that HPA dysregulation, as reflected by an abnormal response to dexamethasone, was a state marker for the depressive syndrome and relatively reliable within individuals.

Cross-Sectional Studies

HPA Axis Abnormalities and Diagnostic Subtypes of Depressive Disorders. The focus of much early research on HPA axis dysregulation in depression involved the use of the DST as a diagnostic marker. This approach acknowledged the heterogeneity of depressive disorders and attempted to use this test as a state marker of clinically identified subtypes. Many problems were encountered in the interpretation of the results from these studies. The topic has been reviewed extensively in the literature (Arana et al., 1985; Carroll, 1985; The APA Task Force on Laboratory Tests in Psychiatry, 1987) and will not be discussed at length here. It was clearly apparent, however, that the frequency of HPA axis dysregulation, as identified by an abnormal DST, varied considerably according to the diagnostic classification system that was used (Davidson, Lipper, Zung, Strickland, Krishnan, & Mahorney, 1984). Although there was a general consensus that the more dysfunctional and episodic forms of depression tended to have a higher frequency of HPA axis dysregulation than the milder or more chronic depressive syndromes, a specific state relationship with a subcategory of depression was not consistently demonstrated.

HPA Axis Abnormalities and Clinical Features of Depression. In addition, the interaction between syndrome severity and diagnostic subtype remains unclear. There continues to be discussion about the validity of several subcategories of depression, such as endogenous or melancholic depressions, with some claiming that they mainly address the severity dimension rather than a categorical distinction. The persistence of multiple definitions for these diagnoses suggests that the question is not yet resolved.

One report indicated that HPA dysfunction was most closely related to those endogenous symptoms that might be expected to be most closely related to hypothalamic dysfunction, such as initial insomnia, agitation, loss of libido, and weight loss, but not to the more cognitive features of paranoia, guilt, worthlessness, helplessness, and suicidal thinking (Miller & Nelson, 1987). Similarly, others have reported a significant relationship between HPA dysregulation and the anxiety and agitation features from the HRSD (Meador-Woodruff, Greden, Grunhaus, & Haskett, 1990; Rubin, Poland, Lesser, Martin, Blodgett, & Winston, 1987).

Severity is certainly confounded with the endogenous/melancholic clinical profile in that the most commonly used rating scales of severity, such as the Hamilton Rating Scale for Depression (HRSD), are heavily weighted by the presence of endogenous clinical features. Recent studies of pituitary peptide response to dexamethasone in depression did not find that severity was closely related with the presence or degree of neuroendocrine disturbance (Akil et al., 1992; Young et al., 1992).

Interestingly, HPA axis dysregulation as a state marker did not appear to distinguish depressions that were grouped according to the most validated categorization of mood disorder, the unipolar/bipolar dichotomy. That is, unipolar and bipolar depressed patients do not appear to have a significant difference in the frequency of this neuroendocrine abnormality (Carroll, Feinberg, Greden, Tarika, Albala, Haskett, et al., 1981). As noted below, however, HPA dysregulation may be a marker of a trait characteristic related to this distinction.

Family Studies

Using the usual definition of a trait marker, namely, that the abnormality is identifiable prior to the appearance of the syndrome or manifest features of the illness and is persistent and relatively unresponsive to changes associated with subsequent fluctuations in state, i.e., episodes and remissions, it has been difficult to identify any trait characteristics of the HPA disturbance.

One finding that suggests the possible existence of a yet undetected trait marker relates to the frequency of HPA axis abnormality in patients categorized according to their family history of depression. A review of results from patients with a family history of depressed first degree relatives revealed a higher frequency of abnormal DST results than patients with no affected relatives (Arana et al., 1985). Another study that examined the DST in normal controls also found that individuals with an abnormal DST had a higher frequency of relatives with evidence of mood disorder (Coryell & Zimmerman, 1987). When the βE response to dexamethasone was examined, a higher frequency of abnormal response was found in patients with either a personal or family history of bipolar disorder compared to those with unipolar disorders (Lahti, Haskett, Murphy-Weinberg, Young, Watson, & Akil, 1991).

Mechanisms of HPA Dysregulation

An important aspect of the complexity of HPA axis regulation involves the existence of multiple negative feedback mechanisms. These systems can be classified according to the anatomical site of the glucocorticoid receptor (anterior pituitary; hypothalamus; other limbic structures); or the response time and mechanism of action (fast rate-sensitive; intermediate and delayed, proportional or level-sensitive) (Keller-Wood & Dallman, 1984; Dallman, Levin, Akana, Jacobson, & Shinsoki, 1987). In addition, there are several distinct glucocorticoid receptor systems in brain and pituitary, which mediate the effects of corticosteroids on mood, behavior, and the underlying neurochemistry, as well as negative feedback on the HPA axis. As a result of animal studies, these glucocorticoid receptors have been classified into Type 1, which can be further divided into mineralocorticoid (MR) and corticosterone-preferring (CR), and Type 2, which is the classical glucocorticoid receptor (GR) (Funder, 1986; Funder, Feldman, & Edelman, 1973; Reul & de Kloet, 1985). The evidence for receptor heterogeneity and the differences in regulation and putative function have been summarized in several recent reviews (DeKloet & Reul, 1987; McEwen, 1987; McEwen, DeKloet, & Rostene, 1986).

In contrast to the elevated urine and plasma corticosteroid levels described above, studies that have examined plasma levels of the pituitary peptides, ACTH and βE/βLPH, did not find consistent evidence of increased or decreased secretion (Fang, Tricou, Robertson, & Meltzer, 1981; Kalin, Weiler, & Shelton, 1982; Nasr, Rodgers, Pandey, Altman, Gaviria, & Davis, 1983; Reus, Joseph, & Dallman, 1982; Risch, 1982). This is intriguing since increased CRH secretion might be expected to increase pituitary peptide levels while elevated circulating corticosteroids would act through normal feedback mechanisms to suppress ACTH and βE. These findings raise the possibility of heterogeneity in the actual mechanisms of HPA axis dysregulation within groups of depressed patients. The complexity of regulation in this neuroendocrine system emphasizes the need to assess its various components.

The characterization of ovine corticotrophin releasing factor (oCRH) (Vale, Spiess, Rivier, & Rivier, 1981) provided an additional strategy for the examination of HPA axis dysregulation at the level of the pituitary. Several groups administered oCRH to assess corticotroph responsivity to stimulation and demonstrated that patients with depression released less ACTH and βE compared to controls (Amsterdam, Maislin, Winokur, Kling, & Gold, 1987; Gold, Loriaux, Roy, Kling, Calabrese, Kellner, et al., 1986; Holsboer, Bardelehen, Gerken, Stalla, & Muller, 1984). The degree to which this hyporesponsiveness was due to feedback effects of elevated basal cortisol levels, adrenocortical hypertrophy, or accentuated fast-feedback by the initial release of ACTH and cortisol continues to be the subject of further studies (Young, Watson, Kotun, Haskett, Greden, Grunhaus, et al., 1990).

Once multiple levels of feedback in the HPA axis were recognized, strategies were sought to distinguish between feedback effects at the pituitary and the brain. Dexamethasone probably exerts its principal feedback effects at the pituitary rather than the brain and, as used in the DST, is acting through delayed or proportional feedback mechanisms. In contrast, animal studies suggested that a rapid, rate sensitive feedback response to an infusion of natural glucocorticoids would provide a means of evaluating the neuronal-hippocampal elements of HPA axis regulation. One study demonstrated that a hydrocortisone infusion exerted a fast feedback effect on βE secretion in control subjects but not in patients with depression (Young, Haskett, Murphy-Weinberg, Watson, & Akil, 1991). It is possible that this reflects a decreased sensitivity to glucocorticoid fast feedback in depressed patients and supported the hypothesis that abnormalities of hippocampal glucocorticoid receptors contribute to the HPA axis dysregulation seen in these patients.

HPA Axis Dysregulation: Depression or Stress?

When addressing the relationship between HPA axis dysregulation and the primary pathogenesis of depression, an important consideration relates to the stress responsiveness of this neuroendocrine system. Since the original proposition of the concept (Selye, 1936), many studies have demonstrated that the HPA axis responds with increased secretory activity to various noxious stimuli, which are commonly referred to as "stress," and include pain, heat, cold, exercise, tissue injury, threat of danger, environmental novelty, etc. Consequently when elevated circulating corticosteroids were first reported in patients with depression, there was concern that this simply represented a nonspecific response to stressors associated with the illness. It was possible that the HPA axis was "simply" reacting to the stress associated with the painful experience of depression, the secondary psychosocial and environmental consequences of the illness, e.g., admission to hospital, strained interpersonal relationships, occupational difficulties; or the nonspecific physiological disturbances that often accompanied depression, e.g., weight loss, sleep deprivation, electrolyte imbalance. Support for a specific relationship between HPA dysfunction and depression did come from the repeated observation that HPA dysregulation was usually more frequent in depressed patients, compared to patients with other severe and distressing psychiatric disorders. Conversely, the demonstration that HPA abnormalities, such as an abnormal DST, could be reproduced in healthy controls subjected to significant weight loss or sleep deprivation suggested to some investigators that the HPA axis disturbances in depression were nonspecific (Mullen, Linsell, & Parker, 1986). The problem with the latter argument was the implicit assumption that the effects of "stress" and depression on the HPA axis were mutually exclusive. The responsivity of the HPA axis to various forms of physical as well as psychological stress does not address the neuroendocrine system's capacity to respond to intrinsic CNS disturbances associated with depression.

Several lines of evidence suggested that the effects of stress often synergise with the stimulatory influences on the HPA axis associated with a depressive disorder. For example, in a study that compared results from a DST performed shortly after hospital admission with repeat tests that were performed 1–2 weeks later, we found that the frequency of an abnormal DST decreased significantly in all diagnostic groups except patients with endogenous depression (Haskett, Zis, Albala, Lohr, & Carroll, 1989). Presumably the reduction in the novelty effect resulting from hospital admission and correction of physiological disturbances secondary to depression had resulted in a decreased HPA axis stimulation due to stress while the specific changes associated with depression remained. It also appears that HPA axis studies in depressed patients, which required blood sampling by repeated venepuncture, found a higher frequency of DST abnormality than studies that utilized an in-dwelling catheter (Young *et al.*, 1992). The latter presumably avoided the acute HPA axis activation due to the stress of venepuncture at the time of the blood sampling. As a result, many reports of HPA axis dysregulation either overestimated the specific HPA effects of depression or claimed to support the lack of specificity of HPA axis changes in depression, because they did not adequately consider the effects of acute stress associated with changes in environment or the actual testing procedure.

Age and the HPA Axis in Depression

Although, as noted previously, no aspects of HPA axis dysfunction have been identified as trait markers for depression, several relatively persistent HPA axis changes in some of these patients can be considered "scars." These are changes in a functional measure that appear for the first time during an episode, do not fully resolve with symptom remission, persist during subsequent episodes of illness and intervals of remission, and may increase in severity with subsequent episodes of illness. Although not yet well defined, this process may provide the most appropriate model for understanding the interactive effects of multiple depressive episodes, environmental stressors, and normal aging on HPA axis function.

In healthy subjects, increased age is not consistently associated with higher circulating cortisol levels, but, in depressed patients, several studies have reported that age is positively correlated with basal and post-dexamethasone plasma cortisol levels (Alexopoulos, Young, Kocsis, Brockner, Butler, & Stokes, 1984; Halbreich, Asnis, Zumoff, & Nathan, 1984; Lewis, Pfohl, Schlechte, & Coryell, 1984; Oxenkrug, Pomara, McIntyre, Branconnier, Stanley, & Gershon, 1983; Tiongco, Hariharan, Haskett, & Greden, 1986; Rubin, Poland, Lesser, Martin, Brodgett, & Winston, 1987). Older patients also tended to have a higher frequency of an abnormal DST result and, when studied during treatment, had slower patterns of normalization (Greden, Flegel, Haskett, Dilsaver, Carroll, Grunhaus, et al., 1986). When resting plasma βE levels were examined in older, depressed women, however, there was a significant negative correlation with age, suggesting that the elevated basal cortisol levels were exerting normal negative feedback on pituitary peptide release (Akil *et al.*, 1992). Since older patients tend to have had more episodes of depression and a greater severity of depression, the interaction between the effects of these variables on HPA regulation and the effect of aging needs further clarification. There is also a report suggesting that this relationship may be more evident in females than males, which raises the question of an interaction in women between aging and the changes in gonadal hormone levels associated with menopause (Young *et al.*, 1992). Considering the preponderance of women affected by depressive illnesses, this is an important topic for future research.

Although important, it remains unclear whether recurrent depressive episodes produce any cumulative changes in the regulatory mechanisms of the HPA axis. Animal studies suggest that repeated or chronic stress does produce sustained dysregulation and it is possible that recurrent episodes of depression, with the associated periods of HPA axis activation, might produce similar neuroendocrine sequelae. The literature contains many reports that animals subjected to chronic or recurrent stress develop adrenal hypertrophy. Whether this is due to increased secretion of ACTH, the physiological stimulus for adrenocortical secretion, or to the release of additional trophic agents such as gamma MSH, remains unclear. In addition, at the level of the pituitary, chronic stress results in an increase in ACTH and βE content while plasma levels of pituitary peptides or corticosteroids are identical to those seen after acute stress (Young & Akil, 1985). Reports that depressed patients had an exaggerated adrenocortical response to synthetic corticotropin (ACTH) and enlarged adrenals when examined with computerized tomographic scanning were consistent with the development of adrenocortical hypertrophy in this patient group (Amsterdam, Marinelli, Arger, & Winokur, 1987; Amsterdam, Winokur, Abelman, Lucki, & Rickels, 1983; Jaeckle, Kathol, Lopez, Meller, & Krummel, 1987; Zis & Zis, 1987). It was also of interest that the increased adrenocortical responsivity to ACTH appeared to persist in some patients after recovery from depression, despite "normalization" of the DST (Amsterdam, Maislin, Winokur, Berwish, Kling, & Gold, 1988; Stokes, Sikes, & Stoll, 1988).

Examination of the effects of repeated or prolonged HPA axis activation on central glucocorticoid receptors in animals may be relevant to understanding the situation in patients with recurrent or chronic depressions. The potential importance of this area is indicated by findings from animal studies, which demonstrate the role of hippocampal and limbic glucocorticoid receptors in negative feedback regulation of the HPA axis (Sapolsky, Krey, & McEwen, 1983a,b, 1984). Chronic stress or chronic corticosterone treatment appears to down-regulate these hippocampal receptors, while it does not affect the number of glucocorticoid receptors in the pituitary, hypothalamus, cortex, and midbrain (Sapolsky, 1986). Rats with down-regulated hippocampal receptors fail to turn off glucocorticoid secretion promptly at the end of a stress. If glucocorticoid treatment is discontinued after several weeks, the receptor loss is reversible and the animals return to normal feedback regulation. When chronic corticosterone treatment is continued for a longer period (e.g, 3 months), there is a permanent loss of glucocorticoid receptors, which is accompanied by actual neuronal loss in the hippocampus. The morphological pattern of this cell loss is typical of that seen in aged rats (Sapolsky et al., 1983).

HPA Axis Stimulation and Psychopathology

CRH Hypersecretion

The possible existence of a direct causal relationship between HPA axis activation and depression has been proposed, but the mechanism and even the direction are not established (Gold, Goodwin, & Chrousos, 1988). The possibility that some of the manifestations of depressive disorder might be a direct result of increased activity in CRH releasing neurons are, in part, based upon the observed effects of intracerebroventricular (ICV) CRH administration to primates. It has been noted that CRH given ICV, not only stimulates the HPA axis, but produces changes in behavior that resemble those seen in depressed patients. They include autonomic arousal, with elevated arterial pressure, heart rate, oxygen consumption, and plasma glucose and catecholamines, as well as altered locomotion, decreased sexual activity, decreased food consumption, and sleep disturbance. This model of CRH overproduction is also indirectly supported by the evidence of elevated CRH levels in the CSF of depressed patients (Nemeroff, Widerlov, Bissette, Walleus, Karlsson, Eklund, et al., 1984), which normalized after successful treatment with electroconvulsive therapy, and the studies showing downregulated CRH receptors in the frontal cortex (Nemeroff, Owens, Bissette, Andorn, & Stanley, 1988) and increased POMC mRNA and corticotroph peptides in the pituitaries of individuals who died by suicide.

Elevated Plasma Cortiscosteroids

An alternative hypothesis involves more peripheral products of the HPA axis. It is clear from the literature that excessive administration of exogenous corticosteroids can produce typical depressive syndromes (Haskett, 1987). In addition, patients with Cushings Disease with primary lesions in the pituitary or the adrenal are frequently noted to have depression (Haskett, 1985). Considering that in these endogenous disorders CRH production is unlikely to be increased, and in patients receiving exogenous corticosteroids, release of the corticotroph peptides is probably inhibited, it is extremely difficult to postulate a unifying hypothesis for the relationship between the products of HPA axis and the development of depression.

Conclusion

The early hopes for the "window into the brain" approach to biological research into depression have required revision. The regulatory mechanisms in the HPA axis and probably other endocrine systems are proving to be very complex. More complicated models that link the HPA dysregulation with depression are being proposed and, as the tools become more sophisticated, are becoming accessible to empirical research. One interesting development is the decrease in the earlier emphasis on separating the influences of stress on the HPA axis from those due to depression and an increase in the search for longitudinal research paradigms. With the increased awareness that mood disorders are a lifelong disease in many patients and the accumulating data for an interaction between exacerbations of depression and environmental stress in some situations, renewed attention to the longitudinal biology of the "stress axis" may provide us with important information about the course of depressive illness and the development of its underlying pathophysiology.

References

Akil H, Haskett RF, Young EA, Grunhaus L, Kotun J, Weinberg V, Greden J, & Watson SJ: Multiple HPA profiles in endogenous depression: Effects of age and sex on cortisol and beta-endorphin. *Biol Psychiat* 1992 (in press).

Alexopoulos GS, Young RC, Kocsis JH, Brockner N, Butler TA, Stokes PE: Dexamethasone suppression tests in geriatric depression. *Biol Psychiat* 1984; 19: 1567–1571.

Amersterdam JD, Maislin G, Winokur A, Berwish N, Kling M, Gold P: The oCRH stimulation test before and after clinical recovery from depression. *J Affect Disord* 1988; 14: 213–222.

Amsterdam JD, Maislin G, Winokur A, Kling M, Gold P: Pituitary and adrenocortical responses to the ovine corticotropin releasing hormone in depressed patients and healthy volunteers. *Arch Gen Psychiat* 1987a; 44: 775–781.

Amsterdam JD , Marinelli DL, Arger P, Winokur A: Assessment of adrenal gland volume by computed tomography in depressed patients and healthy volunteers. *Psychiat Res* 1987b; 21: 189–197.

Amsterdam JD, Winokur A, Abelman E, Lucki I, Rickels K: Cosyntropin ($ACTH_{al-24}$) stimulation test in depressed patients and healthy subjects. *Am J Psychiat* 1983; 140: 907–909.

Arana GW, Baldessarini RJ, Ornsteen M: The dexamethasone suppression test for diagnosis and prognosis in psychiatry. *Arch Gen Psychiat* 1985; 42: 1193–1204.

Bowie PCW, Beaini A: Normalization of the dexamethasone suppression test: A correlate of clinical improvement in primary depressives. *Brit J Psychiat* 1985; 147: 30–35.

Bowie PCW, Beaini AY, Bowie LJ: The prognosis of primary depressive illness: Its relationship to the dexamethasone suppression. *Brit J Psychiat* 1987; 150: 787–790.

Carroll BJ: The dexamethasone suppression test for melancholia. *Brit J Psychiat* 1982; 140: 292–304.

Carroll BJ, Curtis GC, Mendels J: Neuroendocrine regulation in depression, I: Limbic system-adrenocortical dysfunction. *Arch Gen Psychiat* 1976a; 33: 1039–1044.

Carroll BJ, Curtis GC, Mendels J: Neuroendocrine regulation in depression, II: Discrimination of depressed from nondepressed patients. *Arch Gen Psychiat* 1976b; 33: 1051–1058.

Carroll BJ, Feinberg M, Greden JF, Tarika J, Albala AA, Haskett RF, James NM, Kronfol Z, Lohn N, DeVigne JF, Young E: A specific laboratory test for the diagnosis of melancholia. Standardization, validation and clinical utility. *Arch Gen Psychiat* 1981; 38: 15–22.

Coryell W, Zimmerman M: HPA-axis abnormalities in psychiatrically well controls. *Psychiat Res* 1987; 20: 265–273.

Dallman MF, Levin N, Akana SF, Jacobson L, Shinsoki J: In vivo characterization of corticosterone feedback regulation of ACTH secretion. In: Ganong WF, Dallman MF, Roberts JL (Eds.), *The Hypothalamic-Pituitary-Adrenal Axis Revisited.* New York: New York Academy of Science, 1987.

Davidson J, Lipper S, Zung WWK, Strickland R, Krishnan R, Mahorney S: Validation of four definitions of melancholia by the dexamethasone suppression test. *Am J Psychiat* 1984; 141: 1220–1223.

DeKloet ER, Reul JMHM: Feedback action and tonic influence of corticosteroids on brain function: A concept arising from the heterogeneity of brain receptor systems. *Psychoneuroendocrinology* 1987; 12: 83–105.

Fang VS, Tricou BJ, Robertson A, Meltzer HY: Plasma ACTH and cortisol levels in depressed patients: Relation to dexamethasone suppression test. *Life Sci* 1981; 29: 931–938.

Funder JW: Adrenocorticoid receptors in the brain. *Frontiers Neuroendocrinol* 1986; 9: 169–189.

Funder JW, Feldman D, Edelman IS: The roles of plasma binding and receptor specificity in the mineralocorticoid action. *Endocrinology* 1973; 92: 994–1004.

Gold PW, Goodwin FK, Chrousos GP: Clinical and biochemical manifestations of depression: Relation to the neurobiology of stress. *New Engl J Med* 1988; 319: 348–353.

Gold PW, Loriaux DL, Roy A, Kling MA, Calabrese JR, Kellner CH, Nielman LK, Post RM, Pickar D, Gallucci W, Avgerinos P, Paul S, Oldfield EH, Cutler GB, Chrousos GP: Response to corticotropin-releasing hormone in the hypercortisolism of depression and Cushing's disease. *New Engl J Med* 1986; 314: 1329–1335.

Goldberg IK: Dexamethasone suppression test as indicator of safe withdrawal of antidepressant therapy. *Lancet* 1980; i: 376.

Greden JF, Albala AA, Haskett RF, James N, Goodman L, Steiner M, Carroll BJ: Normalization of dexamethasone suppression tests: A probable index of recovery among endogenous depressives. *Biol Psychiat* 1980; 15: 449–458.

Greden JF, DeVigne JP, Albala AA, Tarika J, Buttenheim M, Eiser A, Carroll BJ: Serial dexamethasone suppression tests among rapidly cycling bipolar patients. *Biol Psychiat* 1982; 17: 455–462.

Greden JF, Flegel P, Haskett RF, Dilsaver S, Carroll BJ, Grunhaus L, Genero N: Age effects in serial hypothalamic-pituitary-adrenal monitoring. *Psychoneuroendocrinology* 1986; 11: 195.

Greden JF, Gardner R, King D, Grunhaus L, Carroll BJ, Kronifol Z: Dexamethasone suppression tests in antidepressant treatment of melancholia: The process of normalization and test–retest reproducibility. *Arch Gen Psychiat* 1983; 40: 493–500.

Halbreich U, Asnis GM, Zumoff B, Nathan RS: The effect of age and sex on cortisol secretion in depressives and normals. *Psychiat Res* 1984; 13: 221–229.

Halbreich W, Asnis GM, Shindledecker R, Zumoff B, Nathan RS: Cortisol secretion in endogenous depression, II. Time-related functions. *Arch Gen Psychiat* 1985; 42: 909–914.

Haskett RF: Diagnostic categorization of psychiatric disturbance in Cushing's Syndrome. *Am J Psychiat* 1985; 142: 911–916.

Haskett RF: Psychiatric effects of corticosteroids. In: Burrows GD, Norman TR, Davies B (Eds.), *Drugs in Psychiatry,* Amsterdam: Elsevier, 1987, 237–245.

Haskett RF, Young EA, Greden JF: Natural history of β-endorphin response to dexamethasone during depression. In *Annual Meeting of the Society of Biological Psychiatry.* Program Abstracts, Montreal, Quebec, 1988.

Haskett RF, Zis AP, Albala AA, Lohr NE, Carroll BJ: Comparison of early and delayed inpatient dexamethasone suppression tests. *Psychiat Res* 1989; 27: 161–171.

Holsboer F, Bardelehen U, Gerken A, Stalla G, Muller O: Blunted corticotropin and normal cortisol response to human corticotropin-releasing factor in depression. *N Engl J Med* 1984; 311: 1127.

Jaeckle RS, Kathol RG, Lopez JF, Meller WH, Krummel SJ: Enhanced adrenal sensitivity to exogenous $ACTH_{1\text{-}24}$ stimulation in major depression: Relationship to dexamethasone suppression test results. *Arch Gen Psychiat* 1987; 44: 233–240.

Kalin NH, Weiler SJ, Shelton SE: Plasma ACTH and cortisol concentrations before and after dexamethasone. *Psychiat Res* 1982; 7: 87–92.

Keller-Wood ME, Dallman MF: Corticosteroid inhibition of ACTH secretion. *Endocr Rev* 1984; 5: 1–24.

Lahti AC, Haskett RF, Murphy-Weinberg V, Young EA, Watson SJ, Akil H: Hypothalamic-pituitary-adrenal dysfunction and family subtypes of depression. In *American Psychiatric Association Meeting,* Program Abstract NR74. New Orleans, LA, 1991.

Lewis DA, Pfohl B, Schlechte J, Coryell W: Influence of age on the cortisol response to dexamethasone. *Psychiat Res* 1984; 13: 213–220.

Linkowski P, Mendlewicz J, Leclercq R, Brasseur M, Hubain P, Goldstein J, Copinschi G, Van Cauter E: The 24-hour profile of adrenocorticotropin and cortisol in major depressive illness. *J Clin Endocrinol Metab* 1985; 61: 429.

Matthews J, Akil H, Greden J, Charney D, Weinberg V, Rosenbaum A, Watson SJ: Beta-endorphin/beta-lipotropin-like immunoreactivity in endogenous depression: Effect of dexamethasone. *Arch Gen Psychiat* 1986; 43: 374–381.

McEwen BS: Glucocorticoid-biogenic amine interactions in relation to mood and behavior. *Biochem Pharmacol* 1987; 36(11): 1755–1763.

McEwen BS, DeKloet ER, Rostene WH: Adrenal steroid receptors and actions in the nervous system. *Physiol Rev* 1986; 66: 1121–1188.

Meador-Woodruff J, Greden JF, Grunhaus L, Haskett RF: Severity of depression and hypothalamic-pituitary adrenal axis dysregulation: Identification of contributing factors. *Acta Psychiat Scand* 1990; 81: 364–371.

Miller KB, Nelson JC: Does the dexamethasone suppression test relate to subtypes, factors, symptoms, or severity? *Arch Gen Psychiat* 1987; 44: 769–774.

Mullen PE, Linsell CR, & Parker D: Influence of sleep disruption and calorie restriction on biological markers for depression. *Lancet* 1986; ii: 1051–1055.

Nasr SJ, Rodgers C, Pandey G, Altman EG, Gaviria FM, Davis JM: ACTH and the dexamethasone suppression test in depression. *Biol Psychiat* 1983; 18: 1069–1073.

Nemeroff CB, Owens MJ, Bissette G, Andorn AC, Stanley M: Reduced corticotropin releasing actor finding sites in the frontal cortex of suicide victims. *Arch Gen Psychiat* 1988; 45: 577–579.

Nemeroff CB, Widerlov E, Bissette G, Walleus H, Karlsson I, Eklund K, Kilts CD, Loosen PT, Vale W: Elevated concentrations of CSF corticotropin-releasing-factor-like immunoreactivity in depressed patients. *Science* 1984; 226: 1342–1344.

Oxenkrug GF, Pomara N, McIntyre IM, Branconnier RJ, Stanley M, Gershon S: Aging and cortisol resistance to suppression by dexamethasone: A positive correlation. *Psychiat Res* 1983; 10: 125–130.

Peselow ED, Stanley M, Filippi AM, Barouche F, Goodnick P & Fieve RR: The predictive value of the dexamethasone suppression test. *Brit J Psychiat* 1989; 155: 667–672.

Pfohl B, Sherman B, Schlechte J, Stone R: Pituitary–adrenal axis rhythm disturbances in psychiatric depression. *Arch Gen Psychiat* 1985; 42: 897–903.

Reul JMHM, de Kloet ER: Two receptor systems for corticosterone in rat brain: Microdistribution and differential occupation. *Endocrinology* 1985; 117: 2505–2512.

Reus VI, Joseph MS, Dallman MF: ACTH levels after the dexamethasone suppression test in depression. *N Engl J Med* 1982; 306: 238–239.

Risch SC: Beta-endorphin hypersecretion in depression: Possible cholinergic mechanisms. *Biol Psychiat* 1982; 17: 1071–1079.

Rubin RT, Poland RE, Lesser IM, Martin DJ, Blodgett ALN, Winston RA: Neuroendocrine aspects of primary endogenous depression III. Cortisol secretion in relation to diagnosis and symptom patterns. *Psychol Med* 1987; 17: 609–619.

Sachar EO, Hellman L, Roffward H, Halpern F, Fukushima D, Gallagher D: Disrupted 24-hour patterns of cortisol secretion in psychotic depression. *Arch Gen Psychiat* 1973; 28: 19.

Sapolsky RM: Glucocorticoid toxicity in the hippocampus: Temporal aspects of synergy with kainic acid. *Neuroendocrinology* 1986; 43: 440–444.

Sapolsky R, Krey L, McEwen B: The adrenocortical stress-response in the aged male rat: Impairment of recovery from stress. *Exp Gerontol* 1983a; 18: 55–64.

Sapolsky RM, Krey LC, McEwen BS: Corticosterone receptors decline in a site-specific manner in the aged rat brain. *Brain Res* 1983b; 289: 235–240.

Sapolsky RM, Krey LC, McEwen BS: Glucocorticoid-sensitive hippocampal neurons are involved in terminating the adrenocortical stress response. *Proc Natl Acad Sci* 1984; 81: 6174–6177.

Schlesser MA, Winokur G, Sherman BM: Hypothalamic-pituitary-adrenal axis activity in depressive illness: Its relationship to classification. *Arch Gen Psychiat* 1980; 37: 737–743.

Schweitzer I, Maguire KP, Gee AH, Tiller JWG, Biddle N, & Davies B: Prediction of outcome in depressed patients by weekly monitoring with the dexamethasone suppression test. *Brit J Psychiat* 1987; 151: 780–784.

Selye H: A syndrome produced by diverse noxious agents. *Nature* 1936; 138: 32.

Stokes PE, Pick GR, Stoll PM et al.: Pituitary-adrenal function in depressed patients: resistance to dexamethasone suppression. *J Psychiat Res* 1975; 12: 271–281.

Stokes PE, Sikes CR, Stoll PM: Adrenocortical responsiveness in major depression. *Annual Meeting of Society of Biological Psychiatry.* Program Abstracts No. 103. Montreal, Quebec, 1988.

Tiongco DD, Hariharan M, Haskett RF, Greden JF: Age effects on HPA regulation and dexamethasone plasma levels in normal subjects. In *Annual Meeting of the Society of Biological Psychiatry,* Program Abstracts (pp. 195). Washington, DC, 1986.

Vale W, Spiess J, Rivier C, Rivier J: Characterization of a 41-residue ovine hypothalamic peptide that stimulates secretion of corticotropin and β-endorphin. *Science* 1981; 213: 1394–1397.

Young EA, Akil H: Corticotropin-releasing factor stimulation of adrenocorticotropin and β-endorphin release: Effects of acute and chronic stress. *Endocrin* 1985; 117: 23–30.

Young EA, Watson JS, Kotun J, Haskett RF, Greden JF, Grunhaus L, Murphy-Weinberg V, Vale W, Rivier J, Akil H: β-LPH/β-Endorphin response to low-dose oCRH in endogenous depression: Preliminary studies. *Arch Gen Psychiat* 1990; 47: 449–457.

Young EA, Haskett RF, Murphy-Weinberg V, Watson SJ, Akil H: Loss of glucocorticoid fast feedback in depression. *Arch Gen Psychiat* 1991; 48: 693–699.

Young EA, Kotun J, Haskett RF, Grunhaus L, Greden JF, Watson SJ, & Akil H: Dissociation between pituitary and adrenal suppression to dexamethasone in depression. *Arch Gen Psychiat* 1992 (in press).

Zis K, Zis A: Increased adrenal weight in victims of violent suicide. *Am J Psychiat* 1987; 144: 1214–1216.

Chapter 9

THE MOLECULAR GENETICS OF BIPOLAR DISEASE

Wade H. Berrettini

INTRODUCTION

The development of molecular genetic techniques and relatively complete linkage maps of the human genome has led to the discovery of causative genes for inherited diseases of uncertain pathophysiology, including cystic fibrosis (Riordan et al., 1990), retinoblastoma (Friend et al., 1986), and familial osteoarthritis (Alla-Kokko et al., 1990). The chromosomal locations of genes for numerous other inherited diseases of unknown biochemical etiology have been defined, although the precise genes have not been discovered yet (for review see Human Gene Mapping 10, 1989).

The chromosomal locations of disease genes have been determined using restriction fragment length polymorphism (RFLP) DNA markers and linkage analysis. Although other means of detecting variation in DNA sequence may supplant the RFLP methods in the near future, to date these methods have been widely used with much success. (For a description of the RFLP methodologies, see Gershon et al., 1987.)

This chapter will review some concepts of genetic linkage and the difficulties in conducting linkage studies of complex diseases such as BP illness. Then bipolar (BP) linkage studies will be reviewed, with a clos-

Wade H. Berrettini • Department of Psychiatry and Human Behavior, Jefferson Medical College, Philadelphia, Pennsylvania 19107.

Biology of Depressive Disorders, Part A: A Systems Perspective, edited by J. John Mann and David J. Kupfer. Plenum Press, New York, 1993.

ing commentary on future possibilities. Linkage studies of unipolar illness will not be covered here.

Linkage Concepts

To appreciate how DNA markers have provided new opportunities to understand inherited diseases, a brief review of linkage concepts will be useful. Heritable characteristics (traits) are determined by *genes,* discrete DNA sequences which code for proteins synthesized by the cell's machinery. Genes are arranged in a linear manner on chromosomes. Each individual has a *homologous* pair of every chromosome (except the sex chromosomes), one of maternal origin and one of paternal origin. Thus, for each gene (or genetic locus) there are two copies, maternal and paternal *alleles* (with the exception of those genes found on unpaired sex chromosomes). Alleles for genetic loci that are close to one another on the same chromosome are frequently inherited together. Such loci are said to be linked. During meiosis, the process of *recombination* occurs between homologous pairs of chromosomes (see Fig. 1), so that alleles that are linked are not always inherited together. The greater the distance between two loci, the more frequently recombination will occur between them, and the two alleles will be inherited together only in a random fashion (that is, 50% of the time). The distance between two loci is thus a function of how often the alleles are inherited together. The distance between two loci can be expressed as the recombination fraction, *theta,* the percent of time that recombination between the two loci is observed. If two alleles are inherited together 50% of the time, the two loci are not linked, and $\theta = 0.5$. Two alleles that are inherited together 99 times out of 100 possible meioses are said to be tightly linked ($\theta = 0.01$). One percent θ is roughly equivalent to one million base pairs of DNA, although there are chromosomal regions (hot spots for recombination) where one million base pairs seem to represent a much larger value for θ.

The lod score method (Morton, 1955) can be used to test whether two loci are linked (whether they are close together on the same chromosome). The data for linkage analysis are the observations of the segregation of alleles at the two loci in families. For disease investigations, one locus is a putative illness locus and the other is the marker locus. A lod score compares the probability that the observed segregation of marker and illness in the family would occur in the presence of linkage between the illness gene and the marker versus the probability

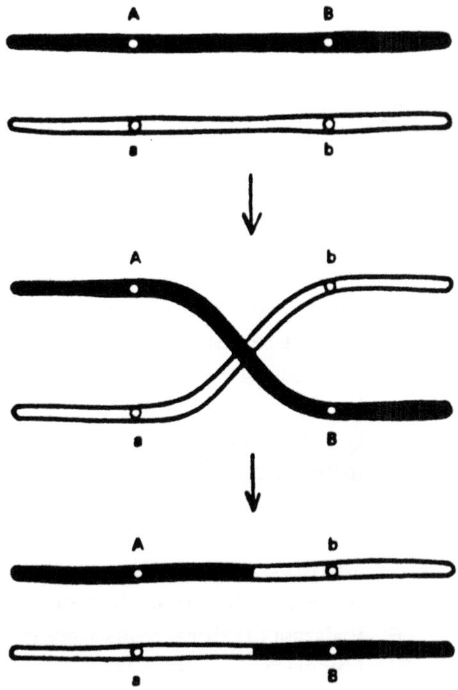

Figure 1. A diagrammatic representation of recombination (crossing over) is shown. Two genetic loci, A and B (with alleles Aa and Bb) are shown. Crossing over between the homologous chromosomes results in recombination between locus A and locus B.

that the observed segregation would occur in the absence of linkage (absence of linkage is $\theta = 0.5$). In the form of an equation we have

$$\text{lod score} = \log_{10} [(P\{\theta < 0.5\})/(P\{\theta = 0.5\})]$$

where P is the probability of observing the segregation of the alleles at the indicated value of theta. In practice, lod scores are calculated for several values of theta between 0 and 0.5, and the value of theta yielding the highest lod score is assumed to be a correct estimate of the distance between the two loci. If the lod score is 3.0 (linkage is 1000 times more likely than nonlinkage), this is considered suggestive evidence for linkage. This should not be confused with a P value of 0.001. A lod score of 3.0 may occur approximately 1 in 20 observations by chance, a P value of 0.05 (Lander, 1988). A lod score more negative than –2.0 is considered as evidence against linkage, and scores between –2.0 and 3.0 indicate a need for more data before a conclusion can be reached. If the sample consists of more than one family, the lod scores for all families can be added to provide a total lod score. For an extensive discussion of linkage see Ott (1985).

The initial successes in localizing causative genes have come with classical Mendelian illnesses such as cystic fibrosis (Riordan et al., 1990). The application of these techniques to complex (non-Mendelian) diseases raises several theoretical and practical difficulties, which develop out of the following characteristics of complex diseases:

1. Reduced penetrance (not everyone who inherits the disease-causing gene develops the illness because of protective genetic or environmental effects);
2. Variable age of onset (a given person may not develop the illness until quite late in life, so that the affection status of a younger person remains uncertain);
3. Phenocopies (individuals without a disease genotype manifest the syndrome from nongenetic causes);
4. Genetic heterogeneity (single mutations at different genetic loci independently may cause clinically indistinguishable disease forms);
5. Polygenic inheritance (individuals may develop the disease from additive effects of several genes, no single one of which may be necessary to cause disease);
6. Epistasis (interactions between alleles at a few loci);
7. Uncertain mode of transmission (segregation analysis is not consistent with a recessive or dominant illness).

The prospects for discovery of causative genes for complex diseases through the application of molecular genetic techniques have been the subject of several analyses (Risch, 1990; Martinez & Golden, 1990; Ott, 1990; Ploughman & Boehnke, 1989). The recent successes with complex illnesses such as Alzheimer's disease (St. George-Hyslop et al., 1987; Van Broeckhoven et al., 1988; Goate et al., 1989) and osteoarthritis (Knowlton et al., 1990) suggest that the difficulties noted above are not insurmountable.

The foregoing discussion is important because psychiatric disorders in general (and BP illness in particular) are complex diseases. In addition to the seven characteristics noted above, BP disease has two other attributes which raise additional problems in conducting molecular genetic searches for causative genes. These include limited (Rice, 1987) diagnostic reliability (even with the most rigorous operationalized criteria) for some affective syndromes (in which a person may not provide reliable history over time) and the uncertain diagnostic boundaries of the inherited spectrum of bipolar disorders. That is, which diagnoses in addition to BPI represent the true inherited spectrum of affective disorders in a given family?

It is worth mentioning the manner in which these difficulties can be addressed in genetic searches for causative genes for BP disease. For example, the unknown penetrance value and mode of transmission can be addressed by analyzing data under several possible penetrance values and modes of transmission. It has been shown that the lod score method is robust to model misspecification, unless gross errors are made (Clerget-Darpoux et al., 1986). Alternatively, data can be collected using affected sibling pair methods (Suarez, 1978) which require no assumptions about the mode of transmission, but are less powerful than the lod score method (Goldstein et al., 1989).

The problems of genetic heterogeneity can be reduced by studying either (1) a single very large family with only one known source of illness, so that strong evidence for linkage can be developed from a single family; or (2) a sufficient number of moderate-sized families with adequate power to detect linkage when only a fraction of the families are linked to any given locus. The power to detect linkage, in the presence of several degrees of heterogeneity, using various small-to-moderate size family structures, has been determined through simulation research with assumptions about the genetic diseases model that may be reasonable for BP illness (Martinez & Golden, 1990). Because so many of the parameters regarding BP disease remain unknown, it is impossible to determine which approach is preferable.

The difficulty imposed by variable age of onset is addressed by adjusting penetrance for age. Thus, a well person in his early 20's provides very little information because the age of risk for BP disease extends into the middle years. Penetrance is maximum for a person in his 60's, as onset of new BP illness in the seventh decade of life is quite unusual and should prompt a search for an organic cause.

The problems of phenocopies and uncertain genetic spectrum for BP disease can be addressed by analyzing the linkage data under a very restrictive affective status model. Assuming that phenocopies occur more frequently for milder forms of the disease, analyzing data with a restrictive affection status model (only BPI for example) will reduce the risk of assigning a person to the ill category when that person does not have the illness gene (a severe error in a linkage study). Because BP disease is clearly characterized by reduced penetrance, within any linked family there will be a number of individuals with the disease gene who remain unaffected. Unfortunately, under reduced penetrance well persons (even when elderly) do not contribute as much to the lod score as do ill persons. Thus, analyzing data under a restrictive model of illness does reduce the amount of linkage information. Investigators also analyze linkage data under broader affection status models to increase linkage information. For the broader models of affection status, data from controlled family studies of BP disease (Gershon et al., 1982) can be used to determine what diagnoses occur with higher frequencies among the first-degree relatives of BP probands. These family studies form a rational foundation for broader affection status models. Thus, if a restrictive affection status model is limited to BPI cases, a broader model may include BPII, unipolar and schizoaffective cases as well. If these diagnoses represent accurately the inherited spectrum of BP illness, then a lod score that is positive with the restrictive affection status model should increase with the broader model, as long as phenocopies do not interfere.

While there may be reasonable solutions to the many problems of a BP disease linkage study, current methods will not allow detection of linkage in the *absence* of one or two *necessary* (but perhaps not sufficient) disease genes within a given family. Even if such necessary disease genes exist within a family, should the illness be very genetically heterogeneous, confirmation of a reported linkage may be quite difficult unless one genetic form of the disease is relatively common.

Bipolar Linkage Studies—X Chromosome

Transmission of BP disease through a gene on the X-chromosome was first proposed by Rosanoff et al. (1935) over 50 years ago. Several decades later Winokur, Clayton, and Reich (1969) reported evidence for X-linkage using genetic marker data as well as family study results. They found a higher morbid risk for female (compared to male) first-degree relatives of BP probands. A lack of apparent male-to-male transmission was also consistent with a dominant X-linked disease. Additionally, they described two BP pedigrees in which color blindness (CB) segregated with affective disorder. Subsequently, many studies of X-linkage in BP illness have been published (see Table 1).

Several years later, Mendlewicz and Fleiss (1974) reported close linkage (lod score = 8.0 at $\theta = 7\%$) of protan and deutan CB to BP illness in 17 pedigrees. They also reported loose linkage ($\theta = 19\%$) of the Xg blood group to BP disease in 23 pedigrees. At the time of their report, these two markers were thought to be loosely linked, so these findings were thought to be internally consistent. It is now known that CB and Xg are at opposite ends of the X-chromosome, making the two findings inconsistent, unless there are separate genes for BP illness on each arm of the X chromosome. Gershon (1991) has analyzed the marker data from these studies, finding a maximum lod score of 3.5 at 4% recombination for CB and Xg, a result that is clearly inconsistent with present

Table 1. Xq28 Linkage Studies of Bipolar Disease[a]

Reference	Marker	θ	Lod Score
Winokur et al. (1969)	Color blindness	0.0	3.3
Mendlewicz and Fleiss (1974)	Color blindness	0.07	8.0
Baron (1977)	Color blindness	0.10	2.3
Mendlewicz et al. (1979)	Color blindness	0.15	1.6[b]
Gershon et al. (1979)	Color blindness	0.0	–6.0
Mendlewicz et al. (1980)	G6PD deficiency	0.06	4.0
Del Zompo et al. (1984)	G6PD deficiency	0.0	1.5
Baron et al. (1987)	CB and G6PD def	0.0	9.2[b]
Mendlewicz et al. (1987)	Factor IX gene	0.11	3.2
Berrettini et al. (1990)	DXS15, DXS52 and F8	0.0	–17
Gejman et al. (1990)	Factor IX	0.0	–9.2
Neiswanger et al. (1990)	DXS52	0.0	–5.5

[a] Published linkage studies of BP disease are listed with the maximum lod score at the relevant value of theta (recombination fraction). Abbreviations are CB, color blindness; F8, factor VIII gene; and G6PD, glucose-6-phosphate dehydrogenase. For negative studies the lod score at the marker used is given.
[b] Heterogeneity detected.

genetic map knowledge. This internal inconsistency, which may have occurred by chance, does raise doubts about the validity of this paper.

Numerous positive and negative linkage studies (see Table 1) of BP pedigrees and genetic markers for this region (Xq28) of the X-chromosome (see Fig. 2 for a genetic map of the Xq28 region) have appeared in the literature over the past ten years (Baron, 1977; Mendlewicz et al., 1979; Gershon et al., 1979; Mendlewicz et al., 1980; Del Zompo et al., 1984; Baron et al., 1987; Mendlewicz et al., 1987; Berrettini et al., 1990; Gejman et al., 1990). Some authors (Mendlewicz et al., 1974; Risch & Baron, 1982; Risch et al., 1986) have suggested that genetic heterogeneity accounts for the negative linkage data and the considerable frequency of apparent male-to-male transmission of illness in BP families.

Some attempts to estimate the fraction of BP illness that is Xq28-linked (estimated at 33% of all BP persons or 40%–90% of BP families without apparent male-to-male transmission) seem to exaggerate this proportion (Risch & Baron, 1982; Risch et al., 1986), especially when one considers how very difficult it is to find multiplex BP families with no apparent male-to-male transmission. It should be noted that several groups (personal communications from Hugh Gurling, John Nurnberger and Bill Byerley) conducting linkage studies of BP disease have had *very limited* success in finding such families despite intensive searches that have yielded more than a dozen large multiplex families with apparent male-to-male transmission.

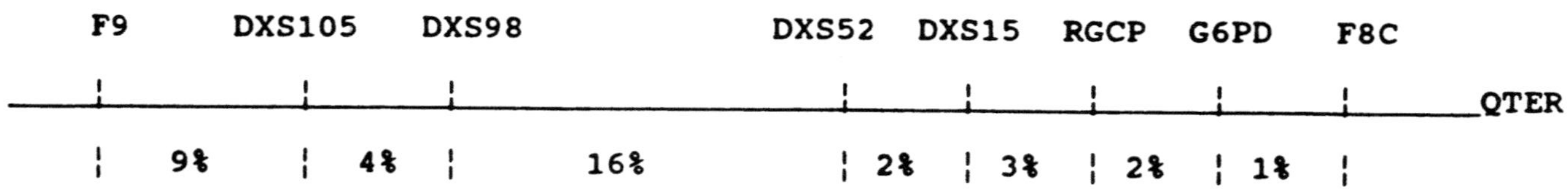

Figure 2. The genetic map of the Xq28 region is redrawn from Berrettini et al. (1990) and is based in part upon Patterson et al. (1987) and Oberle et al. (1985). Abbreviations: F9, Factor IX; F8C, Factor VIII; G6PD, glucose-6-phosphate dehydrogenase; RGCP, red cone pigment and green cone pigment (which constitute the color blindness locus); %, recombination fraction. The DXS numbers refer to anonymous polymorphic loci.

A more recent study (Baron et al., 1987) confirmed that BP illness is linked to Xq28 markers in four Sephardic Israeli pedigrees (lod score = 9.2 at $\theta = 0\%$). The authors noted that the one Ashkenazi family they studied was not linked, suggesting genetic heterogeneity. This study is particularly noteworthy for the robust positive lod scores in the four linked families and for the absence of recombinants, giving rise to the maximum lod score at 0% θ.

All the CB and G6PD studies in which the LOD score exceeds 3.0 (the generally accepted criterion for evidence of linkage) indicate that the disease locus is within 5% recombination of the CB locus (see Table 1 and Fig. 2). The study of Mendlewicz et al. (1987) reporting loose linkage ($\theta = 11\%$) of BP disease to the Factor IX locus may be inconsistent with the previous reports of tight linkage to the CB and G6PD loci (see Fig. 2). The recombination fraction between Factor IX and the CB/G6PD is approximately 30%–35% (Patterson et al., 1987; Oberle et al., 1985). Thus, if the linkage to the Factor IX locus represents the same hypothesized disease gene as the linkages to CB and G6PD loci, then there is an inconsistency of approximately 20% recombination between these positive studies according to the known genetic linkage map of the region (see Fig. 2).

Because the CB and G6PD region is very well mapped, with several highly polymorphic DNA markers (DXS52 [Oberle et al., 1985] and the red and green cone pigment genes [Nathans et al., 1986], for example) within 5% recombination of the CB locus, all BP families without apparent male-to-male transmission can provide linkage data to resolve this issue. If a small proportion (less than 25%) of BP families without male-to-male transmission is linked to Xq28 markers, then a large series (more than 20) of such families (with at least six affected persons per pedigree) is needed to confirm this finding (Martinez et al., 1990). The systematic development of such a large series will require a considerable multicenter effort, such as the NIMH Molecular Genetics Initiative. Should linkage be confirmed, the recent mapping of a GABA receptor subunit gene to this region (Bell et al., 1989) would provide a potentially interesting candidate gene.

Bipolar Linkage Studies—Chromosome 11P15

The report of Egeland et al. (1987) that BP disease was linked to two chromosome 11p15 markers (insulin and Harvey ras oncogene, which are separated by 5% recombination) in the Old Order Amish (OOA) generated considerable excitement, particularly because the tyrosine hydroxylase gene (the enzyme that catalyzes the rate-limiting step in the synthesis of norepinephrine—a logical candidate gene) was localized to this region (O'Malley & Rotwein, 1988). The initial enthusiasm was dampened by several papers (Detera-Wadleigh et al., 1987; Hodgkinson et al., 1987; Byerley & Mellon, 1989) in which linkage was excluded for these markers in other BP pedigrees.

Several years later, Kelsoe et al. (1990) published their reanalysis of the OOA study, in which they confirmed, in a new set of genotypes, the original findings. However, in the intervening years there were two new onsets of affective disorder in the original pedigree. These two new cases reduced the lod score in the original pedigree to less than 3.0. Two new branches of the original pedigree were also defined in the intervening years. These new branches, with multiple affected members, yielded highly negative lod scores for the 11p15 markers that were thought to be linked in the original pedigree. Thus, there are three sets of results, which together substantially reduce the possibility of linkage to the 11p15 region:

1. Several reports excluding numerous other BP pedigrees without a single confirmation by other investigators (Detera-Wadleigh et al., 1987; Hodgkinson et al., 1987; Byerley & Mellon, 1989; Gill et al., 1988; Neiswanger et al., 1990);
2. New onset of disease reduces the lod score of the original pedigree to less than 3.0 (Kelsoe et al., 1990);
3. Exclusion of linkage to 11p15 markers in two new extensions of the original pedigree (Kelsoe et al., 1990).

The original finding had a lod score greater than 4.5, indicating that the probability of this being a chance observation is less than 1 in 200 (O'Malley & Rotwein, 1988). The original genotypes were confirmed by the later study, suggesting that laboratory error is not a likely explanation for the original positive lod score (Kelsoe et al., 1990).

A chance observation may be the best explanation for this unlikely series of events. One point in favor of this explanation is that the OOA study employed multiple markers without a correction. The authors had studied approximately 80 markers before they tested the 11p15 probes. As Ott (1985) has suggested, when the number of markers approaches 100, some correction in the threshold for declaring linkage should be made because of the effect of testing multiple markers. The exact nature of such a correction is difficult to specify, because not all markers represent independent tests of different hypotheses. For example, testing for linkage with Harvey ras oncogene and insulin probes (which are closely linked) does not constitute two independent tests.

A key issue in the 11p15 results is the absence of confirmation in other pedigrees by independent investigators. It remains possible that a rare genetic cause of BP illness may be found in a population isolate that will not be readily evident in a more general gene pool.

Bipolar Linkage Studies—Other Regions

Numerous BP linkage studies have been published (for review see Gershon et al., 1989) using classical markers such as the ABO, Rh, HLA, and Duffy polymorphic loci. Evidence for linkage has not been consistent, in part because these loci (with the exception of the HLA locus) are not highly polymorphic, and lod scores are generally not robust. More recent studies (Cox et al., 1989) are not consistent with close linkage to these markers.

We have conducted systematic linkage mapping on chromosomes 5, 10, 11, 15, and 17 with DNA samples from 15 bipolar pedigrees, consisting of 240 persons, of whom 100 are diagnosed as schizoaffective, BP or recurrent unipolar. Using these diagnoses as an affection status model, linkage (assuming a dominant, 90% penetrance disease model) to 125 DNA markers could be excluded (Berrettini et al., 1991). While there are small regions of these chromosomes where (for lack of sufficiently polymorphic probes) exclusion of linkage is not robust (lod scores more positive than –5), in general our evidence is not consistent with a common genetic cause of BP disease located on these chromosomes. Simulation research indicates that, if 50% of these families were linked to one of the regions studied, linkage would have been detected (Martinez et al., 1990).

A balanced translocation of the end of the long arm of chromosome 11 to the tip of chromosome 9 has been described in a family with BP disease (Smith et al., 1989). The translocation breakpoint is located at 11q22-23. This report suggests that there may be a gene capable of causing BP illness at or near the breakpoint. The translocation produces a dysfunctional gene, giving rise to BP disease. Linkage analysis in our 15 BP families to probes in this region does not provide support for this hypothesis (Berrettini et al., 1991).

Discussion

The reports of BP linkage to two major gene regions (Xq28 and 11p15) remain controversial. The evidence for linkage to 11p15 markers is much less compelling presently, and in the absence of additional linked pedigrees, it will cease to be an actively investigated area of the genome. The question of Xq28 linkage is much more difficult to assess, because of several positive and negative reports from independent investigators (see Table 1). Although there may be inconsistencies with the Xq28 genetic map raising questions about two (Mendlewicz & Fleiss, 1974; Mendlewicz et al., 1987) of the positive reports, there are several other positive papers which suggest that additional pedigrees must be found. Our own results from BP pedigrees without male-to-male transmission suggest that, if Xq28 linkage exists, it does not account for more than 25% of such pedigrees (Berrettini et al., 1990).

Although the field of linkage in psychiatry has seen a series of conflicting reports in its brief lifetime, the prospects for discovery of single major genes for BP disease are excellent. For the first time, the tools are at hand to completely screen the genome for large numbers of multiplex families, using markers for which most families are sufficiently informative to draw conclusions about linkage for the region in question. Given this scenario, if there are single major genes to be discovered in BP disease, it seems likely that at least one will be defined within the next decade.

The considerable clinical heterogeneity of the syndrome may be an indicator of extensive genetic heterogeneity and multifactorial inheritance making linkage difficult to detect. However, clinical heterogeneity does not always indicate genetic heterogeneity. For example Becker and Duchenne muscular dystrophy are quite different clinical syndromes which arise from varying mutations within the same gene (Kunkel et al., 1986).

The failure of segregation analysis to yield unequivocal evidence for a single major locus in BP disease should not deter us from this search. It should be remembered that the multiplex BP families in linkage studies are by no means typical of BP disease. They are remarkable for their high density of illness. Thus, these families may represent a subtype of highly heritable BP disease which may be more genetically homogeneous than the broader spectrum of BP illness on which segregation analysis is based.

In conclusion, these are exciting times for psychiatric geneticists studying BP disease. Large multiplex BP families are needed to take maximum advantage of the molecular biological advances which should allow us to unravel some of the mystery of mood disorder.

References

Ala-Kokko L, Baldwin CT, Moskowitz RW, Prockop DJ: Single base mutation in the type II procollagen gene (COL2A1) as a cause of primary osteoarthritis associated with a mild chrondrodysplasia. *Proc Nat Acad Sci (USA)* 1990; 87: 6565–6568.

Baron M: Linkage between an X-chromosome marker (deutan color blindness) and bipolar affective illness. *Arch Gen Psychiat* 1977; 34: 721–725.

Baron M, Risch N, Hamburger R, Mandel B, Kushner S, Newman M, Drumer D, Belmaker RH: Genetic linkage between X-chromosome markers and bipolar effective illness. *Nature* 1987; 326: 289–292.

Bell MV, Bloomfield J, McKinley M, Patterson MN, Darlison MG, Barnard EA, Davies KE: Physical linkage of a GABA receptor subunit gene to the DXS374 locus in human Xq28. *Am J Human Genet* 1989; 45: 883–888.

Berrettini WH, Goldin LR, Gelernter J, Gejman PV, Gershon ES, Detera-Wadleigh S: X-chromosome markers and manic-depressive illness: Rejection of linkage to Xq28 in nine bipolar pedigrees. *Arch Gen Psychiat* 1990; 47: 366–373.

Berrettini WH: Is investment in molecular genetics worthwhile? *Biol Psychiat* 1991; 30(3): 213–215.

Berrettini WH, Detera-Wadleigh SD, Goldin LR, Martinez M, Hsieh W-T, Hoehe M, Choi H, Muniec D, Coffman D, Ferraro TN, Guroff JJ, Kazuba D, Harris N, Kron E, Nurnberger Jr JI, Alexander R, Gershon ES: Genomic screening for genes predisposing to bipolar disease. *Psychiat Gen* 1991; 2: 41–56.

Byerley W, Mellon C: Mapping genes for manic-depression and schizophrenia with DNA markers. *Trends Neurol Sci* 1989; 12: 46–48.

Clerget-Carpoux F, Bonaiti-Pellie C, Hochez J: Effects of misspecifying genetic parameters in lod score analysis. *Biometrics* 1986; 42: 393–399.

Cox N, Reich R, Rice T, Elston R, Schober J, Keats B: Segregation and linkage analyses of bipolar and major depressive illnesses in multigenerational pedigrees. *J Psychiat Res* 1989; 23: 109–123.

Del Zompo M, Bochetta A, Goldin LR: Linkage between X-chromosome markers and manic-depressive illness: Two Sardinian pedigrees. *Acta Psychiat Scand* 1984; 70: 282–287.

Del Zompo M, Pedditzi M, Ruiu S, Goldin LR, Berrettini WH, Boccheta A:. Association and linkage studies of affective disorders. In: Racagni G, Brunello N, Fukuda T (Eds.), *Biological Psychiatry (Vol. 2). Proceedings of the 5th World Congress of Biological Psychiatry,* Florence, June 9–14, 1991. New York, Elsevier Science Publishing Co., 1991, pp 449–451.

Detera-Wadleigh SD, Berrettini WH, Goldin LR, Boorman D, Anderson S, Gershon ES: Close linkage of c-Harvey-ras-1 and the insulin gene to affective disorder is ruled out in three North American pedigrees. *Nature* 1987; 325: 806–808.

Egeland JA, Gerhard DS, Pauls DL, Sussex JN, Kidd KK, Allen CR, Hostetter AM, Housman DE: Bipolar affective disorder linked to DNA markers on chromosome 11. *Nature* 1987; 325: 783–787.

Friend SH, Bernards R, Rogelj S, Weinberg RA, Rapaport JM, Albert DM, Dryja TP: A human DNA segment with properties of the gene that predisposes to retinoblastoma and osteosarcoma. *Nature* 1986; 323: 643–646.

Gejman PV, Detera-Wadleigh S, Martinez MM, Berrettini WH, Goldin LR,

Gelernter J, Hsieh WT, Gershon ES: Manic-depressive illness not linked to Factor IX in an independent series of pedigrees. *Genomics* 1990; 8: 648–655.

Gershon ES, Targum SD, Matthysee S, Bunney Jr, WE: Color blindness not closely linked to bipolar illness. *Arch Gen Psychiat* 1979; 36: 1423–1430.

Gershon ES, Hamovit J, Guroff JJ, Dibble E, Leckman JF, Sceery W, Targum SD, Nurnberger Jr, JI, Goldin LR, Bunney Jr, WE: A family study of schizoaffective, bipolar I, bipolar II, unipolar and normal control probands. *Arch Gen Psychiat* 1982; 39: 1157–1162.

Gershon ES, Merrill CR, Goldin LR, DeLisi LE , Berrettini WH, Nurnberger Jr, JI: The role of molecular genetics in psychiatry. *Biol Psychiat* 1987; 22: 1388–1405.

Gershon ES, Berrettini WH, Nurnberger Jr, JI, Goldin LR: Genetic studies of affective illness. In: Mann JJ (Ed.), *Models of Depressive Disorders,* New York: Plenum Press, 1989; 109–142.

Gershon ES: Marker genotyping errors in old data on X-linkage in bipolar illness. *Biol Psychiat* 1991; 29: 721–729.

Gill M, McKeon P, Humphries P: Linkage analysis of manic-depression in an Irish family using Hpras 1 and INS DNA markers. *J Med Genet* 1988; 25: 634–637.

Goate AM, Haynes AR, Owen MJ: Predisposing locus for AD on chromosome 21. *Lancet* 1989; i: 352–355.

Goldstein AM, Weeks DE, Cortessis V, Haile RW: Comparison of the affected pedigree member and lod score methods. In: Elston RC, Spence A, Hodge SE, MacCluer J (Eds.), *Multipoint Mapping and Linkage Based Upon Affected Pedigree Members: Genetic Analysis Workshop 6,* New York: Alan R. Liss, 1989, 135–140.

Hodgkinson S, Sherrington R, Gurling HMD, Marchbanks RM, Reeders ST, Mallet J, Petursson H, Brynjolfsson J: Molecular evidence for heterogeneity in manic-depression. *Nature* 1987; 325: 805–807.

Human Gene Mapping 10: Tenth international workshop on human gene mapping (1989). *Cytogenet Cell Genetics* 1989; 51: 1–1147.

Kelsoe JR, Ginns EI, Egeland JA, Gerhard DS, Goldstein AM, Bale SJ, Pauls DL, Long RT, Kidd KK, Conte G, Housman DE, Paul SM: Reevaluation of the linkage relationship between chromosome 11p loci and the gene for bipolar affective disorder in the Old Order Amish. *Nature* 1990; 342: 238–243.

Knowleton RG, Katzenstein PL, Moskowitz RW, Weaver EJ, Malemud CJ, Pathria MN, Jimenez SA, Prockop DJ: Genetic linkage of a polymorphism in the type II procollagen gene (COL2A1) to primary osteoarthritis associated with mild chondrodysplasia. *New Engl J Med* 1990; 322: 525–530.

Kunkel LM, Hejtmancik JF, Kaskey CT: Analysis of deletions in DNA from patients with Becker and Duchenne muscular dystrophy. *Nature* 1986; 322: 73–77.

Lander ES: Mapping complex genetic traits in humans. In: Davies KE (Ed.), *Genome Analysis: A Practice Approach.* Washington, D.C.: IRL Press, 1988; 171–189.

Martinez MM, Goldin LR: Power of the linkage test for a heterogeneous disorder due to two independent inherited causes. *Genet Epidemiol* 1990; 7: 219–230.

Mendlewicz J, Fleiss JL: Linkage studies with X-chromosome markers in bipolar (manic-depressive) and unipolar (depressive) illness. *Biol Psychiat* 1974; 9: 261–294.

Mendlewicz J, Linkowski P, Guroff JJ, Van Praag HM: Color blindness linkage to manic-depressive illness. *Arch Gen Psychiat* 1979; 36: 1442–1447.

Mendlewicz J, Linkowski P, Wilmotte J: Linkage between glucose-6-phosphate dehydrogenase deficiency and manic-depressive psychosis. *Brit J Psychiat* 1980; 137: 337–342.

Mendlewicz J, Simon P, Sevy S, Charon F, Brocas H, Legros S, Vassart G: Polymorphic DNA marker on X-chromosome and manic-depression. *Lancet* 1987; i: 1230–1232.

Morton NE: Sequential tests for the detection of linkage. *Am J Human Genet* 1955; 7: 277–318.

Nathans J, Plantanida TP, Eddy RL, Shows TB, Hogness DS: Molecular genetics of inherited variation in color vision. *Science* 1986; 232: 203–210.

Neiswanger K, Slaugenhaupt SA, Sughes HB, Frank E, Frankel DR, McCarty MJ, Charkravarti A, Zubenko GS, Kupfer DJ, Kaplan BB: Evidence against close linkage of unipolar affective illness to human chromosome 11p markers HRAS1 and INS and chromosome Xq marker DXS52. *Biol Psychiat* 1990; 28: 63–72.

Oberle I, Drayna D, Camerino G, White R, Mandel JL: The telomeric region of the human X-chromosome long arm: Presence of a highly polymorphic DNA marker and analysis of recombination frequency. *Proc Nat Acad Sci USA* 1985; 82: 2824–2828.

O'Malley KL, Rotwein P: Human tyrosine hydroxylase and insulin genes are contiguous on chromosome 11. *Nucleic Acids Res* 1988; 16: 4437–4446.

Ott J: *Analysis of Human Genetic Linkage*. Baltimore: John Hopkins University Press, 1985.

Ott J: Invited editorial: Cutting the Gordian knot in the linkage analysis of complex human traits. *Am J Human Genet* 1990; 46: 219–221.

Patterson M , Kenwrick S, Thibodeau S, Faulk K, Mattel MG, Mattel JF, Davies KE: Mapping of DNA markers close to the fragile site on the human X-chromosome at Xq27.3. *Nucleic Acids Res* 1987; 15: 2639–2651.

Ploughman LM, Boehnke M: Estimating the power of a proposed linkage study for a complex genetic trait. *Am J Human Genet* 1989; 44: 453–451.

Rice J: The estimation of diagnostic sensitivity using stability data: An application to major depressive disorder. *J Psychiat Res* 1987; 21: 337–345.

Riordan JR, Rommens JM, Kerem BS, Alon N, Rozmahel R, Grzelczak Z, Zielenski J, Lok S, Plavsic N, Chou JL, Drumm ML, Iannuzzi MC, Collins FS, Tsue LC: Identification of the cystic fibrosis gene: Cloning and characterization of complementary DNA. *Science* 1990; 245: 1066–1073.

Risch N, Baron M, Mendlewicz J: Assessing X-linked inheritance in bipolar-related major affective disorder. *J Psychiat Res* 1986; 20: 275–288.

Risch N, Baron M: X-linkage and genetic heterogeneity in bipolar related major affective illness: Re-analysis of linkage date. *Ann Human Genet* 1982; 46: 153–166.

Risch N: Genetic linkage and complex diseases with special reference to psychiatric disorders. *Genet Epidemiol* 1990; 7: 3–16.

Rosanoff AJ, Handy L, Plesset IR: The etiology of manic-depressive syndromes with special references to their occurrence in twins. *Am J Psychiat* 1935; 91: 725–740.

Smith M, Wasmuth J, McPherson JD, Wagner C, Grandy D, Civelli O, Potkin S, Litt M: Cosegregation of an 11q22.3-9p22 translocation with affective disorder: Proximity of the dopamine D2 receptor gene relative to the translocation breakpoint. *Am J Human Genet* 1989; 45(suppl): abstract 864.

St. George-Hyslop PH, Tanzi HE, Polinsky RJ, Haines JI, Nee T, Watkins PC, Myers RH, Feldman RG, Pollen D, Drachman D, Growden J, Bruni A, Foncin J-F, Salmon D, Frommett P, Amaducci L, Sorbi S, Piacentini S, Stewart GD, Hobbs WJ, Conneally PM, Gusella JF: The genetic defect causing familial Alzheimer's disease maps on chromosome 21. *Science* 1987; 235: 885.

Suarez BK: The affected sib-pair IBD distribution for HLA linked disease susceptibility loci. *Tissue Antigens* 1978; 12: 87–93.

Van Broeckhoven C, Van Hul W, Backhovens H, Van Camp G, Wehnert A, Stinissen P, Raeymaekers P, De Winter G, Gheuens J, Martin JJ, Vandenberghs A: The familial Alzheimer disease gene is located close to the centromere of chromosome 21. *Am J Human Genet* 1988; 43: a205.

Winokur G, Clayton PJ, Reich T: *Manic-Depressive Illness,* St. Louis: C.V. Mosby Co., 1969, 112–125.

Chapter 10

BRAIN IMAGING STUDIES OF DEPRESSIVE DISORDERS

Harold A. Sackeim and Isak Prohovnik

Introduction

During the last decade, our capacity to obtain in vivo measurement of brain structure and function developed rapidly. Indeed, technological advances in brain imaging often have outpaced their application to problems of psychiatry and neurology, leaving scores of issues to be addressed. This chapter provides an overview of what we have learned from initial studies of brain imaging in affective disorders.

Traditionally, brain imaging studies have focused on either abnormalities in structure or function, with different technologies being used to study each domain. Computerized tomography (CT) and magnetic resonance imaging (MRI) investigations have typically focused on problems of brain structure. Studies using the lateral projection xenon technique to assess regional Cerebral Blood Flow (rCBF), single photon emission computed tomography (SPECT), and positron emission tomography (PET) have concentrated on issues related to brain function. However, increasingly the association of imaging technologies with the distinction between structure and function is becoming blurred. Rapid advances in magnetic resonance spectroscopy (MRS) now allow for use of MRI technology to assess a variety of regional metabolic and bio-

Harold A. Sackeim and Isak Prohovnik • Departments of Biological Psychiatry and Brain Imaging, New York State Psychiatric Institute and Department of Psychiatry, Columbia University, New York, New York 10032.

Biology of Depressive Disorders, Part A: A Systems Perspective, edited by J. John Mann and David J. Kupfer. Plenum Press, New York, 1993.

chemical processes, including measurement of adenosine triphosphate, phosphocreatinine, *N*-acetyl aspartate, GABA, glycogen, lactate, triglycerides, lithium, fluorinated pharmacological agents, and other substances (Luyten & den Hollander, 1987; Keshavan, Kapur, & Pettegrew, 1991). Similarly, techniques traditionally associated with functional imaging may also reveal important anatomical information. For example, the 133-xenon inhalation technique, used to assess regional Cerebral Blood Flow (rCBF), yields a regional index of the amount of perfused gray matter relative to total perfused tissue (gray and white matter). This index has allowed for characterization of hemispheric asymmetry in the normal brain in the relative distributions of gray and white matter (Gur, Packer, Hungerbuhler, Reivich, Obrist, Amarneck, & Sackeim, 1980), as well as assessment of cortical atrophy in patients with Alzheimer's disease (Prohovnik, Smith, Sackeim, Mayeux, & Stern, 1989). Across the technologies used, this chapter outlines the central findings that have been obtained in brain imaging studies of the structural and functional abnormalities in mood disorders.

Abnormalities of Brain Structure

Much of the impetus for the study of structural abnormalities in mood disorders came from neuropathological and brain imaging studies in schizophrenia. A variety of investigations have suggested that subgroups of schizophrenic patients are characterized by enlarged area or volume of the lateral ventricles, by third ventricular enlargement, cortical and/or cerebellar atrophy (sulcal widening), and to a less certain extent, by reduced brain tissue density and disturbances in patterns of anatomic asymmetry (Johnstone, Crow, Frith, Husband, & Kreel, 1976; Weinberger, Torrey, Neophytides, Wyatt, 1979; Weinberger, DeLisi, Perman, Targum, & Wyatt, 1982; see Andreasen, 1989 and Coffman, 1989 for reviews). Despite the uncertain clinical and etiopathological significance of the structural abnormalities in schizophrenia, these findings have been of fundamental cultural and scientific importance. The demonstrations that schizophrenia is associated with abnormalities of brain structure underscored biological perspectives on this disorder in both public and scientific communities and has led to a renewed focus on neuropathological investigation (e.g., Benes & Bird, 1987; Jernigan, Zisook, Heaton, Moranville, Hesselink, & Braff, 1991). Originally, affective disorder patients were included in some brain imaging studies mainly to establish the specificity of findings with respect to schizophrenia (Weinberger et al., 1982). However, the repeated documentation of abnormal brain structure in mood disorders has resulted in this area becoming an independent field of investigation.

Lateral Ventricular Enlargement

The structural abnormality that has received the most investigation is the question of whether mood disorders are associated with lateral ventricular enlargement. Table 1 details the studies that used CT to assess ventricular size in mood disorder patients and at least one comparison or control group. In Table 1, the summary of major findings concerns only the mood disorder patients, and only the positive findings regarding correlates of ventricular size in these patients are presented. In most of these CT studies, the ventricular-to-brain ratio (VBR) was used. This was typically assessed using planimentry to compute the area of the lateral ventricles in the tomographic slice at their largest extent and dividing by the total brain area in the slice, with the result multiplied by 100 (Zatz & Jernigan, 1983).

It is evident that the CT studies have been inconsistent in addressing the basic question of whether mood disorder patients differ from controls in VBR size. While several CT studies have reported larger VBR in mood disorder patients relative to controls, this observation is far from universal, and negative findings were obtained in the several recent investigations (e.g., Dewan, Hadlipur, Lane, Ispahani, Boucher, & Major 1988a; Iacono, Smith, Moreau, Beiser, Fleming, & Lin, 1988; Lewine, Risch, Risby, Stipetic, Jewart, & Eccard, 1991; Schlegel & Kretzschmar, 1987). As Nasrallah et al. (1989) and Schlegel (1991) have commented, the source of these inconsistencies is unclear. Comparisons across the studies are difficult because of variation in the patient populations, the nature of the CT scanning, and in the quantification of ventricular size. It is also noteworthy that with few and contradictory exceptions (e.g., Rieder, Mann, Weinberger, van Kammen, & Post, 1983; Weinberger, DeLisi, Perman, Targum, & Wyatt, 1982), VBR differences between mood disorder and schizophrenic groups have not been observed.

In this context, some central points might be kept in mind. First, whether assessed as a continuous variable or when dichotomously characterized as normal or abnormal, VBR measures are intrinsically insensitive to group differences. Healthy normal controls show marked variability in VBR size, which may be partly under genetic control (Reveley, Reveley, & Murray, 1984; Schulsinger, Parnes, Peterson, Schulsinger, & Teasdale, 1984). Consequently, when VBR is treated as a continuous measure, the magnitude of effects must be large to overcome

Table 1. Computerized Tomography (CT) Studies: Ventricular-to-Brain Ratio (VBR) Values[a]

Study	Mood disorder patients		Normal controls		Schizophrenic patients		Other comparison groups		Major findings in mood disorder patients
	N	VBR	*N*	VBR	*N*	VBR	*N*	VBR	
Van Gall and Becker (1978)	37 UP	?	—	—	45	?	20 OP	?	Higher rate of abnormal VBR than schizophrenic or other psychiatric patients
Jacoby and Levy (1980); Jacoby et al. (1980, 1981)	39 UP, 2BP (manic) all over age 60	?	50 HC	14.2	—	—	40 DP	—	No overall VBR difference with elderly controls and smaller VBR than demented patients; subgroup of 9 mood disorder patients had enlarged VBR; old age, late onset, endogenous features, less anxiety and higher mortality at 2-year follow-up associated with abnormal VBR
Pearlson and Veroff (1981)	16 BP	6.5	—	—	22	7.5	35 OP	3.6	Larger VBR than other psychiatric comparison group
Nasrallah et al. (1982, 1984)	24 BP	7.5	27 MC	4.5	55	8.7	—	—	Larger VBR than normal controls; associated with lower frequency of hospitalizations
Weinberger et al. (1982)	23	3.8	26 MC	2.9	17	6.0	35 SF	5.3	Smaller VBR than schizophrenic or schizophreniform groups; no difference with controls
Kellner et al. (1983)	10	5.5	12 MC	5.5	—	—	—	—	No difference with controls; larger VBR associated with 24-hr urinary free cortisol values
Rieder et al. (1983)	19 BP	5.3	—	—	38	3.7	15 SA	3.9	No differences; tendency for larger VBR compared to schizophrenic and schizoaffective groups
Scott et al. (1983)	10 UP (delusional)	9.4	10 MC	4.2	—	—	—	—	Larger VBR than controls

Targum et al. (1983)	29 UP, 9 BP 20 delusional 18 nondelusional	 5.1 3.6	26 MC	2.9	—	—	—	—	Delusional depressed had larger VBR than nondelusional depressed and control group
Luchins et al. (1984); Meltzer et al. (1984)	14 UP, 6 BP (manic), 2 SA	4.5	62 MC	3.0	45	4.1	—	—	Larger VBR than headache patient control group; associated with psychotic features and low CSF dopamine beta hydroxylase (DBH) activity
Shima et al. (1984)	44 UP, 2 BP	11.2	46 MC	9.1	—	—	—	—	Larger VBR than a combined group of normal controls and headache patients; larger VBR associated with age, single episode, late onset, lack of melancholia, and poor outcome
Pearlson et al. (1984, 1985)	27 BP	6.6	46 HC	4.7	19	6.2	—	—	Larger VBR than healthy controls; larger VBR associated with larger number of hospitalizations, negative symptoms, and persistent unemployment
Dolan et al. (1985)	27 UP, 74 BP	7.3	52 HC	5.6	—	—	—	—	Larger VBR than healthy normal controls; associated with age in both patients and controls
Owens et al. (1985); Johnstone et al. (1986, 1989a)	10 BP inpatients 22 UP & BP outpatients	? ?	—	—	136	11.9	19 OP	?	No VBR differences among the groups; 20% of mood disorder patients had abnormally large VBR; larger VBR in mood disorder outpatients linked to hypothyroidism
Kling et al. (1986)	6 UP	11.1	12 HC	?	6	9.6	—	—	Higher rate of abnormal VBR than in HC; VBR not related to PET measurements of resting regional glucose metabolism
Kolbeinsson et al. (1986)	31 UP, 8 BP, 2 affective other	9.9	16	8.4	—	—	—	—	Larger VBR than healthy normal controls; not related to previous history of ECT

(*Table continues on next page.*)

Table 1. Computerized Tomography (CT) Studies: Ventricular-to-Brain Ratio (VBR) Values[a] (*cont'd*)

	Mood disorder patients		Normal controls		Schizophrenic patients		Other comparison groups		
Study	*N*	VBR	*N*	VBR	*N*	VBR	*N*	VBR	Major findings in mood disorder patients
Sacchetti et al. (1987); Vita et al. (1988)	108	6.1	66	4.7	—	—	—	—	Larger VBR than normal controls; associated with age, age at onset, anxiety, delusions, negative family history of affective disorder, and poor lithium response; in longitudinal subsample, VBR increased at 3-yr follow-up, unlike lack of change in schizophrenic subsample
Schlegel and Kretzschmar (1987); Schlegel et al. (1988, 1989a, 1989b)	33 UP, 22 BP (5 manic), 5 SA	7.4	60 MC	6.9	—	—	—	—	No difference compared to controls; larger VBR associated with symptom severity, negative symptoms, retardation, psychosis, and baseline plasma cortisol levels
Dewan et al. (1988a, 1988b)	26 BP	7.1	22 MC	6.3	—	—	—	—	No difference compred to medical control headache patients
Iacono et al. (1988)	16 UP (psychotic) 18 BP (psychotic)	6.4 6.2	30 MC 44 HC	5.8 6.4	31	6.7	20 SF	5.9	Unipolars had larger VBR than medical controls
Pearlson et al. (1989)	15 UP cognitively impaired 11 UP cognitively normal	12.3 11.0	31 HC	8.6	—	—	13 AD	14.7	VBR increasingly enlarged from normal controls to cognitively intact depressed to cognitively impaired depressed to AD patients

Rossi et al. (1989)	42 UP 30 major depression 12 dysthymia	8.5	37 MC	7.9	36	8.9			Depressed patients intermediate and did not differ from either the schizophrenic or normal group
Abas et al. (1990)	15 UP, 4 BP	?	20 HC	?	—	—	—	—	No VBR difference with controls; larger baseline VBR associated with slowed reaction time during acute phase and with poor performance of difficult cognitive tasks at follow-up during recovery
Andreasen et al. (1990)	26 UP 24 BP	6.2 7.4	75 HC	5.8	108	6.6	—	—	UPs not different from controls, but male BPs had larger VBR than male UPs and male controls
Woods et al. (1990)	9 BP	6.6	—	—	9	10.0	—	—	Schizophrenic group tended to have larger VBR than BP group; at follow-up BPs showed a tendency toward increased ventricular volume; the increase was consistent among schizophrenic patients
Lewine et al. (1991)	19 UP 9 BP	8.9 8.1	91 HC	8.8	45	8.9	—	—	No VBR differences among the groups; higher VBR tended to be associated with lower CSF HVA-HIAA ratio
Van den Bossche et al. (1991)	35 UP	?	10 HC	?	—	—	—	—	No VBR difference between patients or controls or among UPs with major depression with and without melancholia

[a] ?, quantitative mean value not reported in VBR units; UP, unipolar; BP, bipolar; SA, schizoaffective; HC, healthy normal control; MC, medical control; OP, other psychiatric diagnoses; DP, demented patients; SF, schizophreniform; AD, Alzheimer's disease.

this seemingly random natural variation. Furthermore, the choice of control groups is fundamental. It has been suggested that much of the inconsistency in VBR studies of schizophrenia may be more related to variation in the average VBR size of control samples than to variation among samples of schizophrenic patients (cf. Andreasen, Ehrhardt, Swayze, Alliger, Yuh, Cohen, & Ziebell, 1990). In particular, because of exposure to ionizing radiation, many CT studies of mood disorders have used medical populations undergoing CT for diagnostic purposes as the "normal" control sample.

The use of cutoffs to define abnormal VBR enlargement (ventriculomegaly) has related psychometric problems (cf. Harvey, McGuffin, Williams, & Toone, 1990). Most commonly in psychiatric research, an abnormally enlarged VBR is defined as exceeding the control mean by two standard deviations. Indeed, using such an approach it is estimated that between 10% and 30% of mood disorder samples have enlarged VBR (Jeste, Lohr, & Goodwin, 1988). However, this approach to classification is likely to result in a fairly high rate of false negative error relative to false positive error in a sample with truly enlarged VBR. If, for example, mood disorder uniformly resulted in a substantial increase of 0.5 standard deviation in VBR size, far more patients would be "misclassified" as having normal VBR than would be correctly classified as having enlarged VBR. This principle was recently demonstrated in a MRI VBR study of monozygotic twins discordant for schizophrenia (Suddath, Christison, Torrey, Casanova, & Weinberger, 1990). Schizophrenic probands consistently had larger VBRs than their matched discordant twins, discernible by comparative visual inspection. Yet given the background variability in VBR, it is doubtful that probands and their twins could have been distinguished quantitatively had they been treated as independent samples. In essence, the problem here is that in most instances we do not have a way of estimating what the size of a normal VBR should have been for a given patient had they not had mood disorder. Indeed, studies contrasting siblings or twins discordant in their history of mood disorders may provide a means to address this issue.

Use of inversion recovery or T1-weighted sequences with MRI is expected to provide more sensitive morphometric estimates of fluid filled spaces than CT, given MRI's finer spatial and tissue resolution and greater potential for volumetric reconstruction. Table 2 provides a summary of MRI studies of VBR in mood disorder patients. Despite the promise of MRI, it is evident that here too findings have been far from consistent, with the majority of initial studies failing to observe VBR differences between mood disorder patients and controls. Therefore, it

Table 2. Magnetic Resonance Imaging (MRI) Studies: Ventricular-to-Brain Ratio (VBR) Values[a]

Study	Mood disorder patients		Normal controls		Schizophrenic patients		Other comparison groups		Major findings in mood disorder patients
	N	VBR	*N*	VBR	*N*	VBR	*N*	VBR	
Besson et al. (1987)	3 UP, 7 BP	4.5	10 HC	4.3	—	—	—	—	No VBR difference between the groups
Johnstone et al. (1989)	20 BP	?	21 HC	?	21	?	—	—	No VBR difference between the groups
Swayze et al. (1990)	29 BP male 19 BP female	11.8 9.4	28 HC male 19 HC female	9.4 10.2	35 male 17 female	13.4 11.2	—	—	Ventricular enlargement relative to controls observed only in male BP and male schizophrenic subgroups
Lesser et al. (1991)	14 late-onset, psychotic depression (13 UP, 1 BP)	10.0	72 HC	8.5	—	—	—	—	No VBR difference between the groups, despite encephalomalacia in the patients
Rabins et al. (1991)	21 UP	?	14 HC	?	—	—	16 AD	?	Depressed and AD groups were equivalent and by clinical rating had larger lateral ventricles than healthy controls
Coffey (personal communication)	44 UP, 4 BP	21.5^b	76 HC	17.6^b	—	—	—	—	No difference between the groups in mean volume of lateral ventricles or in ratings of ventricular enlargement

[a] ?, quantitative mean value not reported in VBR units; UP, unipolar; BP, bipolar; SA, schizoaffective; HC, healthy normal control; MC, medical control; OP, other psychiatric diagnoses; DP, demented patients; SF, schizophreniform; AD, Alzheimer's disease.
[b] Value is for volume of lateral ventricles

cannot be said with certainty that as a group mood disorder patients manifest lateral ventricular enlargement.

Correlates of Lateral Ventricular Size in Mood Disorder Patients

Given a pattern of both positive and negative findings in clinical samples, it is often useful to examine whether the positive findings are a result of artifactual influences. Alternatively, it is possible that positive findings are genuine, but are specific to patient subtypes, stages of illness, or other clinical features. This latter possibility requires serious attention in the case of mood disorders, given established heterogeneity in phenomenology and in biological disturbances (Vita, Sacchetti, Conte, Alciati, & Pennati, 1985).

Sources of Artifact. The sources of artifact of particular concern in brain imaging studies of mood disorders are the effects of weight loss, comorbid substance and alcohol abuse and dependence, and history of exposure to pharmacological and other somatic treatments. There is evidence from postmortem studies that decreased body weight is associated with decreased brain size (Skullerud, 1985). Furthermore, the sulcal prominence and ventricular enlargement associated with anorexia nervosa may be partially reversible with weight gain and associated metabolic and neuroendocrine normalization (Krieg, Pirke, Lauer, & Backmund, 1988). Alcoholism has been associated with evidence of brain atrophy in pneumoencephalographic (Brewer & Berrett, 1971) and CT studies (Cala & Mastaglia, 1981), with some indication that even heavy, social drinking may have deleterious effects. There has been considerable interest in whether treatment with antidepressant medications, lithium, and particularly ECT may be associated with structural brain changes and, if so, whether these effects are reversible (e.g., Coffey, Figiel, Djang, Sullivan, Herfkens, & Weiner, 1988a; Dolan, Calloway, & Mann, 1985; Kolbeinsson, Arnaldsson, Pëtursson, Skülason, 1986; Rabins, Pearlson, Aylward, Kumar, & Dowell, 1991).

A number of authors have raised the concern that the weight loss associated with the acute expression of major depression and mania or alcohol abuse may be related to findings of ventricular enlargement or sulcal widening (Nasrallah et al., 1989; Hauser, 1991; Schlegel, 1991). Pearlson, Garbacz, Tomkins, Ahn, Gutterman, Veroff, and DePaulo (1984a) excluded subjects who had a recent weight change of 25% or more (gain or loss), as well as heavy use of street drugs or alcohol, and still observed significantly larger VBR in young bipolars relative to matched, healthy controls. In a particularly large series, Dolan,

Calloway, and Mann (1985) excluded patients with a history of alcohol abuse and placed limits on the quantity of alcohol intake in controls. The subgroup of mood disorder patients who were alcohol abstainers had a larger VBR than controls and the abstaining patients did not differ in VBR from patients with mild to moderate alcohol intake.

Weinberger, Torrey, Neophytides, and Wyatt (1979) reported a significant correlation among schizophrenic patients between ventricular size and previous ECT. The issue of whether ECT results in structural brain change has been extensively reviewed (Meldrum, 1986; Siesjö, Ingvar, & Wieloch, 1986; Weiner, 1984), with the predominant view being that ECT, as administered clinically, does not provide the necessary molecular conditions for cellular damage. A number of CT and MRI studies have contrasted mood disorder samples with positive and negative histories of previous ECT and found no differences with respect to measures of cerebral volume and sulcal widening (e.g., Andreasen et al., 1990; Dolan et al., 1985; Kolbeinson et al., 1986; Nasrallah, McCalley-Whitters, & Jacoby, 1982a, Nasrallah, McCalley-Whitters, & Pfohl, 1984; Pearlson et al., 1985; Swayze et al., 1990). CT (Bergsholm, Larsen, Rosendahl, & Holstein, 1989) and MRI (Coffey et al., 1988a; Coffey, Weiner, Djang, Figiel, Soady, Patterson, Holt, Spritzer, & Wilkinson, in press; Scott, Douglas, Whitfield, & Kendell, 1990) prospective, longitudinal, structural imaging studies of patients who have received ECT also have reported negative findings regarding treatment-associated changes. Importantly, Rabin et al. (1991) recently reported that depressed patients who were *subsequently* treated with ECT showed on MRI at pretreatment baseline greater temporal horn abnormality and generally had evidence of greater subcortical pathology than other depressed patients. This type of finding underscores the need for caution in interpreting retrospective associations between types of treatment and evidence of structural change.

There is substantial evidence that lithium, tricyclic antidepressants, and perhaps monoamine oxidase inhibitors can produce fluid changes and alter cerebral microcirculation (e.g., Preskorn, Raichle, & Hartman, 1982; Rubin, Preskorn, & Hartman, 1986). There is the concern that antidepressant or antimanic medications may result in increased extracellular water content, which in turn may result in enlarged fluid-filled spaces (Nasrallah et al., 1989). While there is fairly compelling evidence that neuroleptic exposure is unrelated to evidence of ventricular enlargement in schizophrenia, the possible confounding effects of prior medication history may be more difficult to rule out in mood disorders. As discussed below, it is a reasonable possibility that older depressed patients are particularly likely to manifest structural abnormalities. Samples of medication-naive depressed patients may often be younger, less severely ill, and perhaps less likely to manifest the relevant phenomena. A variety of investigators have examined the role of medication history by correlating the type and length of previous medication exposure with imaging results (e.g., Andreasen et al, 1990; Dewan et al., 1988a; Dolan et al., 1985; Dolan, Calloway, Thacker, & Mann, 1986; Nasrallah et al., 1984; Schlegel & Kretzschmar, 1987). Universally, the results have been negative. At the same time, it must be recognized that retrospective quantification of lifetime medication exposure may have limited reliability and validity.

Clinical and Biological Correlates. *State vs. Trait.* Our understanding of the meaning of lateral ventricular enlargement in mood disorders is highly contingent on whether such enlargement reflects a state, episode, or trait abnormality. Further, if related to clinical state, it would be useful to contrast VBR in bipolar patients during both depressed and manic phases of illness. To date, there has been little investigation of persistence of VBR abnormalities in clinically recovered mood disorder patients. Two longitudinal follow-up studies have been conducted (Vita, Sacchetti, & Cazzullo, 1988; Woods, Yurgelun-Todd, Benes, Frankenburg, & Pope, 1990). In a sample of 21 patients with major affective disorder, VBR showed a marked and consistent increase when patients were rescanned with CT after an average interval of approximately three years. No change in VBR was observed in a comparison group of schizophrenic patients and only slight change in normal controls. Therefore, this study provided the first indication that the VBR abnormality may be progressive in mood disorders, while static in schizophrenia. In contrast, Woods et al. (1990) found an average VBR increase of 25% in a sample of nine young schizophrenics when rescanned at 30 months and an average increase of 11% in nine, young bipolars, when rescanned at 23 months. The change was significant within the schizophrenic group, but not the bipolar group. In both of these studies some patients were clinically improved at the time of second evaluation. Indirectly, these data suggest that VBR is unlikely to decrease with clinical improvement in mood disorders, although more exacting tests need to be conducted. The evidence is too preliminary to resolve the issue of whether a subgroup of patients manifest progressive lateral ventricular enlargement.

Unipolar versus Bipolar Illness. For the most part, studies that have directly compared unipolar and bipolar subsamples in VBR and other measures have failed to observe differences (e.g., Dolan et al., 1985; Schlegel & Kretzschmar, 1987). An important exception is Andreasen et al. (1990), who examined VBR, as assessed by CT, as a function of both polarity and gender. Among male subjects, manic bipolars had the largest VBR relative to unipolar depressed, schizophrenic, and health control groups, with the differences significant for the comparisons with the unipolar and normal control groups. In contrast, among female subjects, bipolars had the smallest VBR, but did not differ from the three remaining diagnostic groups. Similarly, from the same research group, Swayze et al. (1990), using MRI, found a trend suggesting that ventricular enlargement was characteristic of male manic patients, but not females.

It may be noteworthy that some of the strongest findings of VBR abnormality in mood disorders has come from studies focusing on the patients in the manic phase of bipolar illness, often with high or exclusive representation of male patients (e.g., Nasrallah et al., 1982a; Pearlson & Veroff, 1981). This underscores the need to contrast the same patients in depressed and manic phases, at least to rule out the possibility that VBR may covary with phase of illness. Perhaps of more ultimate consequence is the possibility that ventricular enlargement is expressed at different stages of unipolar and bipolar illness. Studies of manic patients have often had considerably younger samples than studies with high representation of unipolar depression (e.g., Andreasen et al., 1990 vs. Dolan et al., 1985). The comparison of bipolar and unipolars has come largely from samples with older patients. Therefore, these negative findings did not address the possibility that ventricular enlargement is typically expressed at different ages in bipolar and unipolar patients.

Age and Age at Onset. Cross-sectional studies suggest that in normal individuals there appears to be little change in VBR between approximately 20 and 60 years of age, while afterwards there is a progressive increase (Zatz, Jernigan, & Ahumada, 1982). This pattern raises both methodological and conceptual concerns. From a methodological viewpoint, the pattern indicates that the association in normals between age and VBR is nonlinear. However, virtually all the studies that have used age as a covariate in comparisons of mood disorder patients with controls or other psychiatric groups have treated this relation in a linear manner. Not only may this have resulted in decreased sensitivity in the comparisons (and higher rates of Type II error), but some of the reported finding may be misleading. For example, Dolan et al. (1985) has been the only investigator to date to examine explicitly interaction effects on VBR between diagnosis of unipolar or bipolar depression and age. This interaction achieved a significance level of 0.18, possibly supporting our suggestion that unipolars and bipolars may manifest ventricular enlargement at different age ranges and a more sensitive data analytic strategy might have revealed a more robust effect. Similarly, the interaction between status as a mood disorder patient (unipolar or bipolar) and age achieved a significance level of 0.14. However, as the authors commented, inspection of the data indicated that it was only in the older age groups that quantitative VBR differences were evident.

Conceptually, there are other suggestions in this literature that emphasize the importance of aging effects. Jacoby and Levy (1980) in a sample of 41 mood disorder patients and 50 healthy controls, all over the age of 60, failed to find a difference in average VBR. They noted, however, that 9 of the patients (22%) have evidence of ventricular enlargement and that this subgroup was characterized by later onset of illness, older age, and higher mortality at 2-year follow-up than the remaining patients. In a recent prospective, longitudinal study of the normal control community sample of Jacoby and Levy (1980), Bird, Levy, and Jacoby (1986) found that ventricular enlargement at baseline was predictive of development of late onset depression. In a similar vein, the MRI work by Coffey and colleagues discussed below has found what may be an extraordinary rate of other types of subcortical pathology in elderly depressed samples (Coffey & Figiel, 1991). The Coffey work has been unusual in its high representation of late onset depressive illness. Some work in schizophrenia has also emphasized the importance of age at onset, albeit with greater deficits reported in the early onset cases. Johnstone, Owens, Bydder, Colter, Crow, and Firth (1989b) found that age at onset was associated with measures of social and intellectual impairment and that it moderated the relations between brain structure measures and the extent of cognitive deficit.

In this light, it is conceivable that a subgroup of patients with late onset depression, typically unipolar, may be especially likely to manifest ventricular enlargement, while ventricular enlargement may also characterize younger bipolar patients. If this were the case, cross-sectional comparison of unipolar and bipolar patients and normal controls in the younger age range would reveal that enlarged VBR is characteristic of bipolar but not unipolar disorder (e.g., Andreasen et al., 1990). At the same time, among older patients there may be no difference between bipolars and unipolars (e.g., Dolan et al., 1985). This type of specification could prove valuable in contributing to our understanding of the etiology of VBR enlargement in mood disorders. Manifestation in younger patients is compatible with the view that ventriculomegaly may reflect neurodevelopmental dysplasia, in which case it should be nonprogressive. Manifestation primarily in older, late onset patients is compatible with an atrophic process, which may or may not be progressive.

Phenomenology. Other than the distinction between psychotic and nonpsychotic illness, few phenomenological features of depression have been found to correlate significantly with VBR or other structural measures in more than one investigation. There have been surprisingly few attempts to associate VBR with overall severity of depressive illness, as assessed by the instruments such as the Hamilton Rating Scale for Depression (HRSD). In the main, such attempts have produced negative results (e.g., Luchins, Levine, and Meltzer, 1984; Van den Bossche, Maes, Brusaard, Schotte, Cosyns, De Moor, & De Schepper, 1991). Schlegel, Maier, Philli, Aldenhoff, Heuser, Kretzschmar, and Benkert (1989a) reported that in 44 medication-free depressed patients, VBR and width of the third ventricle were associated with scores on the Brief Psychiatric Rating Scale, the Bech–Rafaelsen Melancholia Scale, Global Assessment Scale, the Rating for Emotional Blunting, and the Scale for Assessment of Negative Symptoms. At the same time, these structural measures were not associated with HRSD scores or scores on the Hamilton Rating Scale for Anxiety. Item analyses suggested that ventricular size was most associated with retardation-related items. They suggested that scales such as the HRSD may be too multidimensional to reveal consistent effect, particularly since anxiety symptoms and somatic complaints may obscure relations. For the most part, patients subtyping into "endogenous" and "nonendogenous" or "melancholic" and "nonmelancholic" categories has not revealed structural differences (Schlegel, Frommberger, & Buller, 1989b; Van den Bossche et al., 1991). Several investigators have sought associations between VBR or other brain morphometric measures and family history of psychiatric illness (e.g., Dewan et al., 1988b; Dolan et al., 1985, 1986; Nasrallah et al., 1984; Pearlson, Garbacz, Moberg, Ahn, & DePaulo, 1985). No positive findings have been reported.

The significance of delusional or other psychotic symptoms in relation to ventricular enlargement is controversial. Larger VBR in psychotic mood disorder patients was reported by some investigators (Luchins & Meltzer, 1983; Schlegel & Kretzschmar, 1987; Schlegel et al., 1989a; Targum, Rosen, DeLisi, Weinberger, & Citrin, 1983), but not by others (Dewan, Haldipur, Boucher, Ramachandran, & Major, 1988b; Nasrallah et al., 1984; Pearlson et al., 1984a). The reasons for this inconsistency are poorly understood, but it should be noted that in many cases sample sizes have been small. We have already noted that whatever contribution is made by mood disorders to VBR and related measures, that contribution cannot be separated from the large naturally occurring variability. When one adds to this the possibility that effects on VBR may be dependent on age groups, gender, and/or other diagnostic subtyping (unipolar versus bipolar), then power to detect effects was severely compromised in much of this literature.

Cognitive Status. Characteristically, during the acute phase of illness the performance of patients with major depression or mania is reduced on cognitive tests heavily influenced by the adequacy of attention and concentration (e.g., immediate learning and memory) (Sackeim & Steif, 1988). Furthermore, a small subgroup of patients with major depression show more profound cognitive disturbance, with impaired orientation and global performance deficits in the range of demented patients. This constellation in depression is usually referred to as "pseudodementia." In depressed samples not specifically selected for cognitive impairment, there are scattered findings of inverse associations between VBR and specific cognitive measures (Dewan et al., 1988b; Kellner, Rubinow, & Post, 1986; Rabins et al., 1991), as well as a variety of negative findings (Dewan et al., 1988b; Nasrallah et al., 1984; Pearlson et al., 1984a; Schlegel, Maier, Phillipp, Heuser, & Aldenhoff, 1988). Abas, Sahakian, and Levy (1990) reported that degree of residual cognitive impairment following recovery from depression in an elderly sample was positively associated with VBR obtained two years earlier during the acute depressive phase.

Perhaps a more powerful methodology is to contrast structural findings in depressed patients with and without prominent cognitive impairment. Pearlson, Rabins, Kim, Speedie, Moberg, and Burns (1989) identified 15 depressed patients who were cognitively impaired at admission and 11 patients who were unimpaired, using a cutoff of 24 on the Mini-Mental State Examination for this designation. These 26 depressed patients were all older than 60 years and were compared in CT measures to 13 patients with probable Alzheimer's disease (AD) and 31 normal controls. Of the cognitively impaired depressed patients, 11 were reexamined at two year follow-up and only one of this group evidenced cognitive decline. On CT measures of VBR and brain density, the total depressed sample fell intermediate between the AD patients and controls (see Table 1). Specific comparisons indicated that, relative to the normal controls, VBR was larger and brain density was less in the cognitively impaired depressed group and in the AD group. The cognitively-intact depressed group did not differ from normals. Related findings indicated that VBR and brain density measures were associated with neuropsychological performance variables only in the cognitively impaired groups. This study suggests that ventricular enlargement and possibly other structural abnormalities may be particularly manifest in elderly depressed patients with concurrent, reversible, and moderate-to-marked cognitive impairment.

Course of Illness. Many descriptive features regarding the course of affective illness have been correlated with VBR and other brain morphometric measures. With some important exceptions, the findings have been either negative or isolated and unreplicated. It does not appear that duration of affective illness, duration or current episode, number of previous affective episodes, or number of previous psychiatric hospitalizations are related to structural abnormalities (e.g., Roy-Byrne, Post, Kellner, Joffe, & Uhde, 1988). Acute to short-term response to treatment has also shown no relation to VBR (Luchins et al., 1984; Nasrallah et al., 1984). However, Jacoby and Levy (1980) originally observed that treatment response after three months did not distinguish patients with such enlargement at two-year follow-up. Shima, Shikano, Kitamura, Masuda, Tsukumo, and Kanba (1984) reported that larger VBR was associated with poorer clinical outcome nine months after admission in elderly depressed patients. As noted earlier, Bird et al. (1986) suggested that baseline VBR was predictive of the future development of late onset depression.

Biological Correlates. There have been relatively few attempts to examine brain structural findings in the context of other biological abnormalities in mood disorders. With respect to VBR, most attention has focused on the function of the hypothalamic-pituitary-adrenal axis. Dexamethasone suppression test (DST) results were not associated with VBR in a number of studies (Dewan, Haldipur, Boucher, & Major, 1988c; Schlegel & Kretzschmar, 1987; Schlegel, von Bardeleben, Wiedemann, Frommberger, & Holsboer, 1989c; Standish-Barry, Hale, Honig, Bouras, Bridges, & Bartlett, 1985; Targum et al., 1983; Van den Bossche et al., 1991). Kellner, Rubinow, Gold, and Post (1983) reported that VBR was associated with higher 24-hour free urinary cortisol values. Similarly, Schegel et al. (1989c) found that average afternoon plasma cortisol values correlated significantly with VBR. This issue is of particular concern because of the possibility that endogenous hypercortisolemia may result in alterations of brain structure in a fashion similar to that of exogenous steroids (Bentson, Reza,Winter, & Wilson, 1978; Momose, Kiellberg, & Kliman, 1971).

Johnstone, Owens, Crow, Colter, Lawton, and Jagoe (1986) raised the possibility that hypothyroidism may be associated with enlarged VBR in mood disorder patients. However, thyroid-stimulating hormone (TSH) response to thyrotropin-releasing hormone (TRH) was found to be unrelated to VBR (Targum et al., 1983) and T_3 and T_4 levels have also been found to be uncorrelated (Schlegel, 1991). Van den Bossche et al. (1991) were unable to replicate previous findings of a relation between plasma tryptophan and ventricular enlargement (Standish-Barry, Bouras, Hale, Bridges, & Bartlett, 1986). Finally, there are single reports that larger VBR is associated with low plasma dopamine β-hydroxylase activity (Meltzer et al., 1984), increased CSF 5-hydroxyindoleacetic acid (5-HIAA) (Standish-Barry et al., 1986), and an increased ratio of CSF homovanillic acid (HVA) to 5-HIAA (Lewine, Risch, Risby, Stipetic, Jewart, & Eccard, 1991).

To our knowledge, there has been only one study in mood disorder patients that has examined brain structural and functional abnormalities in the same sample (Kling, Metter, Riege, & Kuhl, 1986). PET was used to contrast regional glucose metabolism and CT measurement of VBR and sulcal widening in six chronic schizophrenic patients, six depressed patients, and 12 normal controls. Three of the schizophrenic patients and four of the depressed patients showed evidence of lateral ventricular enlargement and widening of the Sylvian fissure. While overall, the regional metabolic profiles of the three groups could not be distinguished, the patients with evidence of structural abnormality tended to have lower global metabolism. Clearly, additional work of this type is needed with considerably larger samples and with particular attention to partial volume effects (i.e., the spurious reduction of metabolic rates in tissue by inclusion of fluid-filled space).

Other Morphometric and Related Structural Abnormalities

Fewer studies have been conducted on morphometric measures other than lateral ventricular enlargement. Generally, as in the case of lateral ventricular enlargement, either the findings have been inconsistent, suggesting heterogeneity, or there are too few independent replications to state with confidence that a given structural abnormality is an established finding.

The third ventricle has been found to be increased in size in relatively young depressed patients (Dewan et al., 1988a; Schlegel & Kretzchmar, 1987), but negative findings have also been reported (Tanaka et al., 1982). In a recent MRI study of a large sample of elderly depressed patients, Coffey (personal communication) did not observe third ventricle enlargement. A set of studies have examined cerebral sulcal and fissure size. Tanaka, Hazama, Fukuhara, and Tsutsui (1982) described widening of the interhemispheric and Sylvian fissures as particularly characteristic of older depressed patients, whom they felt displayed "atrophy" at an earlier stage than normal controls. Cortical surface atrophy was also described in younger manic patients by Nasrallah, McCalley-Whitters, and Jacoby (1982b). In a large series, Dolan et al. (1986) observed sulcal widening particularly in frontal and temporal areas. Similar results were reported by Kolbeinsson et al. (1986) and Rabins et al. (1991). However, some other studies of sulcal widening in specific cortical areas and other studies using global atrophy measures have been negative (Coffey, personal communication; Dewan et al., 1988a; Iacono et al., 1988; Schegel & Kretzschmar, 1987). A similar pattern pertains to ratings of cerebellar (vermis) atrophy. Both depressed and manic patients have been found to exceed controls in cerebellar atrophy scores (Lippmann, Manshadi, Baldwin, Drasin, Rice, & Alrajeh, 1982; Nasrallah et al., 1982b; Weinberger et al., 1982), and negative results have also been obtained (Pearlson & Veroff, 1981; Yates, Jacoby, & Andreasen, 1987).

More recent investigations have examined the size, volume and/or shape of cerebral structures. Studies of corpus callosum in bipolar patients have reported smaller area (Coffman, Bornstein, Olson, Schwarzkopf, & Nasrallah, 1990) or smaller length (Hauser, Dauphinais, Berrettini, DeLisi, & Gelernter, 1989a), but no abnormalities in unipolar patients (Lammers, Doraiswamy, Husain, Figiel, Lurie, Boyko, Ellinwood, Nemeroff, & Krishnan, 1991). Hauser, Altshuler, Berrettini, Dauphinais, Gelernter, and Post (1989b) found that a sample principally of euthymic bipolar patients had smaller temporal lobe area (relative to cerebrum area) than normal controls. In a related study with somewhat overlapping samples, Altshuler, Contrad, Hauser, Li, Guze, and Denikoff (1991) reported that total temporal lobe volume was reduced in euthymic bipolar patients relative to controls. On the other hand, Johnstone, Owens, Crow, Frith, Alexandropolis, Bydder, and Colter (1989a) did not observe group differences in total brain area, total lateral ventricular area, or total temporal lobe area when contrasting schizophrenic, bipolar, and normal control groups. Coffey (personal communication) in a sample of 48 elderly depressives and 76 normal controls did not observe differences in the volumes of the temporal lobe, the amygdala-hippocampal complex or the cerebral hemisphere, but did find that frontal lobe volume was significantly smaller in the patient group by an average of approximately 7%. This latter finding is of particular interest given the results of functional brain imaging studies suggesting reduced rCBF (Sackeim, Prohovnik, Moeller, Brown, Apter, Prudic, Devanand, & Mukherjee, 1990a) and glucose metabolism (Baxter, Schwartz, Phelps, Mazziotta, Guze, Selin, Gerner, & Sumida, 1989) in frontal cortical areas of depressed patients. Also of relevance to the possibility of reduced frontal volume are recent reports noting decreased centrum semiovale CT attenuation values in the white matter of elderly depressed patients (Pearlson, Rabins, & Burns, 1991) and increased MRI T_1 proton relaxation values in frontal white matter of unipolar patients (Dolan, Poynton, Bridges, & Trimble, 1990). Coffman et al. (1990) observed a trend for reduced frontal area in their MRI study of relatively young, bipolar patients. They also reported that among bipolars a variety of neuropsychological performance measures were inversely related to cerebral, and particularly, frontal area.

Krishman, Doraiswamy, Lurie, Figiel, Husain, Boyko, Ellinwood, and Nemeroff (1991) recently used MRI to examine the size of the pituitary gland in 16 unipolar and 3 bipolar depressed patients who were matched to 19 healthy controls. Pituitary gland length, area, and volume was significantly increased in the depressed patients, an effect that appeared to be particularly pronounced among the elderly. This finding suggests structural disturbance in the neuroendocrine axis in major depression and relations with concurrent neuroendocrine measures are obviously needed.

While this summary has highlighted the inconsistencies in this literature, it should also be evident that there are abundant positive findings regarding morphometric and related structural abnormalities in mood disorders. The frequency of positive findings of abnormality, coupled with negative reports, suggests that the prevalence of structural deficits is, in fact, overrepresented in mood disorders, but that we have yet to identify adequately the subgroups of mood disorder patients most likely to manifest particular abnormalities. In addition to the psychometric issues that we have raised as limiting the power of these types of investigations, there are other broader concerns. Much of the morphometric work has focused on the size of fluid-filled CNS spaces or of fairly large cortical regions. There is little expectation that increased ventricular size in mood disorders is due to hydrocephalus. Rather, the underlying assumption is that dysplasia or atrophic processes result in enlargement of fluid-filled spaces. Therefore, ventricular enlargement would be a secondary and perhaps spatially remote effect of the primary disturbance. Similarly, the morphometric approach to measuring the size of large cortical volumes, such as the frontal or temporal lobes, may be insensitive to pathology that is topographically more discrete within those volumes. With some exceptions, these limitations suggest that morphometric approaches are likely only to provide signals of the existence of neuropathological processes and hints regarding their spatial extent.

Subcortical Encephalomalacia

A consistent finding that has emerged in recent MRI studies is that mood disorder patients have an unusually high prevalence of subcortical hyperintensities (Coffey, Figiel, Djang, Saunders, & Weiner, 1989; Coffey, Figiel, Djang, & Weiner, 1990; Dupont, Jernigan, Butters, Delis,

Hesselink, Heindel, & Gillin, 1990; Figiel, Krishnan, Rao, Doraiswamy, Ellinwood, Nemeroff, Evans, & Boyko, 1991; Lesser, Miller, Boone, Hill-Gutierrez, Mehringer, Wang, & Mena, 1991; Rabins et al., 1991; Swayze et al., 1990; Zubencko, Sullivan, Nelson, Belle, Huff, & Wolf, 1990). The abnormalities appear as areas of increased signal intensity in T2-weighted images and may segregate into three main groupings. Periventricular hyperintensities are observed as a halo or rim adjacent to ventricles that in severe manifestations invade surrounding deep white matter. Single, patchy, or confluent foci may also be observed in subcortical white matter, with or without periventricular findings. Abnormal hyperintensities may also be found in deep gray structures, particularly the basal ganglia, thalamus, and pons. Lucencies in CT examination have also been reported in depression (Coffey, Figiel, Djang, Cress, Saunders, & Weiner, 1988b), but with an expected lower prevalence rate, given lesser sensitivity to this type of pathological process (Awad, Modic, Little, Furland, & Weinstein, 1986a). These abnormalities may be similar to those in Binswanger's disease, a dementing illness that is believed to result from hypoperfusion and small strokes in subcortical watershed areas irrigated by the penetrating medullary arteries (Kinkel, Jacobs, Polanchini, Bates, & Heffner, 1985). The abnormalities have been variously referred to as leukoencephalopathy (Krishnan, Goli, Ellinwood, France, Blazer, & Nemeroff, 1988), leukoaraiosis (Hachinski, Potter, & Merskey, 1987), SAE (subcortical arteriosclerotic encephalopathy; Solomon, Yeates, Burger, & Heinz, 1987), encephalomalacia (Awad, Johnson, Spetzler, & Hodak, 1986b), and UBOs (unidentified bright objects; Swayze et al., 1990). Since the abnormalities in major depression are not restricted to white matter and their etiology is not established, following Awad et al. (1986b), the term encephalomalacia is used here.

In one of the largest prospective series reported to date, all 51 elderly patients (age > 60 years) with major depression, referred for ECT, presented with MRI hyperintensities, with over half rated as moderate to severe and 51% showing lesions of subcortical gray nuclei (Coffey et al., 1989, 1990). The total rate of abnormality and the rate of moderate-to-severe abnormality greatly exceeded that found in a neurologically and medically healthy control sample, with basal ganglia abnormalities most discriminative. Some of the depressed samples studied to date have contained patients with concurrent medical illnesses and without adequate control for factors such as concurrent medication or history of alcohol or other drug abuse. Nonetheless, when depressed patients with preexisting dementia or other neurological conditions were excluded, the rate of encephalomalacia clearly exceeded that of normal controls, despite matching on cerebrovascular risk factors (Coffey et al., 1990). In other studies of elderly healthy controls, when excluding the characteristic halo or caps at the ventricles, approximately 10%–30% presented with MRI white matter abnormalities, with severity typically mild and very low rates of subcortical gray matter abnormalities (Brandt-Zawadski, Fein, Van Dyke, Kierman, Davenport, & De Groot, 1985; Fazekas, Niederkorn, Schmidt, Offenbacher, Horner, Bertha, & Lechner, 1988; Gerard & Weisberg, 1986; Kertesz, Black, Nicholson, Carr, 1988; Mirsen, Lee, Wong, Diaz, Fox, & Hachinski, 1991; Rao, Mittenberg, Bernardin, Haughton, & Leo, 1989).

Two studies have compared rates of encephalomalacia in an elderly depressed sample with that in other neuropsychiatric illnesses. Zubenko et al. (1990) contrasted 67 elderly patients with major depression, 61 patients with dementia (AD and multi-infarct), and 44 healthy controls, using either MRI or CT assessment. Both patient groups markedly exceeded the controls in the prevalence of abnormal findings of cortical atrophy, cortical infarction, gray matter lacunae, and subcortical white matter hyperintensities (MRI) or lucencies (CT). The depressed and demented groups differed only in cortical atrophy scores, with less abnormality in the depressed. The depressed group slightly exceeded the demented group in rates of subcortical white matter abnormalities. Rabins et al. (1991) performed MRI examinations of 21 elderly depressed patients, 16 AD patients, and 14 normal controls. Depressed patients differed from controls in evidence of basal ganglia lesions, white matter hyperintensities, and tended to differ in periventricular hyperintensities. AD patients differed from controls in ratings of white matter hyperintensities, and only tended to differ with respect to basal ganglia lesions and periventricular hyperintensities.

Comparison with the literature suggests that the abnormality rate that has been reported in elderly major depression may equal or exceed that in Alzheimer's disease (Aharon-Peretz, Cummings, & Hill, 1988; Erkinjuntt, Ketonen, Sulkava, Sipponen, Vuorialho, & Iivanainen, 1987; Diaz, Merskey, Hackinski, Lee, Boniferro, Wong, Mirsen, & Fox, 1991; Filley, Davis, Schmitz, Stears, Heaton, Kelly, Culig, & Scherzinger, 1989; Rezek, Morris, & Fulling, 1986) and may be comparable to that in multi-infarct dementia (Aharon-Peretz et al., 1988; Erkinjuntt et al., 1987; Hershey, Modic, Greenough, & Jaffe, 1987). In considerably younger samples, Swayze et al. (1990) found that MRI hyperintensities were more common among bipolar manic patients relative to controls, but schizophrenic patients did not show this abnormality. Likewise, both Dupont et al. (1990) and Figiel et al. (1991) reported that white matter MRI hyperintensities were more common in young bipolars than in controls. Coffey et al. (1989), using only qualitative ratings, also reported high rates of lateral ventricular dilation (69%: 29% moderate to severe) and sulcal prominence (cortical atrophy, 96%: 75% moderate to severe) in their prospective MRI series of elderly depressives. Manifestation of these abnormalities and of encephalomalacia tended to covary. Among demented samples, encephalomalacia and ventricular dilatation have been found to be related (Bradley, Waluch, Brant-Zawadski, Yadley, & Wycoff 1984; Brandt-Zawadski et al. 1985; Hershey et al., 1987). It is commonly hypothesized that periventricular encephalomalacia signals a process that ultimately results in enlarged CSF spaces (Barnes & Enzmann, 1981; Caplan & Schoene, 1978; Hershey et al., 1987).

Clinical and historical features of depressive illness associated with encephalomalacia are not well established. Coffey and colleagues suggested that patients with late onset of illness (onset age ≥ 60 years) and resistance to antidepressant medications are at increased risk. However, the depressed samples reported on in this work comprised consecutive clinical referrals to an ECT service (Coffey et al., 1988b, 1989, 1990). The samples were unusual in that rate of late onset depression was 80% and 86% were said to be medication resistant (Coffey et al., 1988b). The criteria for medication resistance were weak relative to accepted standards for an adequate pharmacotherapy (e.g., Keller, Lavori, Klerman, Adreasen, Endicott, Coryell, Fawcett, Rice, & Hirschfeld, 1986; Sackeim, Prudic, Devanand, Decina, Kerr, & Malitz, 1990b) and included medication intolerant patients. Lesser et al. (1991) reported an excess rate of large subcortical hyperintensities in a comparison of a group of patients with late-onset psychotic depression with normal controls. Coffey et al. (1988b, 1990) have also suggested that patients with encephalomalacia show strong clinical response to ECT, but with enhanced cognitive side effects. However, in much of this work, symptomatic improvement was evaluated globally and not by blind raters using standardized quantitative instruments. There are conflicting reports as to whether the limited encephalomalacia among normal subjects is associated with cognitive impairments (George, deLeon, Kalvin, Rosner, Goodgold, & Chase, 1986; Mirsen et al., 1991; Rao et al., 1989; Steingart, Hachinski, Lau, Fox, Diaz, Cape, Lee, Inzitari, & Merskey, 1987). Concurrent comprehensive neuropsychological evaluation of depressed samples has yet to be conducted. Figiel, Krishnan, and Doraiswamy (1990) suggested that patients with basal ganglia hyperintensities may be particularly likely to develop organic mental syndromes during the course of ECT.

It is evident that elevated rates of encephalomalacia have been observed both in samples of elderly depressed patients who were predominantly unipolar (Coffey et al., 1989; 1990; Lesser et al., 1991; Rabins et al., 1991), and samples of relatively young bipolar patients (Dupont et al., 1990; Figiel et al., 1991; Swayze et al. 1990). Critically, it is unknown whether younger, unipolar samples manifest this abnormality. For the most part, the abnormalities observed in the elderly depressed samples include hyperintensities in subcortical gray matter, particularly the basal ganglia. This particular disturbance may not be observed in younger, bipolar samples (e.g., Figiel et al., 1991).

The pathogenesis of the MRI abnormalities in major depression is unknown. These same abnormalities in other samples have usually been attributed to ischemic cerebrovascular disease (CVD), with the hyperintensities reflecting increased water content in perivascular space, axon and myelin loss, astrocyte proliferation (gliosis), and/or frank infarction (Awad et al., 1986b; Caplan & Schoene, 1978; Román, 1987). Typically, the subcortical white matter most commonly manifesting the increased signal intensities is in watershed areas supplied by small arterioles that branch off long penetrating medullary arteries. These areas receive limited collateral vascular supply, so that they are particularly vulnerable to vascular insult (DeReuk, 1971). In a variety of populations, including major depression and normal controls, encephalomalacia has been found associated with history of hypertension, diabetes, ischemic heart disease, and other cerebrovascular risk factors (Awad, Spetzler, Hodak, Awad, Williams, & Carey, 1987; Coffey et al., 1989; Fazekas et al., 1988; Gerard & Weisberg, 1986; Inzitari, Diaz, Fox, Hachinski, Steingart, Lau, Donald, Wade, Mulic, & Merskey, 1987; Krishnan et al., 1988; Lechner, Schmidt, Bertha, Justich, Offenbacher, & Schneider, 1988). There are some indications that different risk factors are associated with subtype of encephalomalacia, i.e., deep white matter foci versus periventricular (Kertesz et al., 1988). These positive findings indirectly support the vascular hypothesis. Awad et al.'s (1986b) suggestion of "état criblé" is also compatible. Pulsation in ectatic or tortuous blood vessels, rigidified by sclerosis, may mechanically produce increased water-filled, perivascular space or produce these effects and parenchymal atrophy due to ischemic substrate supply. Hachinski et al. (1987) argued that the periventricular white matter effects may be due to leakage of CSF into surrounding tissue. Caps at the frontal horns of the lateral ventricles are common in normal samples, may not be age related, and may be due to interstitial flow into the ventricles, combined with low myelin content and ependymitis granularis in this area (Sze, De Armond, Brant-Zawadzki, Davis, Norman, & Newton, 1986; Young, Randell, Kaplan, Bydder, & Steiner, 1983).

There has been limited histopathological investigation of encephalomalacia and none in patients with major depression. The findings to date are supportive of a vascular etiology in that areas of MRI hyperintensity in neurological and normal samples commonly show arteriolar hyalinization, ectasia, dilated perivascular spaces, gliosis, and/or lacunar infarcts (Awad et al., 1986a; Braffman, Zimmerman, Trojanowski, Gonatas, Hickey, & Schlaepfer, 1988; George et al., 1986; Kirkpatrick & Hayman, 1987). Because of the relatively low mortality rate in the elderly depressed, prospective neuropathological investigation is not likely to be practical, although postmortem MRI and neuropathological investigation may be done concurrently. In contrast, the hypothesis that the encephalomalacia in major depression is due to sclerosis, hypoperfusion, and infarct in subcortical watershed areas may be tested with in vivo perfusion imaging techniques. Studies of the perfusion abnormalities observed with SPECT and PET in stroke samples have consistently shown that the areas of hypoperfusion identified by functional imaging corresponds in location to CT, MRI, or pathology identification of lacunar infarct and related abnormalities. The spatial extent of the functional imaging perfusion abnormalities is typically larger than pathology or structural imaging defined areas of tissue necrosis (Killen, Oster, & Colditz, 1989; Perani, Di Piero, Lucignani, Gilardi, Pantano, Rossetti, Pozzilli, Gerundini, Fazio, & Lenzi, 1988; Seiderer, Krappel, Moser, Hahn, Schmiedek, & Buell, 1989) and remote perfusion abnormalities may be observed (i.e., diaschisis, deafferentation). This indicates that effects of punctate lesions that are smaller in size than the spatial resolution of state-of-the-art functional tomography (~6 mm full width half maximum) may still be observable owing to their association with a larger area of perfusion deficit. No study has examined perfusion in the areas of MRI-defined hyperintensity. One study (Fazekas et al., 188; Fazekas, 1989) used the 133-xenon lateral projection technique to quantify cortical flow in normal subjects categorized by MRI encephalomalacia. Reduced global cortical flow, particularly in the slow compartment (white matter), was observed in subjects with white matter hyperintensities.

Cross-sectional studies of psychiatric, neurologic, and normal samples have all indicated that likelihood and severity of encephalomalacia increases with age (Awad, Spetzler, Hodak, Awad, & Carey, 1986c; Coffey et al., 1989; Fazekas et al., 1988). This fact, the associations with cerebrovascular risk factors, and the hypothesis of underlying CVD suggest that the encephalomalacia and perfusion abnormalities in major depression may be progressive. The only follow-up study of these MRI abnormalities in major depression examined patients prior to and six months following ECT. Blind ratings showed strong test–retest reliability and stability in the occurrence and severity of the abnormalities

(Coffey et al., 1991). The follow-up interval used was probably too short to observe progressive MRI changes; the primary aim was to determine whether ECT resulted in MRI alterations. It is also likely that in vivo perfusion imaging may provide more sensitive measures of progressive effects than MRI. If perfusion abnormalities in deep structures lead to encephalomalacia, they should exist prior to MRI evidence of structural change and their quantification should be more sensitive to progressive change. As described earlier, the possibility that at least of a subgroup of depressed patients are characterized by progressive structural pathology is supported by one of two CT studies of ventricular dilatation (Vita et al., 1988; Woods et al., 1990).

Our research has indicated that some elderly patients with major depression have marked abnormalities in CBF (Sackeim et al., 1990a). We have used the 133-xenon inhalation technique to quantify rCBF in medication-free elderly depressed samples studied prior to, during, and following ECT and rigorously screened for neurological illness or insult. We have reported that pretreatment comparison of resting studies in 41 patients with 40 matched controls revealed a global flow reduction of approximately 15% in cortical CBF (Sackeim et al., 1990a). The magnitude of this deficit was of the same order as in many forms of CVD and was of the same magnitude as we have observed in Alzheimer's disease (Prohovnik, Mayeux, Sackeim, Smith, Stern, & Alderson, 1988; Sackeim, Prohovnik, Mayeux, Moeller, Devanand, Coudreaut, Stern, Brown, & Prudic, 1990c). In applying a new method, the Scaled Subprofile Model (SSM; Moeller, Strother, Sidtis, & Rottenberg, 1987), designed to reveal abnormal regional networks (covariance patterns) in clinical samples, we found that the depressed group had marked abnormality in a specific cortical network, independent of the global flow deficit (Sackeim et al., 1990a). Degree of abnormality in this pattern increased both with patient age and severity of depressive symptoms. Furthermore, in restudying a subgroup two months after the termination of ECT, we have found strong stability in the global and topographic CBF deficits in elderly mood disorder patients, indicating that these abnormalities may be state independent (Sackeim, Prohovnik, Moeller, Maclin, Stade, Devanand, Prudic, & Brown, 1989). These findings are also compatible with the view that a form of CVD may characterize a substantial proportion of the elderly depressed and further highlight the need for concurrent structural and functional imaging studies in this population.

The role of encephalomalacia, cortical perfusion deficits, and the hypothesized underlying CVD process in the pathogenesis of depressive illness is unknown. The fact that a variety of neurological conditions with primarily subcortical involvement (Parkinson's disease;

Huntington's chorea; multiple sclerosis; Binswanger's disease) are associated with high rates of depressive illness supports the possibility that dysfunction in deep gray matter structures or disruption in long communicating white matter tracts can contribute to depression. Indeed, there is suggestive evidence from stroke patients that infarcts in the basal ganglia are more likely to be associated with depression than thalamic lesions, raising possible anatomic specificity (Starkstein, Robinson, Berthier, Parikh, & Price, 1988). Alternatively, it is established that specific reticular and thalamic nuclei, as well as other subcortical regions, modulate global cortical functional activity (e.g., Baron, D'Antona, Pantano, Serdaru, Samson, & Bousser, 1986). In some cases, specific subcortical nuclei have been implicated in the modulation of specific cortical regional networks (e.g., Mesulam, Van Hoesen, Pandya, & Geschwind, 1977; Seltzer & Pandya, 1978). For example, it is known that the medial dorsal thalamic nucleus and the medial pulvinar/lateral posterior thalamic nuclei modulate functional activity in the anterior and posterior portions, respectively, of the prefrontal-parietal network that we have found to be abnormal in elderly depressives (Goldman-Rakic, 1987; Mesulam et al., 1977). Consequently, disruption of specific subcortical nuclei or their ascending pathways may produce cortical dysfunction that, in turn, is responsible for some depressive manifestations. Indeed, cortical disconnection due to encephalomalacia has been suggested as responsible for the dementia of Binswanger's disease (Román, 1987).

Establishing the relations among the structural and vascular cortical and subcortical abnormalities and determining whether, in fact, a form of progressive CVD characterizes a subgroup of the elderly depressed should be a first step in evaluating the relevance of these abnormalities for etiology. In turn, determining the contribution of a possibly progressive CVD process to major depression in the elderly may have important implications for our understanding of phenomenology, prevention, and treatment. For example, there is some evidence that with aging the course of recurrent affective illness may become more virulent, with shorter periods of euthymia between episodes, more abrupt onset of acute symptoms, and greater resistance to somatic treatment (Goodwin & Jamison, 1989; Murphy, 1983; Post, 1972). Progressive cerebrovascular disease could contribute to this pattern. Treatment of depressive illness often involves use of agents with potential cardiovascular and cerebrovascular effects (e.g., TCAs). Conceivably, some of these effects, such as hypotension (Glassman, Roose, Giardiana, & Bigger, 1987) or reductions in cerebral blood flow (Preskorn et al., 1982) or metabolism (Caldecott-Hazard, Mazziotta, & Phelps, 1988), may aggravate a cerebrovascular disease process and perhaps limit efficacy (Georgotas, McCue, & Cooper, 1989; Wehr & Goodwin, 1987) or enhance side effects. Finally, the relations between the perfusion and structural abnormalities with cerebrovascular risk factors may be most critical, suggesting alternative methods of treatment and new approaches to prevention.

Functional Brain-Imaging Studies

Several technologies provide information on regional physiological function. These include the lateral projection, xenon technique to assess rCBF, SPECT, PET, magnetic resonance spectroscopy (MRS), topographic EEG, and magnetoencephalography (MEG). In principle, these approaches allow for quantification of a host of physiological paramters, including concentrations of endogenous and exogenous biochemical substances, various aspects of receptor physiology, brain metabolism, blood flow, and electrical activity. The technologies differ substantially in basic principles of signal acquisition and the types of physiological process they can quantify, spatial and temporal resolution, susceptibility to and the nature of artifacts, and in practical concerns such as radiation exposure and cost. The reader is referred elsewhere for reviews of the basic principles underlying these techniques and of their general application to problems in psychiatry (Andreasen, 1989; Devous, 1989; Morihisa, 1989; Holcomb, Links, Smith, & Wong, 1989).

For the last thirty years, biological hypotheses concerning the pathophysiology of mood disorders has centered on abnormalities in neurochemical transmission, with an emphasis on abnormal levels of neurotransmitters and neuromodulators and alterations in receptor physiology. Brain imaging studies have the potential of advancing our understanding of pathophysiology in two distinct ways. First, in the human much of the investigation of abnormalities in biochemical systems has been indirect, examining peripheral measures in the blood, urine, or CSF to model more central processes. Imaging studies offer the opportunity to examine brain biochemical processes on a regional basis. Examples of attempts to realize this potential in mood disorders include the following. Using PET, Agren, Reibring, Hartvig, Tedroff, Bjurling, Hornfeldt, Andersson, Lundvist, and Langstrom (1991) compared the uptake of L-hyroxytryptophan (5-HTP), the precursor of serotonin, in six patients with major depression and eight normal controls. Two patients were also studied after recovery. Agren et al. (1991) reported that 5-HTP uptake across the blood–brain barrier was reduced in the patients, regardless of clinical state. Mayberg, Robinson, Wong, Parikh, Bolduc,

Starkstein, Price, Dannals, Links, Wilson, Ravert, and Wagner (1988) quantified cortical serotonin-2 receptor binding in the stroke model of secondary depression. Patients with right-hemisphere stroke had greater cortical binding of (3-*N*-^{11}Cmethyl)spiperone in uninjured areas than patients with left-hemisphere stroke or normal controls. Extent of binding in the left hemisphere correlated inversely with depression scores, leading to the suggestion that failure to up-regulate serotonin receptors after stroke may be associated with depression. Using lithium-7 MRS, several groups have measured lithium concentrations both in brain and in peripheral tissue, allowing for investigation of the relations among serum level, brain concentrations, clinical response, and side effects (Gyulai, Wicklund, Greenstein, Bauer, Ciccione, Whybrow, Zimmerman, Kovachich, & Alves, 1991; Komoroski, Newton, Walker, Cardwell, Jagannathan, 1990; Mota de Freitas, Silberberg, Espanol, Dorus, Abraha, Dorus, Elenz, & Whang, 1990).

Second, abnormalities of neurochemical systems are presumably of interest because they alter patterns of excitation and inhibition (tonic or evoked) in distributed neural networks. Consequently, such abnormalities should be associated with disturbed patterns of rCBF and regional metabolic rate (rCMR), since these reflect the activity level or "work load" of neuronal aggregates (Roy & Sherrington, 1890; Sokoloff, 1977). It is commonly argued that there may be heterogeneity in neurochemical abnormalities in mood disorders or that fundamental pathophysiological disturbances involve the dynamic balance of several neurochemical systems (Gerner & Bunney, 1986; Heninger & Charney, 1987). It is conceivable that a diverse set of neurochemical disturbances impinge on the activity level of the same distributed neural network. Therefore, at least conceptually, the study of rCBF and rCMR offers the potential of revealing more consistent patterns of disturbance.

To date, the bulk of functional brain imaging studies in mood disorders have concentrated on examination of rCBF and rCMR for glucose ($rCMR_{glu}$), the focus of this review. It should be recognized that although this area is undergoing rapid change, it is still early in its development. There are inconsistent and, at times, contradictory findings for virtually all the hypotheses that have been raised. Indeed, much of the work so far may be more properly considered exploratory or hypothesis generating than tests of theory. Furthermore, this area is characterized by a set of particular methodological concerns that have yet to be resolved adequately. Before outlining what we believe are the major hypotheses to emerge from this work, we summarize some of the methodological difficulties.

Methodological Issues in Functional Brain Imaging Studies of Mood Disorders

Individual Differences and Sample Size. Part of the mystique of functional brain imaging is the erroneous view that measurement of rCBF or $rCMR_{glu}$ is subject to less noise and a lower magnitude of individual differences than other areas of biological study. Instead, the extent to which normal individuals differ in global cortical or whole brain CBF and CMR is remarkable. Coefficients of variation for global CBF assessed with the xenon inhalation technique in normal samples at rest have typically been in the range of 10%–20%, while this range for $rCMR_{glu}$ typically has been between 15% and 35% (Sackeim et al., 1990a). We determined that the median sample size of mood disorder patients was 14 (range 5–31) in rCBF and $rCMR_{glu}$ in the 13 studies published prior to 1990 (Sackeim et al., 1990a). Even ignoring the possible heterogeneity within patient samples, or attempts to reduce variability by adjusting for individual difference factors that covary with rCBF or $rCMR_{glu}$, in much of the published work statistical power was insufficient to reliably determine whether patients differed from controls in global or topographic effects.

Conditions of Measurement. Regardless of the population, there are a variety of technical factors that can confound brain imaging studies both at the time of data acquisition and at the time of image reconstruction and analysis (e.g., Holcomb et al., 1989). With respect to mood disorders, two issues may be of particular consequence. First, repeat rCBF studies in normal samples at rest have typically found that global CBF, while showing a strong test–retest correlation (in part due to marked individual differences), is typically higher at the first measurement relative to later measurements (Warach, Gur, Gur, Skolnick, Obrist, & Reivich, 1987). This first-measurement effect is traditionally attributed to the anxiety evoked by study participation (e.g., radioisotope exposure, complex laboratory environments). The bulk of the literature has contrasted first measurement at rest in mood disorder patients and controls. It is unknown to what extent the findings of global and topographic CBF and metabolic reductions in patients represent a lack of reactivity to the measurement situation. Furthermore, with recovery, it is possible that alterations in rCBF and $rCMR_{glu}$ partly reflect altered emotional reactivity.

This possibility emphasizes the fact that functional imaging measurements can be exquisitely sensitive to psychological state. This is both a strength and a liability. The second concern also is in this domain.

Resting conditions have varied widely across studies, principally involving issues of whether vision and/or audition is blocked. Regardless of these factors, there is the concern that "resting" conditions augment data variability since no control is imposed on the psychological state of subjects. In addition, the imposition of challenge conditions that intensify activity in a distributed neural system may enhance patient/control differences. For example, Weinberger, Berman, and Zec (1986) found that schizophrenic patients did not differ from controls in prefrontal CBF when studied at rest, but did so under cognitive challenge conditions that raised CBF in this region in normal controls. Two of three attempts to examine mood disorder patients under cognitive challenge conditions found topographic differences relative to normal controls (Cohen, Semple, Gross, Nordahl, King, & Pickar, 1989; Gur, Skolnick, Bur, Caroff, Rieger, Obrist, Younkin, & Reivich, 1984; cf., Warren, Butler, Katholi, McFarland, Crews, & Halsey, 1984). In basic studies of emotion, affective challenge conditions have been very useful in identifying discrete topographic EEG patterning (e.g., Davidson, 1988; Sackeim, 1991). Surprisingly there have yet to be attempts to couple advanced functional imaging techniques with affective challenges to quantify rCBF or rCMR in normal or clinical samples.

Identification of Localized Effects. Brain imaging research presents especially difficult problems in quantification and statistical analysis. With state-of-the-art MRI, PET, or SPECT capable of yielding thousands of data points per measurement, there are obvious issues with respect to data reduction. One approach has been to determine regions of interest by averaging values within a particular area or volume. At the grossest level, these groupings have involved summation within lobes (e.g., Post, DeLisi, Holcomb, Uhde, Cohen, & Buchsbaum, 1987; Weinberger et al., 1986) or brain quadrants (e.g., Gur et al., 1984). Such averaging reduces the spatial resolution of imaging measures. Often, it also involves imposing on the brain a functional organization that may have little physiological validity. An example of this issue arose in some of our work on rCBF in major depression (Sackeim et al., 1990a). A traditional metric has been rCBF or rCMR values in frontal areas, relative to the rest of the brain. There are reports that this frontality ratio or similar anterior-posterior gradients is reduced in major depression (e.g., Buchsbaum, DeLisi, Holcomb, Cappelletti, King, Johnson, Hazlett, Dowling-Zimmerman, Post, Morihisa, Carpenter, Cohen, Pickar, Weinberger, Margolin, & Kessler, 1984). Likewise, we found that the ratio frontal cortex to remaining cortex was reduced in elderly major depressives. However, when applying SSM, a statistical model that attempts to identify intrinsic patterns of regional covariance, it was evident that the frontality ratio did not correspond to SSM's description of the major anterior-posterior network. Implicit in the frontality ratio is the assumption that regions within the frontal cortex have uniform and high covariance and that these regions do not covary with remaining brain regions. In contrast, SSM's detailing of an anterior-posterior gradient indicated considerable variability in the degree to which frontal areas participated and strong participation from some other regions, particularly anterior temporal cortex. Depressed patients did not differ from controls in manifesting SSM's description of the anterior-posterior network. Instead, they showed marked differences in a second pattern that comprised selective frontal, central, superior temporal, and anterior parietal regions. In essence, the topographic organization we impose on the brain through data reduction techniques may or may not correspond to the intrinsic functional organization. Further, the field has yet to reach consensus in resolving these problems of pattern recognition.

Another issue particularly relevant to problems of mood disorders concerns the unit of measurement. A considerable amount of work with functional imaging involves use of semiquantitative techniques. Whereas technologies like xenon-inhalation determination of rCBF may provide full quantification (ml/100 g/min), some applications of PET or SPECT may yield only counts of isotope uptake which cannot be translated into fully quantitative and physiologically meaningful equivalents. Under these circumstances, investigators attempt to discern regional differences and often anchor regional values by comparison to a presumably unaffected region (e.g., a frontal/cerebellar ratio, Kanaya & Yonekawa, 1990). This approach is also widely used to overcome the extensive individual differences in global CBF or CMR, even when full quantification is available. For example, ratios may be derived for counts in frontal regions relative to the whole cortex or the whole brain. This approach has two difficulties. The assumption that a particular anchor region is truly unaffected by illness may be questioned. For instance, as described earlier, in mood disorders there is evidence from structural imaging of cerebellar pathology and Dolan, Bench, Scott, Brown, Friston, & Frackowiak (1991) recently reported findings of elevated CBF in the cerebellar vermis of cognitively impaired depressed patients. Second, such transformation may seriously distort the identification of topographic abnormalities in clinical samples. Take the instance where global CMR is used to normalize regional values. Profound and topographically discrete CMR reductions in a network of brain regions will produce an apparent reduction in global CMR in the clinical sample. Further, regions not participating in the abnormal network will have spuriously elevated ratio values and the conclusion may be drawn that the clinical sample has increased CMR in these areas. Finally, the degree and spatial extent of abnormality in the critical network may be underestimated. These and related analytic problems have been discussed elsewhere (Moeller et al., 1987; Sackeim et al., 1990a).

Central Findings in Functional Brain Imaging Studies of Mood Disorders

The initial questions addressed by functional imaging studies have been (1) whether patients in episodes of major depression or mania have abnormal global CBF or CMR, (2) whether at cortical or subcortical levels there are particular topographic deficits in these populations, and, in particular, whether there are lateralized abnormalities, and (3) the extent to which global and/or regional change is seen with somatic treatment and clinical response.

Global Effects. A number of investigators have reported that global CBF or CMR_{glu} at rest is reduced during episodes of major depression (Baxter, Phelps, Maziotta, Schwartz, Gerner, Selin, & Sumida, 1985; Baxter, Schwartz, Phelps, Mazziotta, Guze, Selin, Gerner, & Sumida, 1989; Devous, 1989; Kanaya, & Yonekawa, 1990; Martinot, Hardy, Feline, Huret, Mazoyer, Attar-Levy, Pappata, & Syrota, 1990; Mathew, Meyer, Francis, Semchuk, Mortel, & Claghorn, 1980; Sackeim et al., 1990a; Warren et al., 1984). Using the carbon 11 (^{11}C) glucose method, Kishimoto, Takazu, Ohno, Yamaguchi, Fujita, Kuwahara, Ishii, Matsushita, Yokoi, and Iio (1987) had intriguing evidence that amino acid pools (aspartic acid, γ-aminobutyric acid, glutamic acid, glutamine) may be globally and markedly reduced in unipolar depression. In small samples, Baxter et al. (1985) found that bipolar, but not unipolar, depressed patients manifested a global CMR_{glu} reduction. On the other hand, Devous (1989) assessing rCBF with SPECT obtained a global reduction in unipolar, but not bipolar, depression. Other investigators have not observed global effects in major depression (Bench, Scott, Brown, Friston, Frackowiak, & Dolan, 1991; Delvenne, Delecluse, Hubain, Schoutens, De Maertelaer, & Mendlewicz, 1990; Gur et al., 1984; Goldstein, Brown, Welch, Marcus, Ewing, & Rosenbaum, 1985; Hagman, Buchsbaum, Wu, Rao, Reynolds, & Blinder, 1990; Kling et al., 1986; Silfverskiöld & Risberg, 1989) and, in a small sample, Rosenberg, Vostrup, Andersen, and Bolwig (1988) reported a global CBF elevation. Buchsbaum, Wu, DeLisi, Holcomb, Kessler, King, Hazlett, Langston, and Post (1986) found a global CMR_{glu}

elevation in 16 bipolar and four unipolar patients who received somatosensory stimulation during tracer uptake (cf., Buchsbaum et al., 1984).

The reasons for these inconsistencies are not clear. In this work, global reductions, no effects, and global elevations have been reported for both CBF and CMR_{glu}, so it does not appear that the findings are particular to one type of physiological measure, or to a particular imaging technology (xenon-RCBF, SPECT, PET). In part, these differences may have more to do with variation in patient characteristics, as this literature includes samples of young outpatient depressives with relatively mild illness and elderly and severely depressed inpatients.

In our work, we have observed what is likely to be the most profound global reduction to be reported in major depression (Sackeim et al., 1990a). The sample was one of the largest that has been studied with functional imaging and was different in other respects. The patients were all referred for treatment with ECT, were relatively elderly (mean age = 60 years), severely depressed (mean HRSD = 32, 39% psychotic depression), and had been washed out of psychotropic medication, other than low doses of a benzodiazepine, for an average of approximately two weeks prior to rCBF measurement. Normal controls were closely matched to the patient group on factors known to influence global CBF (age, sex, end-tidal PCO_2, and hematocrit). We used the xenon-inhalation technique, which provides fully quantitative CBF data, is particularly sensitive to cortical phenomena (and insensitive to subcortical phenomena), but has comparatively poor spatial resolution. On average, gray matter CBF in the cortex was approximately 15% less than that in the normal controls. Hypothetically, the global CBF reduction need not have been truly generalized, but could have been due to reduced flow in specific cortical regions. Indeed, we observed that the depressed group had a specific topographic abnormality. However, using a multivariate technique to tease apart topographic and global effects, we concluded that beyond any regional patterning, there was a significant global cortical reduction in the patients. In subsequent work, we selected and matched 30 depressed patients and 30 normal controls from this sample, and 30 patients with Alzheimer's disease (Sackeim et al., 1990c). Global CBF and metabolic reductions are established phenomena in AD (e.g., Prohovnik et al., 1988). Both patients groups were equivalent in the extent of CBF reductions. The overall magnitude of the global CBF deficit we observed in major depression would appear to be of physiological significance. Such a widespread cortical CBF deficit could result from dysfunction in a variety of subcortical nuclei with widespread cortical projection. It could also result from occlusive cerebrovascular disease.

Considerably less work has been done with the acutely manic, and the practical impediments to enlisting the cooperation of such patients in imaging work should not be underestimated. Baxter et al. (1985) in small samples of patients in mixed or manic states suggested that mania is associated with increased global CMR_{glu}. Repeat studies were also conducted in a rapid-cycling patient who manifested a 36% increase in global CMR_{glu} during hypomania relative to depression. We have reported a similar phenomenon in another rapid-cycling bipolar patient with repeat rCBF assessment (Mukherjee, Prohovnik, Sackeim, Decina, & Lee, 1984). In a sample of 11 medication-free manic patients we also observed evidence of a global CBF enhancement (Sackeim, Prohovnik, Apter, Lucas, Decina, Mukherjee, Prudic, & Malitz, 1987), as did Kishimoto et al. (1987) for the uptake of ^{11}C glucose. Silfverskiöld (1987; Silfverskiöld & Risberg, 1989) could not distinguish a sample of 30 medicated manic patients from normal controls in global CBF, but observed that those manic patients on no or low doses of neuroleptics had higher CBF than those patients receiving higher doses. Therefore, the initial evidence suggests that relative to major depression or euthymia, mania may be associated with increased CBF and CMR.

Topographic Effects: Frontal Cortex. Three major hypotheses have emerged regarding the regional distribution of abnormalities in flow and metabolism in major depression. One current hypothesis is that patients in episode of major depression evidence reduced activity in frontal cortical regions, particularly dorsolateral prefrontal cortex (e.g., Baxter et al., 1989). A related view is that this disturbance is more manifest in the left hemisphere than the right hemisphere. A third hypothesis is that depressed patients have reduced flow and metabolism in the basal ganglia, particularly in the caudate.

Investigation of frontal flow and metabolic deficits in depression was first stimulated by reports of hypofrontality in schizophrenic patients (Ingvar & Franzen, 1974). Buchsbaum et al. (1984, 1986) reported that both schizophrenic and bipolar depressed patients had lower anterior-posterior gradients in CMR_{glu} than controls. Baxter et al. (1985) found that the global CMR_{glu} decrease in bipolars was greatest in frontal cortex. Baxter et al. (1989) contrasted subgroups with unipolar depression, bipolar depression, bipolar mania, obsessive-compulsive disorder with secondary depression, obsessive-compulsive disorder without depression, and normal controls. The depressed groups all manifested reduced metabolism in the dorsolateral prefrontal cortex. This was most consistent for left-side measurements, and the ratio of metabolism in left dorsolateral prefrontal cortex relative to the total left hemisphere correlated significantly with HRSD scores. Bipolar manic patients differed from the bipolar depressed in this ratio for the left side, but not for the right side. Hurwitz et al. (1990) contrasted small samples of unipolar depressed and controls and also reported anterior frontal hypometabolism, but with a greater effect on the right side. Martinot et al. (1990) found that 10 depressed patients (7 bipolar, three unipolar) evidenced significant frontal hypometabolism, which was more accentuated on the left side. Cohen et al. (1989) studying depressed patients during the performance of an attentional task obtained a metabolic reduction in medial frontal cortex. Mayberg, Starkstein, Sadzot, Preziosi, Andrezejewski, Dannals, Wagner, and Robinson (1990) comparing depressed and nondepressed Parkinson's patients found that the former were characterized by lower caudate and lower orbital inferior frontal lobe metabolism. In contrast, another group of reports indicate either a general absence of topographic abnormalities in depressed samples studied at rest (Gur et al., 1984; Warren et al., 1984; Goldstein et al., 1985; Kling et al., 1986; Silfverskiöld & Risberg, 1989) or an absence of hypofrontal effects in the context of other topographic deficits (Devous, 1989).

As with the issue of the global flow and metabolism reduction, there are a sufficient number of positive reports of frontal reductions in major depression to give credence to a frontal deficit hypothesis. Furthermore, a number of the studies that did not observe this phenomenon were often either small in size, impressionistic, or included medicated patients. Our findings on this issue take a somewhat different direction. As indicated, using a traditional data analytic approach, we replicated the finding that flow in frontal cortex is reduced relative to the rest of the cortex (Sackeim et al., 1990a). However, in our hands, this particular metric did not receive empirical validation. Using SSM, an anterior-posterior gradient did contribute prominently to the rCBF profiles of both depressed patients and normal controls. The topography of this network departed in important ways from a pattern equally weighting all frontal cortex. Most critically, depressed patients and controls did not differ in the empirically determined anterior-posterior gradient. Selective frontal regions played a critical role in the network that distinguished patients and controls, but so did selective central, superior temporal, and anterior parietal regions. The method used to identify this abnormal network was blind to whether subjects were depressed patients or normal controls. Nonetheless, patients manifested flow reduction across this pattern and the extent of this topographic abnormality correlated significantly with HRSD scores and patient age. In subsequent work, we have found that exclusive manifestations of this topographic abnormality also distinguishes elderly depressives from AD patients (Sackeim et al., 1990c).

The spatial distribution of our topographic abnormality does not conform to that of the major cerebral arteries, making a primary cerebrovascular pathology unlikely. The distribution encompasses broad areas of polymodal association cortex, with the posterior regions corresponding to the inferior parietal lobule (supramarginal and angular gyri) and the superior temporal sulcus. These areas have strong reciprocal connectivity with prefrontal cortex and appear to form a parallel distributed network involved in arousal, attention, motivation, as well as representational cognitive functions (e.g., Goldman-Rakic, 1987; Mountcastle, 1978). Therefore, our findings are in basic agreement with those of others that some selective frontal regions may have reduced functional activity during episodes of major depression. They suggest, however, that these deficits are only a portion of the mosaic, an important part of a more widespread cortical network.

Topographic Effects: Lateral Asymmetries. Both psychophysiological studies in normals (Davidson, 1988) and studies of secondary depression and mania following brain insult (Robinson & Starkstein, 1989) suggest that a dimension of lateralization is critical in the regulation of mood (see Sackeim, 1991 for a review). A group of imaging studies have emphasized left-sided flow and metabolic reductions as being particularly common or pronounced in major depression. Mathew et al. (1980) reported that left-hemisphere CBF was significantly reduced in major depression, with a smaller effect on the right side. Baxter et al. (1985) suggested that some unipolar patients may have a striking asymmetry in frontal cortex, with lower right-side metabolism. As noted earlier, in a subsequent study, Baxter et al. (1989) found that the left-sided dorsolateral frontal metabolism was reduced in unipolar, bipolar, and secondary depression and that metabolic rate in this region covaried with depression severity. Delevenne et al. (1990) reported that CBF was primarily reduced in the left hemisphere of bipolar and endogenous depressives. A similar effect was reported by Kanaya and Yonekawa (1989) and Martinott et al. (1991) claimed to replicate the finding of a specific left prefrontal metabolic reduction in the depressed state. In contrast, this pattern of relatively reduced flow or metabolism in the left hemisphere, and particularly in prefrontal regions, has not been observed by others (e.g., Devous, 1989). Indeed, Uytdenhoef, Portelange, Jacquy, Charles, Linkowski, and Medlewicz (1983) reported that unipolar depressives manifested higher relative left frontal and right posterior (CBF) when compared to euthymic bipolars and normal controls. In a small sample, Hurwitz et al. (1990) reported that significant metabolic reductions in unipolar depressives were most prominent in right, anterior frontal regions. Post et al. (1987), examining temporal lobe glucose utilization, found that maximal metabolism was reduced in primary mood disorders, with the effect significant on the right side and marginal on the left side.

At present, the weight of the evidence suggests that there may be greater relative involvement of left hemisphere regions in the flow and metabolic reductions in major depression. Considerable caution is necessary, however, since few investigators conducted the types of analyses that are traditionally required in laterality research to establish group differences in asymmetry (e.g., Davidson, 1988). The hypotheses being offered require demonstration of group by hemisphere interactions. Perhaps because of limitations of sample size and the notorious lack of power of interaction effects, most investigators have used independent contrasts of depressed and control groups in left and right side regions. The fact that significant differences may emerge for contrasts on one side and not for contrasts on the other is very weak evidence of group differences in asymmetry.

If, in fact, left prefrontal flow and metabolism is particularly reduced in major depression, there will be a compelling parallel to the work by Robinson and others on secondary depression acutely following brain insult. A variety of studies have suggested that left-sided insult is more likely to result in depressive manifestations than right-side insult, with some evidence indicating that the more anterior the insult within the left hemisphere, the more likely this effect is observed (Robinson & Starkstein, 1989). However, considerable behavioral and neuropsychological evidence has implicated disturbance of right hemisphere in major depression and psychophysiological studies have tended to emphasize right frontal activation as accompanying depressed states in normals (see Sackeim, 1991 for a review). Integration of these lines of evidence may be problematic. One possibility is that bilateral frontal flow and metabolic reduction is a trait characteristic in major depression. Hypothetically, during the depressed state there is a right-sided frontal increase, perturbing right hemisphere function (e.g., Sackeim, Greenberg, Weiman, Gur, Hungerbuhler, & Geschwind, 1982). Consequently, cross-sectional studies during the acute phase would most likely identify left-side abnormalities as most prominent in major depression.

Topographic Effects: Basal Ganglia. A small group of studies have examined rCMR in subcortical structures of depressed patients. To date the findings have been relatively consistent and suggest reduced activity in the basal ganglia. Baxter et al. (1985) reported that unipolar depressives had reduced caudate/hemisphere metabolic ratios. Buchsbaum et al. (1986), in a sample composed primarily of bipolar depressives, reported relatively decreased basal ganglia metabolism, particularly in the caudate. Hagman et al. (1990) contrasted eight depressed women, eight bulimic women, and eight normal control women. Decreased basal ganglia metabolism differentiated the depressed women from the other two groups. Mayberg et al. (1990) noted that depressed Parkinsonism patients had both lower caudate metabolism and lower metabolism in orbital inferior frontal cortex. This pattern distinguished the depressed patients from Parkinsonism patients with other behavioral disturbances, including dementia. They suggested that disruption in basal ganglia and inferior frontal pathways may be critical to manifestations of depression. Martinot et al. (1991) did not observe abnormal striatal metabolism in medicated depressed patients.

These findings may be of particular note, given the evidence emerging in structural imaging studies that elderly major depressives have a particularly elevated rate of basal ganglia hyperintensities on T2-weighted MRI examination. Furthermore, there is some evidence that basal ganglia stroke is more likely to result in mood disturbance than thalamic lesions (Starkstein et al., 1988) and a particularly high prevalence of mood disorders in patients with basal ganglia disease has been documented (Folstein, Peyser, Starkstein, & Folstein, 1991; Mayeux, Stern, Rosen, & Leventhal, 1981).

Treatment and Recovery Effects. Relatively little is known about the fate of the flow and metabolic disturbances with somatic treatment and/or clinical response, as most investigations have focused only on the acute phase of illness. Longitudinal investigation is particularly important with functional imaging techniques, given their sensitivity to alterations in psychological state.

The majority of treatment studies have concerned ECT samples. Shortly following seizure termination there is often marked reduction in CBF and CMR (Ackermann, Engel, & Baxter, 1986; Prohovnik, Sackeim, Decina, & Malitz, 1986; Rosenberg et al. 1988; Sackeim et al., 1987; Sackeim et al., 1989; Silfverskiöld, 1987; Silfverskiöld & Risberg, 1989; Volkow, Bellar, Mullani, Jould, & Dewey, 1988). The laterality of this effect depends on electrode placement, with bilateral ECT producing bilateral reductions, particularly in anterior cortex, and right unilateral ECT resulting in more pronounced reductions over the stimulated hemisphere (Prohovnik et al., 1986; Sackeim et al., 1989; Silfverskiöld 1987). Silfverskiöld (1987) demonstrated that such effects are not obtained with sham ECT (anesthesia alone). In preliminary studies of major depression, we have been able to separate the effects of ECT technique from those associated with clinical recovery. Decreased manifestation of a regional covariance pattern, comprised almost exclusively of frontal areas, was associated with superior clinical response to either bilateral or right unilateral ECT (Sackeim et al., 1989). This effect indicated that CBF reductions in selective frontal regions was associated with superior clinical outcome. The effect was observed both in analyses of acute changes in CBF (30 min prior to ECT compared to 1 hour post ECT) and analysis of subacute changes (pretreatment versus several days post-ECT course, while maintained medication free). The changes in CBF associated with good clinical outcome did not result in change in the specific topographic structure that was abnormal in these patients at pretreatment baseline. Indeed, in clinical responders, the global CBF deficit was deepened further at measurements taken during and a few days following the ECT course. Repeat studies of these patients two months following the termination of ECT, now with the confound of continuation pharmacotherapy, indicated that both the global flow deficit and the particular topographic abnormality showed strong stability (Sackeim et al., 1989). This area of research has indicated that, at least in the short-term, ECT results in diminished CBF and CMR. There is also some evidence that the extent and regional distribution of this effect is associated with clinical outcome. Since the bulk of findings in major depression indicate that during the acute phase such patients have global or topographic CBF and CMR reductions, it would not appear that this form of treatment exerts its action by reversing the baseline abnormalities. It appears, therefore, that some of these abnormalities may be state independent. It should be noted that Mukherjee (1989) found that clinical outcome in mania was also associated with the magnitude of acute and subacute CBF reductions.

Animal research has demonstrated that antidepressant medications may have substantial effects on rCBF and rCMR, with common findings of regional reductions in activity following chronic administration (e.g., Caldecott-Hazard, 1988; Preskorn et al., 1982). In clinical research, there is some evidence that abnormal patterns of asymmetry may normalize with antidepressant medication treatment and clinical response (Baxter et al., 1989; Martinot et al., 1990) and there is a suggestion that reduced caudate metabolic activity may also normalize (Baxter et al., 1985). However, recent studies have claimed that global cortical and frontal hypometabolism may persist despite clinical improvement (Hurwitz et al., 1990; Martinot et al., 1990). For instance, in comparing medicated patients while depressed and following clinical response, Martinot et al.

(1990) reported normalization of the asymmetry in prefrontal cortical metabolism, but persistence of frontal and global cortical hypometabolism. Clearly, more definitive research needs to be done. Nonetheless, the work conducted with responders to antidepressant medications also suggests that some of the CBF and CMR abnormalities in major depression may be state independent.

Conclusion

Several generalizations may be drawn from this complex literature. Sufficient evidence has accrued from structural and functional imaging studies to conclude that mood disorder samples are likely to evidence both structural pathology and central dysregulation of blood flow and metabolism. However, we require better characterization of the nature of these abnormalities, the clinical features associated with their manifestation, and prospective, longitudinal investigation of their course. Simultaneous study of structural and functional deficits is an obvious need, particularly since there is evidence from both domains that a subgroup of patients may have state-independent deficits, compatible with, if not, suggestive of occult cerebrovascular disease. Furthermore, the possibility that at least some of the functional deficits are state independent raises a serious complication. Surely, the state of depression is associated with altered metabolic demand in critical networks. Teasing apart the contributions of state and trait will require serious attention in the years ahead.

Acknowledgments

Preparation of this chapter was supported in part by grants MH35636 and MH44779 from the National Institute of Mental Health, grant AG05433 from the National Institute of Aging, and an Established Investigator Award from the National Alliance for Research in Schizophrenia and Depression.

References

Abas, MA, Sahakian BJ, Levy R: Neuropsychological deficits and CT scan changes in elderly depressives. *Psychol Med* 1990; 29: 507–520.

Ackermann RF, Engel J, Baxter J: Positron emission tomography and autoradiographic studies of glucose utilization following electroconvulsive seizures in humans and rats. *Ann NY Acad Sci* 1986; 462: 263–269.

Agren H, Reibring L, Hartvig P, Tedroff J, Bjurling P, Hornfeldt K, Andersson Y, Lundvist H, Langstrom B: Low brain uptake of L-[11C]-hydroxytrotophan in major depression: A positron emission tomography study on patients and healthy volunteers. *Acta Psychiat Scand* 1991; 83: 449–455.

Aharon-Peretz J, Cummings JL, Hill MA: Vascular dementia and dementia of the Alzheimer type: cognition, ventricular size, and leuko-araiosis. *Arch Neurol* 1988; 45: 719–721.

Altshuler LL, Conrad A, Hauser P, Li XM, Guze BH, Denikoff K: Reduction of temporal lobe volume in bipolar disorder: a preliminary report of magnetic resonance imaging. *Arch Gen Psychiat* 1991; 48: 482–483.

Andreasen NC: Nuclear magnetic resonance imaging. In: Andeasen NC (Ed.), *Brain Imaging: Applications in Psychiatry*, Washington, DC: American Psychiatric Press, 1989, 67–121.

Andreasen NC, Swayze V, 2d, Flaum M, Alliger R, Cohen G: Ventricular abnormalities in affective disorder: clinical and demographic correlates. *Am J Psychiat* 1990; 147: 893–900.

Awad IA, Johnson PC, Spetzler RF, Hodak JA: Incidental subcortical lesions identified on magnetic resonance imaging in the elderly. II. Postmortem pathological correlations. *Stroke* 1986a; 17: 1090–1097.

Awad IA, Modic M, Little JR, Furlan AJ, Weinstein M: Focal parenchymal lesions in transient ischemic attacks: correlation of computed tomography and magnetic resonance imaging. *Stroke* 1986b; 17: 399–403.

Awad IA, Spetzler RF, Hodak JA, Awad CA, & Carey R. (1986c). Incidental subcortical lesions identified on magnetic resonance imaging in the elderly. I. Correlation with age and cerebrovascular risk factors. *Stroke, 17,* 1084–1089.

Awad IA, Spetzler RF, Hodak JA, Awad CA, Williams F. Jr, & Carey R. (1987). Incidental lesions noted on magnetic resonance imaging of the brain: prevalence and clinical significance in various age groups. *Neurosurgery, 20,* 222–7.

Barnes, D.M., Enzmann DR: The evaluation of white matter disease as seen on computed tomography. *Radiology* 1981; 138:379–383.

Baron JC, D'Antona R, Pantano P, Serdaru M, Samson Y, Bousser MG: Effects of thalamic stroke on energy metabolism of the cerebral cortex. A positron tomography study in m an. *Brain* 1986; 109: 1243–1259.

Baxter LR, Phelps ME, Maziotta JC, Schwartz JM, Gerner RH, Selin CE, Sumida RM: Cerebral metabolic rates for glucose in mood disorders. *Arch Gen Psychiat* 1985; 42: 441–447.

Baxter LR, Jr, Schwartz JM, Phelps ME, Mazziotta JC, Guze BH, Selin CE, Gerner RH, Sumida RM: Reduction of prefrontal cortex glucose metabolism common to three types of depression. *Arch Gen Psychiat* 1989; 46: 243–250.

Bench CJ, Scott LC, Brown RG, Friston KJ, Frackowiak RSJ, Dolan RJ: Regional cerebral blood flow in depression determined by positron emission tomography. *J Cerebral Blood Flow Metab* 1991; 11 Supp 2: S654.

Benes FM, Bird ED: An analysis of the arrangement of neurons in the cingulate cortex of schizophrenic patients. *Arch Gen Psychiat* 1987; 44: 608–615.

Bentson J, Reza M, Winter J, Wilson G: Steroids and apparent cerebral atrophy on computed tomography scans. *J Comput Assist Tomogr* 1978; 2: 16–23.

Bergsholm P, Larsen JL, Rosendahl K, Holsten F: Electroconvulsive therapy and cerebral computed tomography. A prospective study. *Acta Psychiat Scand* 1989; 80: 566–572.

Bird JM, Levy R, Jacoby RJ: Computed tomography in the elderly: Changes over time in a normal population. *Br J Psychiat* 1986; 148: 80–85.

Bradley WG, Waluch V, Brant-Zawadski M, Yadley RA, Wycoff RR: Patchy, periventricular white matter lesions in the elderly: A common observation during NMR imaging. *Nonivas Med Imag* 1984; 1: 35–41.

Braffman BH, Zimmerman RA, Trojanowski JQ, Gonatas NK, Hickey WF, Schlaepfer WW: Brain MR: Pathologic correlation with gropss histopathology. 2. Hyperintense white-matter foci in the elderly. *AJNR* 1988; 9: 629–636.

Brewer C, Barrett L: Brain damage due to alcohol consumption: An airencephalographic, psychometric, and electroencephalgrahic study. *Br J Addict* 1971; 66: 170–182.

Buchsbaum MS, DeLisi LE, Holcomb HH, Cappelletti J, King AC, Johnson J, Hazlett E, Dowling-Zimmerman S, Post RM, Morihisa J, Carpenter W, Cohen R, Pickar D, Weinberger DR, Margolin R, Kessler RM: Anteroposterior gradients in cerebral glucose use in schizophrenia and affective disorders. *Arch Gen Psychiat* 1984; 41: 1159–1166..

Buchsbaum MS, Wu J, DeLisi LE, Holcomb H, Kessler R, Johnson J, King AC, Hazlett E, Langston K, Post RM: Frontal cortex and basal ganglia metabolic rates assessed by positron emission tomography with [18F]2-deoxyglucose in affective illness. *J Affective Disord* 1986; 10: 137–152.

Cala LA, Mastaglia FL: Computerized tomography in chronic alcoholics. *Alcoholism*, 1981; 5: 283–294.

Caldecott-Hazard S, Mazziotta J, Phelps M: Cerebral correlates of depressed behavior in rats, visualized using 14C-2-deoxyglucose autoradiography. *J Neurosci* 1988; 8: 1951–1961.

Caplan LR, Schoene WC: Clinical features of subcortical arteriosclerotic encephalopathy (Binswanger disease). *Neurology* 1978; 28: 1206–1215.

Coffey CE, Figiel GS: Neuropsychiatric significance of subcortical encephalomalacia. In: Carroll, BJ, Barrett JE (Eds.), *Psychopathology and the Brain*, New York: Raven, 243–264.

Coffey CE, Figiel GS, Djang WT, Sullivan DC, Herfkens RJ, Weiner RD: Effects of ECT on brain structure: A pilot prospective magnetic resonance imaging study. *Am J Psychiat* 1988a; 145: 701–706.

Coffey CE, Figiel GS, Djang WT, Cress M, Saunders WB, Weiner RD: Leukoencephalopathy in elderly depressed patients referred for ECT. *Biol Psychiat* 1988b; 24: 143–161.

Coffey CE, Figiel GS, Djang WT, Saunders WB, Weiner RD: Subcortical white matter hyperintensity on magnetic resonance imaging: Clinical and neuroanatomic correlates in the depressed elderly. *J Neuropsychiat Clin Neurosci* 1989; 1: 135–144.

Coffey CE, Figiel GS, Djang WT, Weiner RD: Subcortical hyperintensity on magnetic resonance imaging: a comparison of normal and depressed elderly subjects. *Am J Psychiat* 1990; 147: 187–189.

Coffey CE, Weiner RD, Djang WR, Figiel GS, Soady SAR, Patterson LJ, Holt PD, Spritzer CE, Wilkinson WE: Brain anatomic effects of ECT: a prospective magnetic resonance imaging study. *Arch Gen Psychiat* 1991; 48: 1013–1021.

Coffman JA: Computed tomography. In Andreasen NC (Ed.), *Brain Imaging: Applications in Psychiatry*, Washington, DC: American Psychiatric Press, 1–66.

Coffman JA, Bornstein RA, Olson SC, Schwarzkopf SB, Nasrallah HA: Cognitive impairment and cerebral structure by MRI in bipolar disorder. *Biol Psychiat* 1990; 27: 1188–1196.

Cohen RM, Semple WE, Gross M, Nordahl TE, King AC, Pickar D, Post RM: Evidence for common alterations in cerebral glucose metabolism in major affective disorders and schizophrenia. *Neuropsychopharmacology* 1989; 2: 241–254.

Davidson RJ: Cerebral asymmetry, affective style, and psychopathology. In: Kinsbourne M (Ed.), *Cerebral Hemisphere Function in Depression*, Washington, DC: American Psychiatric Press, 1988, 1–22.

Delvenne V, Delecluse F, Hubain PP, Schoutens A, De Maertelaer V, Mendlewicz J: Regional cerebral blood flow in patients with affective disorders. *Br J Psychiat* 1990; 157: 359–365.

De Reuck J, Crevits L, DeCoster W, Sieben G, vander Eecken H: Pathogenesis of Binswanger chronic progressive subcortical encephalopathy. *Neurology* 1980; 920–928.

DeReuk J: The human periventricular arterial blood supply and the anatomy of cerebral infarctions. *Eur Neurol* 1971; 5: 321–334.

Devous MD Sr: Imaging brain function by single-photon emission computer tomography. In: Andreasen NC (Ed.), *Brain Imaging: Applications in Psychiatry*, Washington DC: American Psychiatric Press, 1989, 147–234.

Dewan MJ, Haldipur CV, Lane EE, Ispahani A, Boucher MF, Major LF: Bipolar affective disorder. I. Comprehensive quantitative computed tomography. *Acta Psychiat Scand* 1988a; 77: 670–676.

Dewan MJ, Haldipur CV, Boucher MF, Ramachandran T, Major LF: Bipolar affective disorder. II. EEG, neuropsychological, and clinical correlates of CT abnormality. *Acta Psychiat Scand* 1988b; 77: 677–682.

Dewan MJ, Haldipur CV, Boucher M, Major LF: Is CT ventriculomegaly related to hypercortisolemia? *Acta Psychiat Scand* 1988c; 77: 230–231.

Diaz JF, Merskey H, Hachinski VC, Lee DH, Boniferro M, Wong CJ, Mirsen TR, Fox H: Improved recognition of leukoaraiosi and cognitive impairment in Alzheimer's disease. *Arch Neurol* 1991; 48: 1022–1025.

Dolan RJ, Calloway SP, Mann AH: Cerebral ventricular size in depressed subjects. *Psychol Med* 1985; 15: 873–878.

Dolan RJ, Calloway SP, Thacker PF, Mann AH: The cerebral cortical appearance in depressed subjects. *Psychol Med* 1986; 16: 775–779.

Dolan RJ, Poynton AM, Bridges PK, Trimble MR: Altered magnetic resonance white-matter T1 values in patients with affective disorder. *Br J Psychiatry* 1990; 157: 107–110.

Dolan RJ, Bench CJ, Scott LC, Brown RG, Friston KJ, Frackowiak RSJ: Regional cerebral blood flow in 'depressive pseudodementia' determined by positron emission tomography. *J Cerebral Blood Flow Metab* 1991; 11 Suppl 2: S17.

Dupont RM, Jernigan TL, Butters N, Delis D, Hesselink JR, Heindel W, Gillin JC: Subcortical abnormalities detected in bipolar affective disorder using magnetic resonance imaging: Clinical and neuropsychological significance. *Arch Gen Psychiat* 1990; 47: 55–59.

Erkinjuntt T, Ketonen L, Sulkava R, Sipponen J, Vuorialho M, Iivanainen M: Do white matter changes on MRI and CT differentiate vascular dementia from Alzheimer's disease? *J Neurol Neurosurg Psychiat* 1987; 50: 37–42.

Fazekas F: Magnetic resonance signal abnormalities in asymptomatic individuals: their incidence and functional correlates. *Eur Neurol* 1989; 29: 164–168.

Fazekas F, Niederkorn K, Schmidt R, Offenbacher H, Horner S, Bertha G, Lechner H: White matter signal abnormalities in normal individuals: Correlation with carotid ultrasonography, cerebral blood flow measurements, and cerebrovascular risk factors. *Stroke* 1988; 19: 1285–1288.

Figiel GS, Krishnan KR, Rao VP, Doraiswamy M, Ellinwood EH, Nemeroff CB, Evans D, Boyko O: Subcortical hyperintensities on brain magnetic resonance imaging: a comparison of normal and bipolar subjects. *J Neuropshychiat Clin Neurosci* 1991; 3: 18–22.

Figiel GS, Krishnan KR, Doraiswamy PM: Subcortical structural changes in ECT-induced delirium. *J Geriatr Psychiat Neurol* 1990; 3: 172–176.

Filley CM, Davis KA, Schmitz SP, Stears JC, Heaton RK, Kelly J, Culig KM, Scherzinger AL: Neuropsychological performance and magnetic resonance imaging in Alzheimer's disease and normal aging. *Neuropsychiat, Neuropsychol Behav Neurol* 1989; 2: 81–92.

Folstein SE, Peyser CE, Starkstein SE, Folstein MF: Subcortical triad of Huntington's disease: a model for a neuropathology of depression, dementia, and dyskinesia. In: Caroll BJ, Barrett JE (Eds.), *Psychopathology and the Brain,* New York: Raven, 1991, 65–75.

Georgotas A, McCue RE, Cooper TB: A placebo-controlled comparison of nortriptyline and phenelzine in maintenance therapy of elderly depressed patients. *Arch Gen Psychiat* 1989; 46: 783–786.

Gerard G, Weisberg LA: MRI periventricular lesions in adults. *Neurology* 1986; 36: 998–1001.

Gerner RH, Bunney Jr, WE: Biological hypotheses of affective disorders. In: Berger PA, Brodie KH (Eds.), *Biological Psychiatry, American Handbook of Psychiatry,* New York: Basic Books, 1986, Vol. 8, 265–301.

Glassman AH, Roose SP, Giardiana EV, Bigger Jr. JT: Cardiovascular effects of tricyclic antidepressants. In: Meltzer HY (Ed.), *Psychopharmacology: The Third Generation of Progress*, New York: Raven, 1987, 1437–1442.

Goldman-Rakic PS: Circuitry of the prefrontal cortex and the regulation of behavior by representational knowledge. In: Plum F, Mountcastle V (Eds.), *Handbook of Physiology, Sec. 1: The Nervous System, Vol. 5: Higher Cortical Function*, Bethesda, MD: American Physiology Society, 1987, 373–417.

Goldstein PC, Brown GG, Welch KMA, Marcus A, Ewing JR, Rosenbaum G: Age-related decline of rCBF in schizophrenia and major affective disorder. *J. Cereb Blood Flow Metab* 1985; 5, Suppl 1: S203–204.

Goodwin FK, Jamison KR: *Manic-Depressive Illness*. New York: Oxford University Press, 1989.

Gur RC, Packer IK, Hungerbuhler JP, Reivich M, Obrist WD, Amarnek WS, Sackeim HA: Differences in the distribution of gray and white matter in human cerebral hemispheres. *Science* 1980; 207: 1226–1228.

Gur RE, Skolnick BE, Gur RC, Caroff S, Rieger W, Obrist WD, Younkin D, Reivich M: Brain function in psychiatric disorders. II. Regional cerebral blood flow in medicated unipolar depressives. *Arch Gen Psychiat* 1984; 41: 695–699.

Gyulai L, Wicklund SW, Greenstein R, Bauer MS, Ciccione P, Whybrow PC, Zimmerman J, Kovachich G, Alves W: Measurement of tissue lithium concentration by lithium resonance spectroscopy in patients with bipolar disorder. *Biol Psychiat* 1991; 29: 1161–1170.

Hachinski VC, Potter P, Merskey H: Leuko-araiosis. *Arch Neurol* 1987; 44: 21–23.

Hagman JO, Buchsbaum MS, Wu JC, Rao SJ, Reynolds CA, Blinder BJ: Comparison of regional brain metabolism in bulimia nervosa and affective disorder assessed with positron emission tomography. *J Affective Disord* 1990; 19: 153–162.

Harvey I, McGuffin P, Williams M, Toone BK: The ventricle-brain ratio (VBR) in functional psychoses: An admixture analysis. *Psychiatry Res* 1990; 35: 61–69.

Hauser P, Dauphinais ID, Berrettini W, DeLisi LE, Gelernter J: Corpus callosum dimensions measured by magnetic resonance imaging in bipolar affective disorder and schizophrenia. *Biol Psychiat* 1989a; 26: 659–668.

Hauser P, Altshuler LL, Berrettini W, Dauphinais ID, Gelernter J, Post RM: Temporal lobe measurement in primary affective disorder by magnetic resonance imaging. *J Neuropsychiat Clin Neurosci* 1989b; 1: 128–134.

Hauser P: Magnetic resonance imaging in primary affective disorder. In: Hauser P. (Ed.), *Brain Imaging in Affective Disorders*, Washington, DC: American Psychiatric Press, 1991, 25–61.

Heninger GR, Charney DS: Mechanisms of action of antidepressant treatments: Implications for the etiology and treatment of depressive disorders. In: Meltzer HY (Ed.), *Psychopharmacology: The Third Generation of Progress*, New York: Raven, 535–544.

Hershey LA, Modic MT, Greenough PG, Jaffe DF: Magnetic resonance imaging in vascular dementia. *Neurology* 1987; 37: 29–36.

Holcomb HH, Links J, Smith C, Wong D: Positron emission tomography: Measuring metabolic and neurochemical characteristics of the living nervous system. In: Andreasen NC (Ed.), *Brain Imaging: Applications in Psychiatry,* Washington, DC: American Psychiatric Press, 235–370.

Hurwitz TA, Clark C, Murphy E, Klonoff H, Martin WR, Pate BD: Regional cerebral glucose metabolism in major depressive disorder. *Can J Psychiat* 1990; 35: 684–688.

Iacono WG, Smith GN, Moreau M, Beiser M, Fleming JA, Lin TY: Ventricular and sulcal size at the onset of psychosis. *Am J Psychiat* 1988; 145: 820–824.

Ingvar DH, Franzen G: Abnormalities of cerebral blood flow distribution in chronic schizophrenia. *Acta Psychiatr Scand* 1974; 50: 425–462.

Inzitari D, Diaz F, Fox A, Hachinski VC, Steingart A, Lau C, Donald A, Wade J, Mulic H, Merskey H: Vascular risk factors and leukoaraiosis. *Arch Neurol* 1987; 44: 42–47.

Jacoby RJ, Levy R: Computed tomography in the elderly, 3: Affective disorder. *Br J Psychiatry* 1980; 136: 270–275.

Jacoby RJ, Levy R, Dawson JM: Computed tomography in the elderly. I: The normal population. *Br J Psychiat* 1980; 136: 249–255.

Jacoby RJ, Levy R, Bird JM: Computed tomography and the outcome of affective disorder: A follow-up study of elderly patients. *Br J Psychiat* 1981; 139: 288–292.

Jernigan TL, Zisook S, Heaton RK, Moranville JT, Hesselink JR, Braff DL: Magnetic resonance imaging abnormalities in lenticular nuclei and cerebral cortex in schizophrenia. *Arch Gen Psychiat* 1991; 48: 881–890.

Jeste DV, Lohr JB, Goodwin FK: Neuroanatomical studies of major affective disorders: A review and suggestions for future research. *Br J Psychiat* 1988; 153: 444–459.

Johnstone EC, Crow TJ, Frith CD, Husband J, Kreel L: Cerebral ventricular size and cognitive impairment in chronic schizophrenia. *Lancet* 1976; 2: 924–926.

Johnstone EC, Owens DG, Crow TJ, Colter N, Lawton CA, Jagoe R: Hypothyroidism as a correlate of lateral ventricular enlargement in manic-depressive and neurotic illness. *Br Psychiat* 1986; 148: 317–321.

Johnstone EC, Owens DG, Crow TJ, Frith CD, Alexandropolis K, Bydder G, Colter N: Temporal lobe structure as determined by nuclear magnetic resonance in schizophrenia and bipolar affective disorder. *J Neurol Neurosurg Psychiat* 1989a; 52: 736–741.

Johnstone EC, Owens DG, Bydder GM, Colter N, Crow TJ, Frith CD: The spectrum of structural brain changes in schizophrenia: age of onset as a predictor of cognitive and clinical impairments and their cerebral correlates. *Psychol Med* 1989b; 19: 91–103.

Keller MB, Lavori PW, Klerman GL, Andreasen NC, Andicott J, Coryell W, Fawcett J, Rice JP, Hirschfeld RM: Low levels and lack of predictors of somatotherapy and psychotherapy received by depressed patients. *Arch Gen Psychiatry* 1986; 43: 458–466.

Kellner CH, Rubinow DR, Gold PW, Post RM: Relationship of cortisol hypersecretion to brain CT scan alterations in depressed patients. *Psychiat Res* 1983; 8: 191–197.

Kellner CH, Rubinow DR, Post RM: Cerebral ventricular size and cognitive impairment in depression. *J Affect Disord* 1986; 10: 215–219.

Kertesz A, Black SE, Tokar G, Benke T, Carr T, Nicholson L: Periventricular and subcortical hyperintensities on magnetic resonance imaging. *Arch Neurol* 1988; 45: 404–408.

Keshavan MS, Kapur S, Pettegrew JW: Magnetic resonance spectroscopy in psychiatry: potential, pitfalls, and promise. *Am J Psychiatry* 1991; 148: 976–985.

Killen AR, Oster G, Colditz GA: An assessment of the role of 123I-*N*-isopropyl-*p*-iodoamphetamine with single-photon emission computed tomography in the diagnosis of stroke and Alzheimer's disease. *Nucl Med Commun* 1989; 10: 271–284.

Kinkel WR, Jacobs L, Polanchini I, Bates V, Heffner RR: Subcortical arteriosclerotic encephalopathy (Binswanger's disease). *Arch Neurol* 1985; 42: 951–959.

Kirkpatrick JB, Hayman LA: White-matter lesions in MR imaging of clinically healthy brains of elderly subjects: Possible pathologic basis. *Radiology* 1987; 162: 509–511.

Kishimoto H, Takazu O, Ohno S, Yamaguchi T, Fujita H, Kuwahara H, Ishii T, Matsushita M, Yokoi S, Iio M: 11C-glucose metabolism in manic and depressed patients. *Psychiat Res* 1987; 22:81–88.

Kling AS, Metter EJ, Riege WH, Kuhl DE: Comparison of PET measurement of local brain glucose metabolism and CAT measurement of brain atrophy in chronic schizophrenia and depression. *Am J Psychiat* 1986; 143: 175–180.

Kolbeinsson H, Arnaldsson OS, Pétursson H, Skúlason S: Computed tomographic scans in ECT-patients. *Acta Psychiat Scand* 1986; 73: 28–32.

Komoroski RA, Newton JE, Walker E, Cardwell D, Jagannathan NR: In vivo NMR spectroscopy of lithium-7 in humans. *Magn Reson Med* 1990; 15: 347–356.

Krieg J-C, Pirke K-M, Lauer C, Backmund H: Endocrine, metabolic, and cranial computed tomographic findings in anorexia nervosa. *Biol Psychiat* 1988; 23: 377–387.

Krishnan KR, Doraiswamy PM, Lurie SN, Figiel GS, Husain MM, Boyko OB, Ellinwood EH Jr, Nemeroff CB: Pituitary size in depression. *J Clin Endocrinol Metab* 1991; 72: 256–259.

Krishnan KR, Goli V, Ellinwood EH, France RD, Blazer DG, Nemeroff CB: Leukoencephalopathy in patients diagnosed as major depressive. *Biol Psychiat* 1988; 23: 519–522.

Lammers CH, Doraiswamy PM, Husain MM, Figiel GS, Lurie SN, Boyko OB, Ellinwood EH, Nemroff CB, Krishnan KRR: MRI of corpus callosum and septum pellucidum in depression. *Biol Psychiat* 1991; 29: 300–301.

Lesser IM, Miller BL, Boone KB, Hill-Gutierrez E, Mehringer CM, Wang K, Mena I: Brain injury and cognitive function in late-onset psychotic depression. *J Neuropsychiat Clin Neurosci* 1991; 3: 33–40.

Lewine RR, Risch SC, Risby E, Stipetic M, Jewart RD, Eccard M: Lateral ventricle-brain ratio and balance between CSF HVA and 5-HIAA in schizophrenia. *Am J Psychiat* 1991; 148: 1189–1194.

Luchins DJ, Levine RJ, Meltzer HY: Lateral ventricular size, psychopathology and medication response in psychoses. *Biol Psychiatry* 1984; 19: 29–44.

Luchins DJ, Metlzer HY: Ventricular size and psychosis in affective disorder. *Biol Psychiat* 1983; 18: 1197–1198.

Luyten PR, den Hollander JA: Observations of metabolites in the human brain by MR spectroscopy. *Radiology* 1986; 161: 795–798.

Martinot JL, Hardy P, Feline A, Huret JD, Mazoyer B, Attar-Levy D, Pappata S, Syrota A: Left prefrontal glucose hypometabolism in the depressed state: a confirmation. *Am J Psychiat* 1990; 147: 1313–1317.

Mathew RJ, Meyer JS,Francis DJ, Semchuk KM, Mortel K, Claghorn JL: Cerebral blood flow in depression. *Am J Psychiat* 1980; 137: 1149–1450.

Mayberg HS, Robinson RG, Wong DF, Parikh R, Bolduc P, Starkstein SE, Price T, Dannals RF, Links JM, Wilson AA, Ravert HT, Wagner HN: PET imaging of cortical S2 serotonin receptors after stroke: lateralized changes and relationship to depression. *Am J Psychiat* 1988; 145: 937–943.

Mayberg HS, Starkstein SE, Sadzot B, Preziosi T, Andrezejewski PL, Dannals RF, Wagner HN Jr, Robinson RG: Selective hypometabolism in the inferior frontal lobe in depressed patients with Parkinson's disease. *Ann Neurol* 1990; 28: 57–64.

Mayeux R, Stern Y, Rosen J, Leventhal J: Depression, intellectual impairment, and Parkinson's disease. *Neurology* 1981; 31: 645–650.

Meldrum BS: Neuropathological consequences of chemically and electrically induced seizures. *Ann NY Acad Sci* 1986; 462: 186–193.

Meltzer HY, Tong C, Luchins DJ: Serum dopamine-β-hydroxylase activity and lateral ventricular size in affective disorders and schizophrenia. *Biol Psychiat* 1984; 19: 1395–1402.

Mesulam M-M, Van Hoesen GW, Pandya DN, Geschwind N: Limbic and sensory connections of the inferior parietal lobule (Area PG) in the rhesus monkey: A study with a new method for horseradish peroxidase histochemistry. *Brain Res.* 1977; 136: 393–414.

Mirsen TR, Lee DH, Wong CJ, Diaz JF, Fox H, Hachinski VC, Merskey H: Clinical correlates of white-matter change on magnetic resonance imaging scans of the brain. *Arch Neurol* 1991; 48: 1015–1021.

Moeller JR, Strother SC, Sidtis JJ, Rottenberg DA: Scaled subprofile model: A statistical approach to the analysis of functional patterns in positron emission tomographic data. *J Cereb Blood Flow Metab* 1987; 7: 649–658.

Momose KJ, Kielberg RN, Kliman B: High incidence of cortical atrophy of the cerebral and cerebellar hemispheres in Cushing's disease. *Radiology* 1971; 99: 341–348.

Morihisa JM: Computerized EEG and evoked potential mapping. In: Andreasen NC (ED.), *Brain Imaging: Applications in Psychiatry,* Washington DC: American Psychiatric Press, 123–145.

Mota de Freitas D, Silberberg J, Espanol MT, Dorus E, Abraha A, Dorus W, Elenz

E, Whang W: Measurement of lithium transport in RBC from psychiatric patients receiving lithium carbonate and normal individuals by 7Li NMR spectroscopy. *Biol Psychiat* 1990; 28: 415–424.

Mountcastle V: Brain systems for directed attention. *J R Soc Med* 1978; 71: 14–27.

Mukherjee S: Mechanisms of the antimanic effect of electroconvulsive therapy. *Convulsive Ther* 1989; 5: 227–243.

Mukherjee S, Prohovnik I, Sackeim HA, Decina P, Lee C: Regional cerebral blood flow in a rapid-cycling bipolar patient. *Presented at the Annual Meeting of the American Psychiatric Association*, 1984.

Murphy E: The prognosis of depression in old age. *Br J Psychiat* 1983; 142: 111–119.

Nasrallah HA, McCalley-Whitters M, Jacoby CG: Cerebral ventricular enlargement in young manic males. A controlled CT study. *J Affective Disord* 1982a; 4: 15–19.

Nasrallah HA, McCalley-Whitters M, Jacoby CG: Cortical atrophy in schizophrenia and mania: A comparative study. *J Clin Psychiat* 1982b; 43: 439–441.

Nasrallah HA, McCalley-Whitters M, Pfohl B: Clinical significance of large cerebral ventricles in manic males. *Psychiat Res* 1984; 13: 151–156.

Nasrallah HA, Coffman JA, Olson SC: Structural brain-imaging findings in affective disorders: An overview. *J Neuropsychiatry Clin Neurosci* 1989; 1: 21–26.

Pearlson GD, Veroff AE: Computerized tomographic scan changes in manic-depressive illness. *Lancet* 1981; 2: 470.

Pearlson GD, Garbacz DJ, Tompkins RH, Ahn HS, Gutterman DF, Veroff AE, DePaulo JR: Clinical correlates of lateral ventricular enlargement in bipolar affective disorder. *Am J Psychiat* 1984; 141:253–256.

Pearlson GD, Garbacz DJ, Moberg PJ, Ahn HS, DePaulo JR: Symptomatic, familial, perinatal, and social correlates of computerized axial tomography (CAT) changes in schizophrenics and bipolars. *J Nerv Ment Dis* 1985; 173: 42–50.

Pearlson GD, Rabins PV, Kim WS, Speedie LJ, Moberg PJ, Burns A: Structural brain CT changes and cognitive deficits in elderly depressives with and without reversible dementia ("pseudodementia"). *Psychol Med* 1989; 19: 573–584.

Pearlson GD, Rabins PV, Burns A: Centrum semiovale white matter CT changes associated with normal aging, Alzheimer's disease and late life depression with and without reversible dementia. *Psychol Med* 1991; 21: 321–328.

Perani D, Di Piero V, Lucignani G, Gilardi MC, Pantano P, Rossetti C, Pozzilli C, Gerundini P, Fazio F, Lenzi GL: Remote effects of subcortical cerebrovascular lesions: A SPECT cerebral perfusion study. *J Cereb Blood Flow Metab* 1988; 8: 560–567.

Post F: The management and nature of depressive illnesses in late life: A follow-through study. *Br J Psychiat* 1972; 121: 393–404.

Post RM, DeLisi LE, Holcomb HH, Uhde TW, Cohen R, Buchsbaum MS: Glucose utilization in the temporal cortex of affectively ill patients: Positron emission tomography. *Biol Psychiat* 1987; 22: 545–553.

Preskorn SH, Raichle ME, Hartman BK: Antidepressants alter cerebrovascular permeability and metabolic rate in primates. *Science* 1982; 217: 250–252.

Prohovnik I, Sackeim HA, Decina P, Malitz S: Effects on regional cerebral blood flow: Modality and therapeutic outcome. *Ann NY Acad Sci* 1986; 462: 249–262.

Prohovnik I, Mayeux R, Sackeim HA, Smith G, Stern Y, Alderson PO: Cerebral perfusion as a diagnostic marker of early Alzheimer's disease. *Neurology* 1988; 38: 931–937.

Prohovnik I, Smith G, Sackeim HA, Mayeux R, Stern Y: Gray-matter degeneration in presenile Alzheimer's disease. *Ann Neurol* 1989; 25: 117–124.

Rabins PV, Pearlson GD, Aylward E, Kumar AJ, Dowell K. Cortical magnetic resonance imaging changes in elderly inpatients with major depression. *Am J Psychiat* 1991; 148: 617–620.

Rao SM, Mittenberg W, Bernardin L, Haughton V, Leo GJ: Neuropsychological test findings in subjects with leukoaraiosis. *Arch Neurol* 1989; 46: 40–44.

Reveley AM, Reveley MA, Murray RM: Cerebral ventricular enlargement in nongenetic schizophrenia: a controlled twin study. *Br J Psychiat* 1984; 144: 89–93.

Rieder RO, Mann LS, Weinberger DR, van Kammen DP, Post RM: Computed tomographic scans in patients with schizophrenia, schizoaffective, and bipolar disorder. *Arch Gen Psychiat* 1983; 40: 735–739.

Robinson RG, Starkstein SE: Mood disorders following stroke: New findings and future directions. *J Geriatr Psychiat* 1989; 22: 1–15.

Román GC: Senile dementia of the Binswanger type. A vascular form of dementia in the elderly. *J Am Med Associ* 1987; 258: 1782–1788.

Rosenberg R, Vostrup S, Andersen A, Bolwig T: Effect of ECT on cerebral blood flow in melancholia assessed with SPECT. *Convulsive Ther* 1988; 4: 62–73.

Rossi A, Stratta P, Di Michele V, Bolino F, Nistico R, de Leonardis R, Sabatini MD, Casacchia M: A computerized tomographic study in patients with depressive disorder: A comparison with schizophrenic patients and controls. *Acta Psychiat Belg* 1989; 89: 56–61.

Roy CS, Sherrington CS: On the regulation of the blood supply of the brain. *J Physiol* 1890; 11: 85–108.

Roy-Byrne PP, Post RM, Kellner CH, Joffe RT, Uhde TW: Ventricular-brain ratio and life course of illness in patients with affective disorder. *Psychiatry Res* 1988; 23: 277–284.

Rubin EH, Preskorn SH, Hartman BK: The role of the central adrenergic system in the regulation of the cerebromicrocirculation: Basic and clinical implications. *Ann NY Acad Sci* 1986; 462: 224–231.

Sacchetti E, Vita A, Calzeroni A, Invernizzi G, Cazzullo CL: Neuromorphological correlates of schizophrenic disorders: Focus on cerebral ventricular enlargement. In: Cazzullo GL, Invernizzi G, Sacchetti E, Vita A (Eds.), *Etiopathogenetic Hypotheses of Schizophrenia: The Impact of Epimediological, Biochemical, and Neuromorphological Studies,* Lancaster, England: MTP Press, 1987, 67–91.

Sackeim HA: Emotion, disorders of mood, and hemispheric functional specialization. In: Carroll BJ, Barrett JE (Eds.), *Psychopathology and the Brain,* New York: Raven, 1991, 209–242.

Sackeim HA, Greenberg MS, Weiman AL, Gur RC, Hungerbuhler JP, Geschwind

N: Hemispheric asymmetry in the expression of positive and negative emotions: Neurologic evidence. *Arch Neurol* 1982; 39: 210–218.

Sackeim HA, Prohovnik I, Apter S, Lucas L, Decina P, Mukherjee S, Prudic J, Malitz S: Regional cerebral blood flow in affective disorders: Relations to phenomenology and effects of treatment. In: Takahashi R, Flor-Henry P, Gruzlier J, Niwa S (Eds.), *Cerebral Dynamics, Laterality and Psychopathology,* New York: Elsevier, 1987, 477–492.

Sackeim HA, Steif BL, The neuropsychology of depression and mania. In: Georgotas A, Cancro R (Eds.), *Depression and Mania,* New York, Elsevier, 1988, 265–289.

Sackeim HA, Prohovnik I, Moeller JR, Macklin E, Stade N, Devanand DP, Prudic J, Brown RP: ECT-induced alterations in functional neural networks: Evidence from studies of rCBF and EEG. *Abstracts of panels and posters presented at the Annual Meeting of the American College of Neuropsychopharmacology,* 1989, 91.

Sackeim HA, Prohovnik I, Moeller JR, Borwn RP, Apter S, Prudic J, Devanand DP, Mukherjee S: Regional cerebral blood flow in mood disorders. I. Comparison of major depressives and normal controls at rest. *Arch Gen Psychiat* 1990a; 47: 60–70.

Sackeim HA, Prudic J, Devanand DP, Decina P, Kerr B, Malitz S: The impact of medication resistance and continuation pharmacotherapy on relapse following electroconvulsive therapy in major depression. *J Clin Psychopharmacol* 1990b; 10: 96–104.

Sackeim HA, Prohovnik I, Mayeux R, Moeller JR, Devanand DP, Coudreaut M, Stern Y, Brown R, Prudic J: Abrnormal cortical neural networks in major depression and Alzheimer's disease: specificity and relations to clinical state. *Biol Psychiat* 1990c; 27: 41A.

Schlegel S: Computed tomography in affective disorders. In: Hauser P (Ed.), *Brain Imaging in Affective Disorders,* Washington DC: American Psychiatric Press, 1991, 1–24.

Schlegel S, Kretzschmar K: Computed tomography in affective disorders, Part I: ventricular and sulcal measurements. *Biol Psychiat* 1987; 22: 4–14.

Schlegel S, Maier W, Philipp M, Aldenhoff JB, Heuser I, Kretzschmar K, Benkert O: Computed tomography in depression: Association between ventricular size and psychopathology. *Psychiat Res* 1989a; 29: 221–230.

Schlegel S, Frommberger U, Buller R: Computerized tomography (CT) in affective disorders: relationship with psychopathology. *Psychiat Res* 1989b; 29: 271–272.

Schlegel S, von Bardeleben U, Wiedemann K, Frommberger U, Holsboer F: Computerized brain tomography measures compared with spontaneous and suppressed plasma cortisol levels in major depression. *Psychoneuroendocrinology* 1989c; 14: 209–216.

Schulsinger F, Parnes J, Peterson ET, Schulsinger H, Teasdale TW, Mednick SA, Moller L, Silverton L: Cerebral ventricular size in offspring of schizophrenic mothers: a preliminary study. *Arch Gen Psychiat* 1984; 41 602–606.

Scott AI, Douglas RH, Whitfield A, Kendell RE: Time course of cerebral magnetic resonance changes after electroconvulsive therapy. *Br J Psychiat* 1990; 156: 551–553.

Seiderer M, Krappel W, Moser E, Hahn D, Schmiedek P, Buell U: Detection and quantification of chronic cerebrovascular disease: comparison of MR imaging, SPECT, and CT. *Radiology* 1989; 170: 545–548.

Seltzer B, Pandya DN: Afferent cortical connections and architectonics of the superior temporal sulcus and surrounding cortex in the rhesus monkey. *Brain Res* 1978; 149: 1–24.

Siesjö BK, Ingvar M, Wieloch T: Cellular and molecular events underlying epileptic brain damage. *Ann Ny Acad Sci* 1986; 462: 207–223.

Silfverskiöld P: *Depression and Mania: Clinical and Neurophysiological Aspects and Effects of ECT,* Lund, Sweden: University of Lund, 1987.

Silfverskiöld P, Risberg J: Regional cerebral blood flow in depression and mania. *Arch Gen Psychiat* 1989; 46: 253–259.

Skullerud K: Variations in the size of the human brain: Influence of age, sex, body length, body mass index, alcoholism, Alzheimer's changes, and cerebral atherosclerosis *Acta Neurol Scand (Suppl)* 1985; 102: 1–94.

Sokoloff L: Relation between physiological function and energy metabolism in the central nervous system. *J Neurochem* 1977; 29:13–26.

Standish-Barry HM, Hale AS, Honig A, Bouras N, Bridges PK, Bartlett JR: Ventricular size, the dexamethasone suppression test and outcome of severe endogenous depression following psychosurgery. *Acta Psychiat Scand* 1985; 72: 166–171.

Starkstein SE, Robinson RG, Berthier ML, Parikh RM, Price TR: Differential mood changes following basal ganglia vs. thalamic lesions. *Arch Neurol* 1988; 45: 725–730.

Steingart A, Hachinski V, Lau C, Fox AJ, Diaz F, Cape R, Lee D, Inzitari D, Merskey H: Cognitive and neurologic findings in subjects with diffuse white matter lucencies on computed tomographic scans (leuko-araiosis). *Arch Neurol* 1987; 44: 32–35.

Suddath RL, Christison GW, Torrey EF, Casanova MF, Weinberger DR: Anatomical abnormalities in the brains of monozygotic twins discordant for schizophrenia. *N Engl J Med* 1990; 322: 789–794.

Swayze VW 2d, Andreasen NC, Alliger RJ, Ehrhardt JC, Yuh WT: Structural brain abnormalities in bipolar affective disorder. Ventricular enlargement and focal signal hyperintensities. *Arch Gen Psychiat* 1990; 47: 1054–1059.

Sze G, De Armond SJ, Brant-Zawadzki M, Davis RL, Norman D, Newton TH: Foci of MRI signal (pseudolesions) anterior to the frontal horns: Histologic correlations of a normal finding. *Am J Roent* 1986; 147: 331–337.

Tanaka Y, Hazama H, Fukuhara T, Tsutsui T: Computerized tomography of the brain in manic-depressive patients—a controlled study. *Folia Psychiat Neurol Jpn* 1982; 36: 137–143.

Targum SD, Rosen LN, DeLisi LE, Weinberger DR, Citrin CM: Cerebral ventricular size in major depressive disorder: association with delusional symptoms. *Biol Psychiat* 1983; 18: 329–336.

Uytdenhoef P, Portelange P, Jacquy J, Charles G, Linkowski P, Mendlewicz J:

Regional cerebral blood flow in lateralized hemispheric dysfunction in depression. *Br J Psychiat* 1983; 143: 128–132.

Van den Bossche B, Maes M, Brussaard C, Schotte C, Cosyns P, De Moor J, De Schepper A: Computed tomography of the brain in unipolar depression. *J Affect Disord* 1991; 21: 67–74.

Vita A, Sacchetti E, Conte G, Alciati A, Pennati A: Heterogeneity of major affective disorders. Biological and clinical evidence. *Encephale* 1985; 11: 71–77.

Vita A, Sacchetti E, Cazzullo CL: A CT follow-up study of cerebral ventricular size in schizophrenia and major affective disorder. *Schizophrenia Res* 1988; 1: 165–166.

Volkow ND, Bellar S, Mullani N, Jould L, Dewey S: Effects of electroconvulsive therapy on brain glucose metabolism: A preliminary study. *Convul Ther* 1988; 4:199–205.

Von Gall M, Becker H: Zur Anwendung der Comptertomographie (CT) in der klinischen Psychiatrie. *Fortschr Neurol, Psychiat* 1978; 17: 381–386.

Warach S, Gur RC, Gur RE, Skolnick BE, Obrist WD, Reivich M: The reproducibility of the 133-Xe inhalation method technique in resting studies: Task order and sex related effects in healthy young adults. *J Cereb Blood Flow Metab* 1987; 7: 702–708.

Warren LR, Butler RW, Katholi CR, McFarland CE, Crews EL, Halsey Jr JH: Focal changes in cerebral blood flow produced by monetary incentive during a mental mathematics task in normal and depressed subjects. *Brain and Cognition* 1984; 71–85.

Wehr TA, Goodwin FK: Can antidepressants cause mania and worsen the course of affective illness? *Am J Psychiat* 1987; 144: 1403–1412.

Weinberger DR, Torrey EF, Neophytides AN, Wyatt RJ: Lateral cerebral ventricular enlargement in chronic schizophrenia. *Arch Gen Psychiat* 1979; 36: 735–739.

Weinberger DR, DeLisi LE, Perman GP, Targum S, Wyatt RJ: Computed tomography in schizophreniform disorder and other acute psychiatric disorders. *Arch Gen Psychiat* 1982; 39: 778–783.

Weinberger DR, Berman KF, Zec RF: Physiologic dysfunction of dorsolateral prefrontal cortex in schizophrenia: II. Regional cerebral blood flow evidence. *Arch Gen Psychiat* 1986; 43: 114–124.

Weiner RD: Does ECT cause brain damage? *Behav Brain Sci* 1984; 7: 1–53.

Woods BT, Yurgelun-Todd D, Benes FM, Frankenburg FR, Pope HG Jr: Progressive ventricular enlargement in schizophrenia: Comparison to bipolar affective disorder and correlation with clinical course. *Biol Psychiat* 1990; 27: 341–352.

Yates WR, Jacoby CG, Andreason NC: Cerebellar atrophy in schizophrenia and affective disorder. *Am J Psychiat* 1987; 144: 465–467.

Zatz LM, Jernigan TL: The ventricular brain ratio on computed tomography scans: validity and proper use. *Psychiat Res* 1983; 8:207–214.

Zatz LM, Jernigan TL, Ahumada AJ: Changes on computed cranial tomography with aging: intracranial fluid volume. *Am J Neuroradiol* 1982; 3: 1–11.

Zubenko GS, Sullivan P, Nelson JP, Belle SH, Huff FJ, Wolfe GL: Brain imaging abnormalities in mental disorders of late life. *Arch Neurol* 1990; 47: 1107–1111.

INDEX

GPSR Compliance
The European Union's (EU) General Product Safety Regulation (GPSR) is a set of rules that requires consumer products to be safe and our obligations to ensure this.

If you have any concerns about our products, you can contact us on

ProductSafety@springernature.com

In case Publisher is established outside the EU, the EU authorized representative is:

Springer Nature Customer Service Center GmbH
Europaplatz 3
69115 Heidelberg, Germany

www.ingramcontent.com/pod-product-compliance
Ingram Content Group UK Ltd.
Pitfield, Milton Keynes, MK11 3LW, UK
UKHW020238160926
475610UK00001B/43
* 9 7 8 1 4 7 5 7 9 5 0 0 4 *